Quechua Indian from the Peruvian Andes who was born and spent all his life at an altitude of 4000 m. In such native highlanders, the carotid bodies are enlarged and show prominence of the dark variant of chief cells containing methionine- and leucine-enkephalins. These big carotid bodies are associated with a diminished ventilatory drive in response to hypoxia.

Donald Heath · Paul Smith

Diseases of the Human Carotid Body

With 136 Figures

Springer-Verlag
London Berlin Heidelberg New York
Paris Tokyo Hong Kong
Barcelona Budapest

Donald Heath DSc, MD, PhD, FRCP, FRCPEd, FRCPath
George Holt Professor of Pathology, University of Liverpool, Honorary Consultant
Pathologist, Royal Liverpool University Hospital, Department of Pathology, Duncan
Building, Royal Liverpool University Hospital, Liverpool L69 3BX, UK

Paul Smith BSc, PhD
Senior Lecturer, Department of Pathology, Duncan Building, Royal Liverpool
University Hospital, Liverpool L69 3BX, UK

Front cover illustration: Latex cast of the glomic vasculature in a normal carotid body.
The main glomic artery arises from the external carotid artery. A smaller, subsidiary
glomic artery arises from a separate location in front of it. Both vessels divide into three
branches. The large convoluted vessels at the upper pole are glomic veins.

British Library Cataloguing in Publication Data
Heath, Donald, *1928–*
Diseases of the human carotid body.
I. Title   II. Smith, Paul, *1944–*
616.8

Library of Congress Cataloging-in-Publication Data
Heath, Donald.
Diseases of the human carotid body/Donald Heath. Paul Smith. p. cm.
Includes index.

ISBN-13: 978-1-4471-1876-3   e-ISBN-13: 978-1-4471-1874-9
DOI: 10.1007/978-1-4471-1874-9

1. Carotid body—Pathophysiology.   2. Carotid body—Histopathology.
I. Smith, Paul, 1944–   . II. Title.
[DNLM]: 1. Carotid Body—pathology.   WL 102.9 H437d]
RC409.H43 1992
612.4′92—dc20
DNLM/DLC   91-4837
for Library of Congress   CIP

2128/3830-543210 Printed on acid-free paper

This book is dedicated to
Christine Smith
and to the memory of
Florence Heath

Preface

Ever since its discovery in 1742 the carotid body has remained an organ of mystery. Originally described as a ganglion, it was subsequently regarded as a gland, chromaffin paraganglion and non-chromaffin paraganglion. In 1928 it was shown to be a chemoreceptor with close associations with the function of baroreception in the adjacent carotid sinus and perhaps within its own substance. These discoveries led physiologists to embark on a series of elegant experimental studies on a number of animal species which have, however, so far failed to identify the transducer for detection of changes in tension of arterial blood gases or the mechanism of chemoreception. Pathologists on the other hand have largely ignored the carotid body, restricting their interest to its tumour, the chemodectoma. A remarkable disparity in knowledge of the organ has resulted, with most information being available on the physiology of chemoreceptor tissue in laboratory animals. In contrast, there has been sparse interest and awareness of the pathology in man of this nodule of tissue lying in the carotid bifurcation whose functional activity is suggested by the high blood flow it receives, and its rich content of biogenic amines and a wide variety of peptides.

This book is an attempt to redress this unsatisfactory situation. During the last few years our understanding of the detailed histology and ultrastructure of the human carotid body has improved. It has become apparent that its structure does not remain static throughout life but changes with age. In later life lymphocytic infiltration of the glomic tissues occurs and frequently results in aggregations which may constitute a distinct pathological entity, chronic carotid glomitis, reminiscent of autoimmune disease. At the other end of life the histology of the developing carotid body in fetuses and neonates is now being documented, and changes in the glomus of the infant have been reported in diseases characterised by hypoxia and in the sudden infant death syndrome. The content and distribution of peptides in the adult human carotid body have been established. It is feasible that some of them may be concerned with functions other than chemoreception and in this respect the reported rôle of the carotid body in sodium metabolism is considered in this book. The histological features of the carotid sinus and glomic arteries subserving baroreception are described and the structural alterations of the human carotid body in systemic hypertension are described. Consideration of all of these matters make this volume the first text available on the pathology of diseases of the human carotid body. We believe it will be of interest and value to pathologists, physiologists and clinicians.

We thank Miss Linda Byron and Miss Jill Shaw for their expertise and patience in typing the manuscript. Mr Alan Williams and Dr David Williams were most generous in providing much time and skill in their preparation of the photomicrographs and line diagrams. Much of the work described in this book is the result of happy collaboration over many years with valued colleagues. Some of our early studies were carried out with physicians and pathologists such as Emeritus Professor Peter Harris or Dr Christopher Edwards, who are established authorities in cardiopulmonary medicine and pathology, respectively. We have also been privileged to work in the Andes with members of a distinguished group of Peruvians including Professor Javier Arias-Stella and Dr Hever Krüger. Finally, we acknowledge our research students over the years en route to their PhDs, who helped to collect data on the carotid bodies and then joined us in enthusiastic and often heated discussions on the results. In particular we recall with great pleasure our association with Dr Ross Jago, Mr Gerard Hurst and Dr Qamar Khan.

Liverpool, 1991   D.H.
                                                                                        P.S.

Acknowledgements

The following colleagues have kindly provided us with clinical and pathological material from which the figures indicated have been prepared, or were co-authors in papers from which the illustrations have been taken:

Professor J. Arias-Stella
Figs. 21.3, 21.4, 21.5, 21.6

Mrs H. Burnett
Figs. 8.2, 8.6

Dr Y. Castillo
Figs. 21.3, 21.4, 21.5, 21.6

Dr R. Drewe
Figs. 9.5, 18.5

Dr C. Edwards
Figs. 2.5, 2.6, 16.6, 16.7, 21.3, 21.4, 21.5, 21.6

Mr R. Fitch
Figs. 9.1, 9.2, 9.6, 9.7, 9.8, 9.9, 21.1, 21.2

Mrs H. Flenley
Figs. 12.2, 12.3

Dr R. Galloway
Fig. 19.14

Dr J. Gosney
Figs. 8.2, 8.6

Professor P. Harris
Figs. 2.5, 2.6, 9.1, 9.2, 20.2, 20.3, 21.1, 21.2, 21.3, 21.4, 21.5, 21.6

Mr G. Hurst
Figs. 4.2, 4.6, 9.5, 9.6, 9.7, 9.8, 9.9, 14.2, 14.3, 18.5

Dr R. Jago
Figs. 2.1, 2.2, 2.7, 2.8, 9.3, 9.4, 10.3, 10.8, 16.2, 16.8

Dr Q. Khan
Figs. 2.3, 2.4

Dr H. Krüger
Figs. 21.3, 21.4, 21.5, 21.6

Mr P. Lowe
Figs. 20.6, 20.7

Mr D. Moore
Figs. 9.6, 9.7, 9.8, 9.9

Dr E. Weitzenblum
Figs. 9.6, 9.7, 9.8, 9.9

Dr M. Winson
Figs. 20.2, 20.3, 20.4, 20.5

We are indebted to the Editors of the following journals and Publishers for permission to reproduce the illustrations listed below which were previously published by them:

British Journal of Diseases of the Chest
Figs. 14.2, 14.3

Cardiovascular Research
Figs. 16.2, 16.6, 16.7, 16.8, 20.4, 20.5

Cell and Tissue Research
Figs. 8.2, 8.6

Edward Arnold
Figs. 2.8, 3.6, 10.6, 16.3, 16.4, 17.1, 18.4, 19.1, 20.4, 20.5

Histopathology
Figs. 9.5, 18.5

Journal of Clinical Pathology
  Figs. 2.3, 2.4

Journal of Comparative Pathology
  Figs. 9.1, 9.2, 21.1, 21.2

Journal of Laryngology and Otology
  Figs. 20.6, 20.7

Journal of Pathology
  Figs. 2.1, 2.2, 2.7, 2.8, 9.3, 9.4, 9.6,
  9.7, 9.8, 9.9, 10.3, 10.8, 20.2, 20.3,
  21.3, 21.4, 21.5, 21.6

Liverpool University Press
  Figs. 12.2, 12.3, 16.2, 16.9, 17.3, 17.4,
  17.6, 18.1, 18.3, 18.6

Thorax
  Figs. 2.5, 2.6

Contents

1 The Nature of the Carotid Body

Situated at the bifurcation of the common carotid artery on both sides of the neck is a small nodule of tissue called the carotid body. It comprises multiple clusters of specialised cells which are rich in biogenic amines and peptides. These cells have a most intimate and complex relationship with minute blood vessels which form a network arising from the glomic arteries, which themselves originate from the bifurcation of the common carotid artery (Fig. 1.1). These small nodules of tissue comprise the main chemoreceptor in man, although there are subsidiary contributions also from the aortic bodies. The carotid body has a complex nerve supply, derived from the glosso-pharyngeal and sympathetic nerves and its cells come into intimate contact with the nerve fibrils. The origin of the internal carotid artery shows abrupt thinning of its media, often with some dilatation, to form the carotid sinus, which functions as the major baroreceptor of the body (Figs. 1.1 and 1.2). There are, however, histological data to suggest that the carotid body itself is directly sensitive to the level of intravascular pressure within the glomic arteries. Understanding of the physiology and pathology of the human carotid body remains incomplete and controversial. A plethora of nerves, biogenic amines and peptides is found in and around its chief cells and yet the transducer for chemoreception has still not been identified. Furthermore, there is the intriguing possibility that the carotid bodies have additional endocrine functions unrelated to chemoreception.

Discovery of the Carotid Body

In 1742 Haller demonstrated nerve twigs leaving what he considered to be a small, hitherto undescribed, ganglion in the carotid bifurcation (Adams 1958). This discovery of a "ganglion minutum"

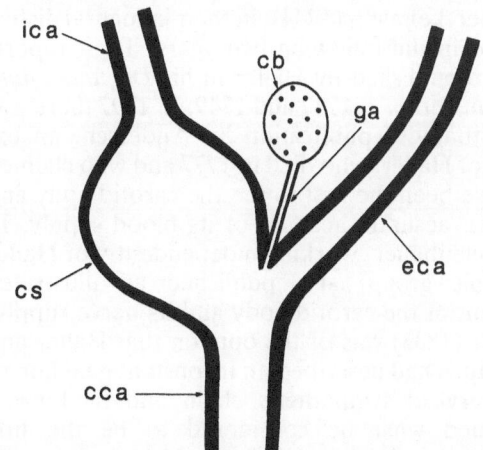

Fig. 1.1. Diagrammatic representation of the carotid bifurcation to show the carotid body and the carotid sinus and their vascular connections. The common carotid artery (*cca*) divides into internal (*ica*) and external (*eca*) branches. The first part of the internal carotid artery is dilated to form the carotid sinus (*cs*) subserving baroreception. The carotid body (*cb*, stippled) is supplied by blood through one or more glomic arteries (*ga*) which usually arise from the bifurcation. The various features, such as the length of the glomic artery and the thinness and degree of dilatation of the carotid sinus have been deliberately exaggerated to show the different components clearly.

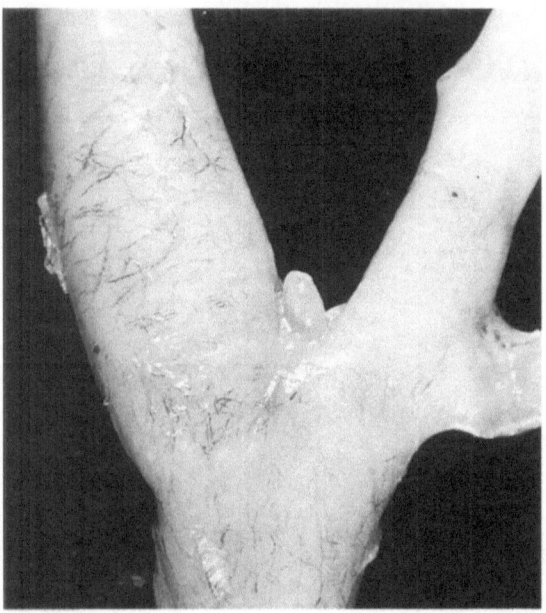

Fig. 1.2. Left carotid bifurcation from a man of 50 years who died from polycystic disease of the kidneys. Reference to Fig. 1.1 allows easy recognition of the carotid body and sinus and the other components of the bifurcation. In reality the glomic arteries are very short and embedded in the fibroelastic tissue of Mayer's ligament situated at the base of the carotid body. The carotid sinus is dilated compared to the origin of the external carotid artery. The carotid body weighed 6.9 mg (× 3.5).

was referred to by both of his pupils, Taube (1743) and Berckelmann (1744), in their inaugural dissertations in the following two years. Their reports were republished by Haller in his *Disputationum anatomicarum* in 1747 and 1749. In 1797 there was a posthumous publication by Andersch, an ex-pupil of Haller, who died in 1777 and who claimed to have been the first to see the carotid body and give an accurate account of its blood supply. In 1772 Neubauer, working independently of Haller and his group, also published an illustrated account of the carotid body and its nerve supply. Mayer (1833) was of the opinion that Haller and his pupils had described an inconstant ganglion of the cervical sympathetic chain and he himself provided what he considered to be the first description of the carotid body. He noted that the carotid body was attached to the wall of the carotid bifurcation by a ligamentous strand of tissue. It is now known that this strand encloses the glomic artery carrying the blood supply to the carotid body and it is called to this day "Mayer's ligament". It is of interest that he was probably the first to recognise that the carotid body is supplied by the glossopharyngeal nerve.

Glomic Tissue

Thus, by 1833 the stage was set for further investigation by anatomists, embryologists, physiologists and pathologists. Since that time the carotid body has been studied intensively, but there is still much controversy as to its nature and function. It comprises a complex network of arteries, capillaries and veins that make it into a highly vascular ball of tissue. The Latin word for a rolled-up body or ball is "glomus" (plural "glomera") and this term is classically used in descriptive histology to indicate such a spherical conglomeration of cells and blood vessels. Thus, in its diminutive form, it is used to describe the renal glomerulus. Care must be taken not to confuse the term used in this sense with the glomus found in the subcutaneous tissue of the fingers and toes that comprises a collection of arteriovenous anastomoses, which are not a feature of the human carotid body. Throughout this book we use the terms glomus and glomic tissue in a non-specific sense, implying merely a sphere of vascular tissue exhibiting an intimate association of cells and small blood vessels.

The Basic Histological Unit

The cellular parenchyma of the carotid body consists of two types of cell. One has a large round nucleus and copious cytoplasm containing vesicles rich in biogenic amines and peptides; it is termed the chief cell. The other is elongated and ensheathes nerve axons, and resembles the Schwann cell; it is called the sustentacular cell. These cells are arranged in clusters with chief cells at the centre and sustentacular cells at the periphery. The clusters are grouped together to form lobules that are themselves embedded in a fibrous stroma through which run numerous nerve fibrils and small blood vessels.

The carotid body is not alone in showing this basic microscopic structure. There are glomera widely scattered throughout the body, showing the same basic histological unit of rounded chief and elongated sustentacular cells embedded in a richly vascular background. Kjaergaard (1973) concluded that there is no histological or ultrastructural criterion by which the carotid body can be distinguished from other glomera and thus it appears to be part of a generalised system. It is, however, prudent to bear in mind that similarity in histological appearance does not mean that these tissues have the same function throughout the body. Only the carotid bodies and the aortic glomera, for example, have been shown to be chemoreceptors.

Glenner and Grimley (1974) made a study of neoplasms arising in this generalised system and they too found that these tumours shared the same histological features. This supported the concept that these widespread tissues are closely related. On the basis of their studies these authors believe that the glomera of this system fall into four groups. We may consider these here to set the carotid body in its background of this widely dispersed system.

Branchiomeric. This is the group which includes the carotid body. Initially it was thought that the glomera were derived from the mesoderm in the baroreceptive vascular zones. In other words it was considered that they were actually derived from cells in the branchial arches and could thus be termed branchiogenic. As such, an extremely intimate connection between the glomera and the respective vascular zone had to be presupposed (Kjaergaard 1973). Later it became apparent that the glomera were laid down at the site of the branchial arches rather than being derived from them. Hence they are correctly termed branchiomeric. As we describe below, the glomera are thought to be derived from cells, probably neurectodermal, which migrate along the nerves corresponding to the branchial arches. In the case of the carotid body, the migration would be along the glossopharyngeal nerve to the third branchial arch. It would seem that the nervous connections predominate over the vascular.

This group also includes jugulotympanic glomic tissue in the adventitia of the bulb of the jugular vein immediately beneath the bony floor of the middle ear or along the tympanic branch of the glossopharyngeal nerve. This tissue is called the glomus jugulare. Others in the group are the aortico-pulmonary bodies, some of which have been described just above the ductus arteriosus in the angle between it and the descending portion of the arch of the aorta. Others lie near the root of the innominate artery, on the anterolateral aspect of the left part of the aortic arch, and near the origin of the left coronary artery. One of the branchiomeric glomera is found on the bifurcation of the pulmonary trunk, where it has acquired the spurious designation of the "glomus pulmonale" (see Chapter 6). Orbital glomera have been described in both man and chimpanzee. Laryngeal glomera have been described just above the anterior end of the vocal cords and between the thyroid and cricoid cartilages (Glenner and Grimley 1974).

Intravagal. Groups of glomic cells are found interior to the perineurium, just beneath the nerve sheath or between nerve fibres. One such glomus has been described within or below the ganglion nodosum of the vagus nerve, where it is termed the "glomus intravagale". Intravagal glomera are not related to arteries but cannot be distinguished from other nodules of glomic tissue elsewhere in the body on the grounds of histology, ultrastructure or cytochemistry.

Aortico-sympathetic. These are associated with segmental ganglia of the sympathetic chain and collateral ganglia. A member of this group, especially prominent in the fetus and newborn, is the organ of Zuckerkandl.

Visceral-autonomic. These are associated with the blood vessels of viscera. Such glomera have been reported in the interatrial septum of the heart, in the hilus of the liver, in the wall of the urinary bladder and in association with the mesenteric vessels. Cell groups in association with the duodenal wall have been considered as further candidates for inclusion in this group.

Embryology

Above, we point out that the carotid body is branchiomeric rather than branchiogenic. In other words it is laid down in the third branchial arch rather than being derived from its tissues. It develops next to the third arch artery, which will come to form the initial portion of the internal carotid artery. The primordium of the carotid body can first be recognised in the human embryo at about 6 weeks of gestation (12.5 mm crown-rump length) as a condensation of undifferentiated cells in the adventitia of what is to be the internal carotid artery (Boyd 1937). Soon after, the local mesodermal thickening becomes localised to the medial aspect of the third arch artery near the bifurcation of the future common carotid artery.

The contribution made by this condensation of cells to the formation of the carotid body has not been easy to establish. Many believe that the nervous connections predominate over the vascular so that the glomus is thought to be derived from cells that migrate to this original mesodermal thickening along the nerves corresponding to the branchial arch. The question of whether glomus cells arise from mesenchyme or from neural crest cells that migrate there has proved difficult to answer because embryologists have not been able to identify primordial glomus cells with certainty (McDonald 1981). It was soon recognised that a branch of the glossopharyngeal nerve was associated with the primordium of the carotid body from its earliest stages. The question was whether the nerve was a source of neuroblasts, had an inductive effect on mesenchymal cells, or was not involved at all in the development of the carotid body (McDonald 1981). Boyd (1937) concluded that glomus cells differentiate from mesenchymal cells. In contrast, Smith (1924) had earlier come to the conclusion that glomus cells probably develop from neuroblasts. De Winiwarter (1939) also ascribed a neurogenic origin to the glomus cells, claiming migration of neuroblasts along the arterial wall from the nodose ganglion of the vagus nerve as well as from the petrosal ganglion of the glossopharyngeal nerve. Other authors attempted to combine the hypotheses, believing that the original mesenchymal condensation was responsible for the formation of the stroma of the organ, while neuroblasts from both cranial nerves and adjacent sympathetic or parasympathetic nerves or ganglia led to the development of the glomus cell. The association between the developing carotid body and migrating sympathetic neuroblasts may have an alternative explanation. Using fluorescence microscopy, Korkala and Hervonen (1973) found small catecholamine-containing cells in the carotid body and superior cervical ganglion primordium of an 8-week (28 mm) human fetus and there was in addition a cord of fluorescent cells that linked the carotid body and the ganglion, suggesting that glomus cells are derived from the sympathetic anlage. However, the possibility has to be kept in mind that sympathetic neuroblasts migrating along the cord into the carotid body might lead to the development there of sympathetic ganglion cells and not glomus cells. Zak and Lawson (1982) reviewed the various theories of development that have been advanced and concluded that only the suggested contribution from pharyngeal endoderm referred to later in this chapter has been eliminated as a factor in the embryogenesis of the carotid body.

In conclusion, it seems likely that neural crest cells that migrate along the glossopharyngeal nerve to the third branchial arch early in development of the carotid body may subsequently differentiate into both chief cells and sustentacular cells. Finally it has to be kept in mind that the segment of the third branchial arch artery where the primordial mesodermal condensation forms gives rise to the carotid sinus as well as to the origin of the internal artery. Both carotid body and carotid

sinus receive their sensory innervation from the carotid sinus branch of the glossopharyngeal nerve. Hence, from the outset, chemoreception and baroreception are closely linked.

Chromaffin Paraganglia

The embryological events just described lead to the formation of tissue whose nature has defied categorisation by anatomists. At the turn of the century Kohn (1900) developed the concept that the carotid body and the system of the glomera throughout the body shared, with the adrenal medulla, the characteristic of staining with potassium dichromate and formed a system of chromaffin paraganglia. However, for glomera to be classified as sympathetic paraganglia they have to meet certain strict criteria. Thus, the glomic cells must be similar in nature to the postganglionic sympathetic ganglion cells, their main innervation must be sympathetic efferent nerve fibres, the cells must secrete catecholamines, and they must be chromaffin (Kjaergaard 1973). These criteria are not met by the carotid body and the glomera, which cannot thus legitimately be regarded as chromaffin paraganglia.

Non-chromaffin Paraganglia

The concept of the carotid body as a chromaffin paraganglion was discarded largely because the carotid body and other glomera were found to be non-chromaffin except to a minute degree (see Chapter 3). To meet this objection Watzka (1930/31) introduced the concept of "non-chromaffin paraganglia", as the parasympathetic homologue. The strict criteria to be met here are that the cells must be homologous to the postganglionic parasympathetic ganglion cells, their main innervation must be parasympathetic efferent nerve fibres, the cells must secrete an acetylcholine-like substance and the cells must be non-chromaffin (Kjaergaard 1973). There is little doubt that acetylcholine is present in the carotid body, but its location is a subject of considerable debate (McDonald 1981). However, while there is a parasympathetic element in the glossopharyngeal nerve, its afferent sensory supply to the carotid body far exceeds the efferent. Hence, once again the criteria for non-chromaffin paraganglia are not met by the glomera.

To the naked eye the human carotid body is undeniably non-chromaffin. The technique consists of mixing 100 ml of a 5% (w/v) solution of potassium dichromate with 7 ml of a 5% (w/v) solution of potassium chromate (Kiernan 1981). The blocks of tissue to be stained are fixed in this mixture for 24 hours and then washed in running water for approximately 8 hours. They are then dehydrated, cleared and embedded in paraffin-wax in the usual manner. In human carotid bodies there is frequently no stain at all discernible to the naked eye, or at best a very pale reaction which is difficult to distinguish from the non-specific background staining of other tissue (Pryse-Davies et al. 1964). However, in reality the "non-chromaffin" carotid body includes chromaffin granules that can be detected at the ultrastructural level and they represent the "dense-core vesicles" (see Chapter 17) which contain catecholamines. If one accepts the suggestion of Böck and Gorgas (1976) that the term "chromaffin" should mean the capability to synthesise and store catecholamines or indolamines on this basis, the distinction between chromaffin and non-chromaffin referred to above ceases to exist. From a practical standpoint the human carotid body is non-chromaffin only in that the reaction is too insensitive to label the cells containing biogenic amines which are visible either macroscopically or on light microscopy.

The Carotid Body as a Chemoreceptor

The problem of the nature and designation of the carotid body is made even more complex because this particular glomus of the system is adjusted to become the undoubted major chemoreceptor of the human body, although the aortic bodies play a subsidiary rôle in this respect. The demonstration by de Castro (1928) of the afferent nature of the nerve supply of the carotid body from the glossopharyngeal first indicated that it was, at least in part, a sensory organ. His interpretation of its histological structure as being consistent with the monitoring of blood gases was soon confirmed by the studies of Heymans and his colleagues (1930), who showed that the carotid body could detect changes in arterial gas tensions and pH.

The Carotid Body as an Endocrine Gland

In the light of recent observations it is a matter of considerable interest that, when the first microscopic study of the carotid glomus was performed in 1862 by Luschka, he thought on histological grounds that it was a gland. He assumed that the carotid body had developed from the pharyngeal endoderm and this view appeared to be supported by the studies of Stieda (1881) until Jacoby (1896) showed that Stieda had confused the primordium of glomic tissue with that of the inferior parathyroid gland. Luschka's concept of the carotid body as an endocrine gland seemed to be dead.

However, over the past 20 years a series of observations has revived new interest in old ideas. It is clear that the chief cell of the carotid body is an archetypal APUD (amine precursor uptake and decarboxylation) cell. Such cells have the capacity for the formation, storage and secretion of peptides of low molecular weight. Ultrastructurally they are characterised by dense-core vesicles. Chief cells of the carotid body contain the peptides methionine- and leucine-enkephalins in the proteinaceous cores of the vesicles. Vasoactive intestinal polypeptide and substance-P-like immunoreactivity are to be found localised in the nerve fibrils of the carotid body. Pearse (1969) gave an overall non-committal designation of "glomin" to describe a putative endocrine of glomic tissue, but he did not offer a suggestion as to its action. Certainly it is apparent that not all functions of the carotid body are related to chemoreception. Several years ago we found that the carotid bodies are enlarged in cases of systemic hypertension with left ventricular hypertrophy (Edwards et al. 1971; see Chapter 14). Subsequently a reflex effect on sodium secretion, originating in the carotid body, was discovered (Honig 1983). Possibly there is endocrine activity in the carotid bodies concerned with sodium metabolism that provides an unsuspected link with renal function (see Chapter 15).

It is apparent from this account that the nature of the carotid body is complex and its classification as a tissue highly controversial. Pathologists have tended to accept the designation of this group of histologically identical tissues as the non-chromaffin paraganglionic system and of its tumours as non-chromaffin paragangliomas. While such terminology cannot be defended on purist grounds, it has the power of generally accepted usage.

References

Adams WE (1958) The comparative morphology of the carotid body and carotid sinus. Thomas, Springfield, IL

Andersch CS (1797) Tractatio anatomico-physiologica de nervis humani corporis aliquibus. Andersch EP (ed) A Fasch, Regiomonti, Chap VI, pp 132–133

Berckelmann MLR (1744) De nervorum in arterias imperio, Inaug Diss Gottingae. A Vandenhoeck, Gottingae

Böck P, Gorgas K (1976) Catecholamines and granule content of carotid body type I cells. In: Coupland RE, Fujita T (eds) Chromaffin, enterochromaffin and related cells. Elsevier Scientific Publishing Company, Amsterdam, Chap 23, pp 355–374

Boyd JD (1937) The development of the human carotid body. Contr Embryol Carnegie Inst 26:1–31

de Castro F (1928) Sur la structure et l'innervation du sinus carotidien de l'homme et des mammifères. Nouveaux faits sur l'innervation et la fonction du glomus caroticum. Études anatomiques et physiologiques. Trav Lab Rech Biol 25:331–380

De Winiwarter H (1939) Origine et développement du ganglion carotidien. Appendice: participation de l'hypoblaste à la constitution des ganglions craniens. Arch Biol (Paris) 50:67–94

Edwards C, Heath D, Harris P (1971) The carotid body in emphysema and left ventricular hypertrophy. J Pathol 104:1–13

Glenner GG, Grimley PM (1974) Tumors of the extra-adrenal paraganglion system (including chemoreceptors). Armed Forces Institute of Pathology, Washington, DC

Haller A (1747, 1749) Disputationum anatomicarum selectarum. A Vandenhoeck, Gottingae. Vol II, pp 939–951 (reprint of Taube's Inaugural Dissertation); vol IV, pp 425–445 (reprint of Berckelmann's Inaugural Dissertation)

Heymans C, Bouckaert JJ, Dautrebande L (1930) Sinus carotidien et réflexes respiratoires; influences respiratoires réflexes de l'acidose, de l'alcalose, de l'anhydride carbonique, de l'ion hydrogène et de l'anoxémie. Sinus carotidiens et échanges respiratoires dans les poumons et au dela des poumons. Arch Int Pharmacodyn Ther 39:400–448

Honig A (1983) Role of the arterial chemoreceptors in the reflex control of renal function and body fluid volumes in acute arterial hypoxia. In: Acker H; O'Regan RG (eds) Physiology of the peripheral arterial chemoreceptors. Elsevier Biomedical Press, Amsterdam/New York, pp 395–429

Jacoby M (1896) Ueber die Entwickelung der Nebendrüsen der Schildrüse und der Carotidendrüse. Anat Anz 12:152–157

Kiernan JA (1981) Histological and histochemical methods: theory and practice. Pergamon Press, Oxford

Kjaergaard J (1973) Anatomy of the carotid glomus and carotid glomus-like bodies (non chromaffin paraganglia). F.A.D.L.'s Forlag, Copenhagen

Kohn A (1900) Ueber den Bau und die Entwicklung der sog. Carotisdrüse. Arch Mikrosk Anat 56:81–148

Korkala O, Hervonen A (1973) Origin and development of the catecholamine-storing cells of the human fetal carotid body. Histochemie 37:287–297

Luschka H (1862) Ueber die drüsenartige Natur des sogenannten Ganglion intercaroticum. Arch Anat Physiol Lpz 405–414

Mayer AFJK (1833) Ueber ein neuentdecktes Ganglion im Winkel der aüssern und innern Carotis, be'im Menschen und den Säugethieren (Ganglion intercaroticum). Notiz Geb Natur Heilk (L. von Froriep) 36:8–9

McDonald DM (1981) Peripheral chemoreceptors. Structure–function relationships of the carotid body. In: Hornbein TF (ed) Regulation of breathing. Marcel Dekker, New York, pp 256–257

Neubauer JE (1772) Descriptio anatomica nervorum cardiacorum. Sectio 1. De nervo intercostali cervicali, dextri imprimis lateris. Fleischer, Frankfurt Leipzig, p 222

Pearse AGE (1969) The cytochemistry and ultrastructure of polypeptide hormone-producing cells of the APUD series and the embryologic, physiologic, and pathologic implications of the concept. J Histochem Cytochem 17:303–313

Pryse-Davies J, Dawson IMP, Westbury G (1964) Some morphologic, histochemical and chemical observations on chemodectomas and the normal carotid body, including a study of the chromaffin reaction and possible ganglion cell elements. Cancer 17:185–202

Smith C (1924) The origin and development of the carotid body. Am J Anat 34:87–131

Stieda L (1881) In: Untersuchungen über die Entwickelung der Glandula thymus, Glandula thyreoidea und Glandula carotica. Engelmann, Leipzig, p 34

Taube HWL (1743) De vera nervi intercostalis origine, Inaug Diss Gottingae. A Vandenhoeck, Gottingae

Watzka M (1930/31) Ueber die Verbindungen inkretorischer und neurogener Organe. Verh Anat Ges 39:185–190

Zak FG, Lawson W (1982) Embryology. In: The paraganglionic chemoreceptor system. Springer Verlag, New York, pp 133–141

2 Size, Weight and Anatomical Variation

The classic carotid body is a spherical or ovoid structure situated between the arms of the carotid bifurcation and arising from its angle. When all adherent adipose tissue has been removed, the carotid body proves to be reddish-brown or tan in colour. It is attached to its artery of origin by a short stalk, the ligament of Mayer, whose rôle in the discovery of the organ has already been described in Chapter 1. This ligament contains the glomic artery providing the glomus with its blood supply. The carotid body is thus a pedunculated structure, although sometimes its lower pole extends downwards to cover the stalk and create the impression that it is sessile. There is considerable variation in both the macroscopic form and location of the carotid bodies and it is important for the pathologist to be well acquainted with this so that anatomical variation may be distinguished from disease with confidence at post mortem.

Anatomical Variation in Form

Ovoid

In most cases the carotid bodies on both sides are single and ovoid, sometimes with a pointed apex and situated at the bifurcation (see Fig. 1.2). They are usually sessile, with the glomic artery hidden beneath them. Occasionally they are pedunculated, arising several millimetres along the glomic artery. This simple appearance is the classic form of the normal carotid body, and its prevalence has been found to be between 83% and 95% in two large series comprising together 250 cases at necropsy (Smith et al. 1982; Khan et al. 1988; see Table 2.1).

Table 2.1. Variation in form of carotid bodies

Form	Right (%)	Left (%)
Data from Khan et al. (1988), 100 subjects		
Single ovoid	83	86
Double	7	6
Bilobed (including V form)	9	7
Leaf shaped	1	1
Data from Smith et al. (1982), 150 subjects		
Single ovoid	95	93
Double	4	3
Bilobed	1	4

Double

In an appreciable minority, found to range between 3% and 7% (Table 2.1) double carotid bodies were found on one or other side, and they occurred in various places. Usually both members of the pair arose at the same site, such as the carotid bifurcation or external carotid artery (Fig. 2.1). Sometimes, however, they occurred at different sites such as one carotid body at the bifurcation and a second supplied by a glomic artery arising from the carotid sinus. The combined weight of double carotid bodies was similar to that of normal single ones, because both members of the pair were roughly half the normal size.

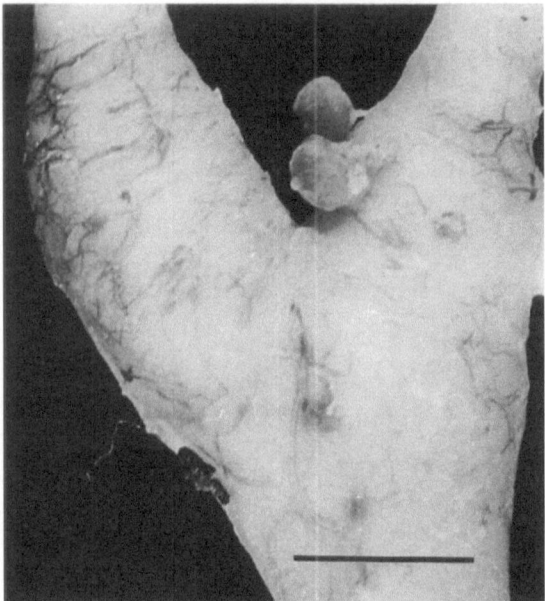

Fig. 2.1. Double carotid body. Left carotid bifurcation from a woman of 51 years who died from acute pancreatitis. Two rounded carotid bodies arise from separate regions of the external carotid artery. Scale line = 0.5 cm.

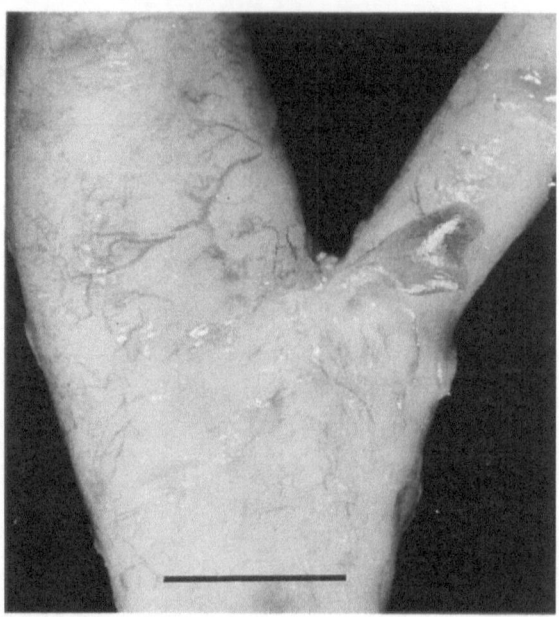

Fig. 2.3. V-shaped left carotid body (weighing 7 mg) arising from the bifurcation in a woman of 57 years dying from congestive cardiac failure caused by mitral stenosis. Scale line = 0.5 cm.

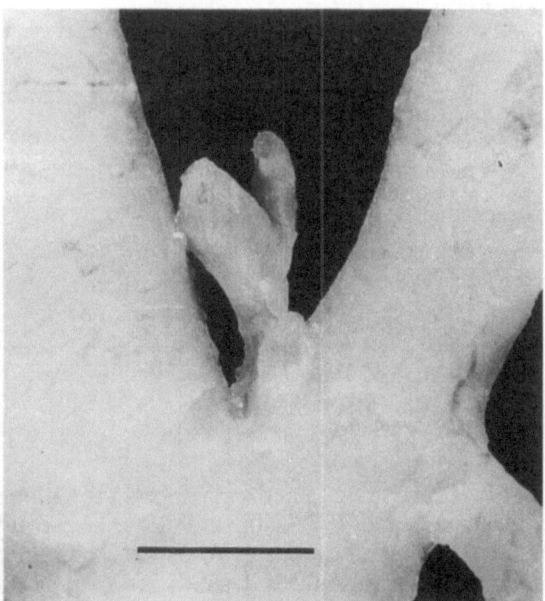

Fig. 2.2. Bilobed carotid body from a man of 65 years who died from carcinoma of the bronchus. There are two distinct lobes which fuse at their base to share a common origin from the bifurcation. Scale line = 0.5 cm.

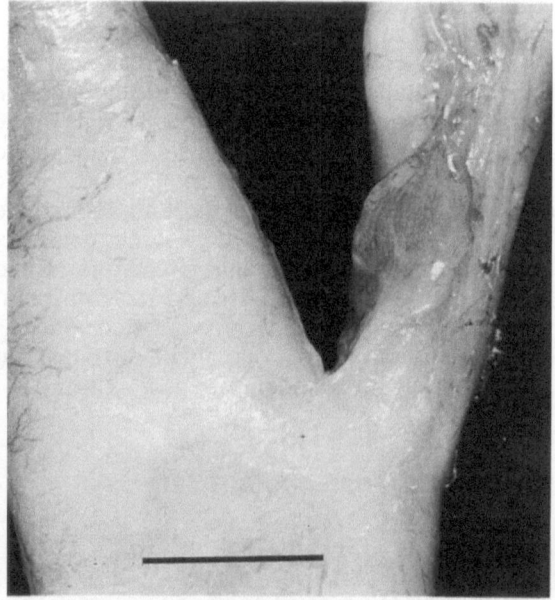

Fig. 2.4. Leaf-shaped left carotid body (weighing 14 mg) lying over left external carotid artery in a woman of 66 years with acute myeloid leukaemia. Scale line = 0.5 cm.

Bilobed

In up to 9% of cases studied (Table 2.1) the carotid bodies were bilobed, two distinct apical lobes fusing at the base to share a common origin from a single glomic artery (Fig. 2.2). Bilobed carotid bodies were usually found at or near the bifurcation but occur at any site where single bodies are found. In some instances the carotid bodies were V-shaped (Fig. 2.3), but this may be regarded as a variant of the bilobed variety rather than as a distinct form, the two lobes being fused to form a common base.

Leaf-shaped

Rarely the carotid bodies on one or other side assumed an expanded, leaf-shaped appearance (Fig. 2.4).

Variation in Location

Normal carotid bodies show great variation not only in their form but also in their location (Table 2.2). In 86% to 88% of cases studied carotid bodies lay over and slightly posteriorly to the carotid bifurcation supplied by glomic artery or arteries arising from that great vessel. Much less commonly their arterial supply came from the external carotid artery or carotid sinus, so that the carotid bodies were situated over these major arteries rather than over the bifurcation. Rarely the glomic arteries arose from the ascending pharyngeal arteries.

Table 2.2. Variation in location of carotid bodies

Location	Right (%)	Left (%)
Data from Khan et al. (1988), 100 subjects		
Carotid bifurcation	86	87
External carotid artery	7	9
Carotid sinus	5	2
Ascending pharyngeal artery	2	2
Data from Smith et al. (1982), 150 subjects		
Carotid bifurcation	88	87
External carotid artery	5	5
Carotid sinus	5	5
Ascending pharyngeal artery	1	3

Anatomical Variation and Pathological Abnormality

It is important that the pathologist should be able to distinguish between anatomical variation and

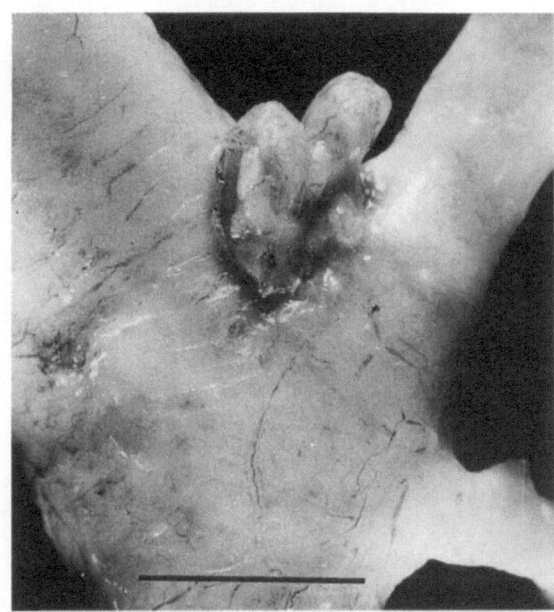

Fig. 2.5. Left carotid body (combined carotid body weight 55.8 mg) from a man aged 58 years with pronounced left ventricular hypertrophy caused by systemic hypertension. It has an apparently coarse nodular appearance produced mainly by enlargement of a bilobed variant. Scale line = 0.5 cm.

disease in carotid bodies dissected out at necropsy. In general the often striking appearances illustrated so far in this chapter may be readily recognised for what they are, namely normal variants of no pathological significance.

The important criterion for macroscopic abnormality of a carotid body is an increase in its weight. When this is associated with an anatomical variation this may lead to difficulty in interpretation. One such case that we examined was a man of 58 years who had pronounced left ventricular hypertrophy caused by systemic hypertension. The left carotid body was enlarged, the combined carotid body weight being 55.8 mg. Naked-eye examination at first suggested that the glomus was involved by a process of coarse nodularity but closer analysis suggested that the appearances were in fact produced by a bilobed variant with remaining nodules of adipose tissue (Fig. 2.5).

Diseased carotid bodies are unequivocally enlarged but they usually retain a smooth, ovoid appearance (Fig. 2.6). When the enlargement is due to chronic hypoxaemia, the carotid bodies have a cyanotic hue (Fig. 2.6). Such enlarged ovoid glomera will show the histological features of carotid body hyperplasia. Sometimes pathologically enlarged carotid bodies show fine nodularity (Fig. 2.7) in contrast to the apparent coarse

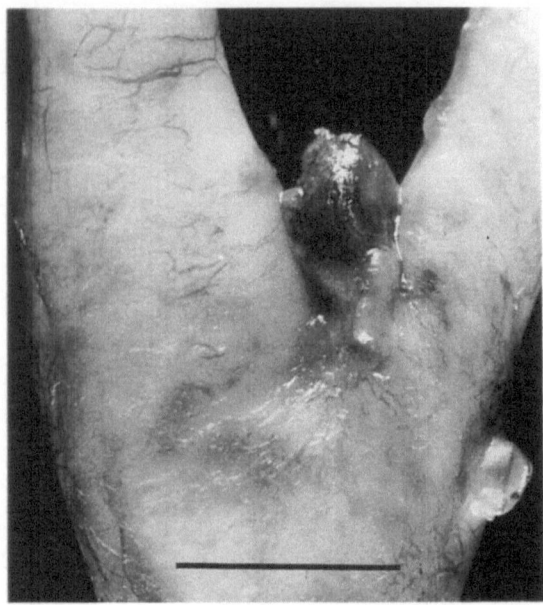

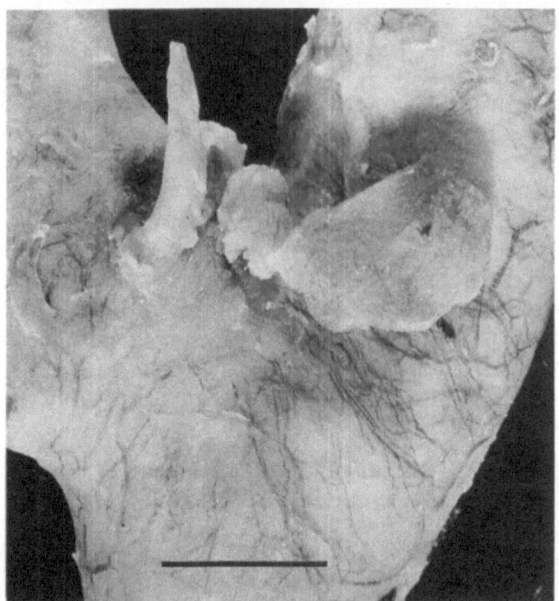

Fig. 2.6. Left carotid body (combined carotid body weight 43.9 mg) from a man aged 62 years with panacinar emphysema and biventricular hypertrophy. It is large and ovoid with a smooth, cyanotic appearance. Scale line = 0.5 cm.

Fig. 2.7. Right carotid bifurcation of a man of 53 years with systemic hypertension and left ventricular hypertrophy. The carotid body arises from the bifurcation and is enlarged with fine nodularity. Histological examination showed that it was hyperplastic. Scale line = 0.5 cm.

nodularity referred to above, which we believe is due to anatomical variation (Fig. 2.5).

In summary, when carotid bodies are examined at necropsy they may present a considerable range of macroscopic appearances. Some of these are anatomical variants but others are of pathological importance and it is necessary to distinguish between the two. The classic carotid body is a simple ovoid but it has several anatomical variants that are, in descending order of frequency, the bilobed (including the V-form), the double and the leaf-shaped. The hallmark of pathological abnormality on macroscopic appearance is enlargement, which indicates histological hyperplasia. Usually the enlarged glomus retains its ovoid appearance, with a smooth surface, but sometimes it assumes a fine nodularity. In cases of chronic obstructive lung disease it is cyanotic. Finally, it has to be kept in mind that anatomical variants themselves may become hyperplastic so that enlargement of a bilobed or multilobed carotid body may lead to the macroscopic appearance of coarse nodularity.

Weight

From these observations on anatomical variation and pathological abnormality it is clear that the weight of the carotid bodies is of interest to pathologists for its increase is one indication of hyperplasia. It is, of course, greatly increased when the organ becomes neoplastic but then the change is so pronounced as to make the measurement superfluous. When the carotid bodies are dissected prior to weighing, it is important that every fragment of adherent connective and adipose tissue should be removed. Such retained fragments of tissue could constitute a significant proportion of the weight of this small organ. Careless dissection can easily transform a normal carotid body into an apparently hyperplastic one. In spite of these potential shortcomings, weighing is the most convenient and speedy measurement for routine use in the post-mortem room and the laboratory, and its inherent small inaccuracies have to be accepted as the price for this.

The mean weight of both carotid bodies combined is up to 30 mg, in subjects free from cardiopulmonary disease. There is considerable variation in weight in different subjects. From an unselected group of 42 cases the mean weight of the left carotid body was 11.3 mg and that of the right was 12.9 mg, with a mean combined weight of 24.2 mg (Heath et al. 1970). This series included several examples of carotid body hyperplasia associated with chronic obstructive lung disease and systemic hypertension and hence these values are slightly higher than those found in subsequent studies. In the weights from 13 cases selected to be free from cardiopulmonary disease, the combined carotid body weight ranged from 6.2 to 34.4 mg, with a mean of 21.1 mg (Edwards et al. 1971).

In the later study of 57 cases with normal carotid bodies, the mean combined weight was 18.0 mg with a range of 8.5–31 mg (Smith et al. 1982). From these data the upper limit of normal can be estimated by calculating the limits of probability within which 95% of the population as a whole might be expected to fall. These limits are 6.8–29.2 mg, so that 30 mg may be regarded as the upper limit of normal. Although the mean weights of hyperplastic carotid bodies, as determined by their histological appearance, are considerably in excess of normal, their 95% probabilities overlap the normal range completely. Thus the weight of the carotid bodies is a measurement which can be obtained rapidly but does not always distinguish between normal and hyperplastic organs. A combined weight exceeding 30 mg usually denotes hyperplasia but a weight less than this value does not necessarily indicate normality (Table 2.3). In order to identify hyperplasia more precisely, histological assessment is necessary.

Table 2.3. Weights of normal and hyperplastic carotid bodies from 98 cases coming to necropsy (Data from Smith et al. 1982)

Group	No. of cases	Range of 95% probability		Mean (mg)
		Lower	Upper	
Control	57	7	30	18
Hypoxaemia	5	4	99	52
Systemic hypertension	36	5	54	30

There is no increase in size and weight in the carotid bodies with age in healthy subjects living at sea level. This is in sharp contrast to the progressive increase in weight of these organs with age which occurs in the native Quechua highlanders of the Andes exposed to hypobaric hypoxia throughout their lives (see Chapter 11).

Size

In a series of 42 unselected cases Heath et al. (1970) found the average height, breadth and depth of the carotid body of the human adult to be 3.3, 2.2 and 1.7 mm, respectively. This series included some hyperplastic carotid bodies as well as normal ones, so normal values are likely to be slightly lower than this. Much earlier, Sato (1932) had provided data on the dimensions of the carotid bodies of Japanese adults (Table 2.4). His figures are certainly lower than ours but the measurements were made from fixed material, subsequently paraffin-embedded and sectioned serially.

Table 2.4. Dimensions of the adult human carotid body based on fixed material from Japanese subjects (after Sato 1932)

Age (years)	Dimensions (mm)		
16	0.73	0.96	0.23
19	1.25	2.62	2.40
24	1.57	1.68	1.72
25	2.55	1.12	1.68
32	2.97	1.56	1.48
39	2.34	1.92	0.84
42	1.40	2.80	1.76
53	1.06	2.21	1.44
62	2.52	1.50	0.98

Hence the tissues would have undergone shrinkage due to fixative and subsequent processing and the true dimensions of human carotid bodies are probably somewhat larger than he indicated. The classic figures for the dimensions of the human carotid body are those given by Luschka (1862). These are 5 (average) to 7 mm (maximum) by 2.5 (average) to 4 mm (maximum) by 1.5 mm. Adams (1958) notes that figures given by later authorities such as Schaper (1892), Kohn (1900) and Mönckeberg (1905) are in fact those of Luschka. These figures seem rather high to us. Indeed they resemble those of hyperplastic carotid bodies. Thus in the series referred to above (Heath et al. 1970) the biggest carotid body found occurred in a case of panacinar emphysema complicated by hypoxaemia and measured 6.2 mm × 3.3 mm × 2.5 mm.

Volume

Even the carotid bodies of infants and small children may be dissected out and weighed accurately, although this requires more time than is needed for the same procedures in adults, which can be carried out rapidly in the post-mortem room. The carotid bodies of some neonates and of laboratory animals used in research are not suitable for simple dissection and weighing because they are too small. Under these circumstances the more time-consuming techniques of tissue morphometry have to be employed to determine the volume of the carotid body. One almost hesitates to mention this because there appears to be a widely held misapprehension amongst pathologists that the preparation of serial sections and the application of morphometric techniques are required for the examination of all carotid bodies, a mistaken view that has undoubtedly held back the study of the pathology of the human carotid body and deprived medicine of much information on chemoreceptors in heart and lung disease.

A common method used to determine the volume of the carotid body is to apply Simpson's rule as quoted by Dunnill (1968). By this rule the volume, V, of an organ is given by:

$$V = \frac{h}{3}\left[(A_0 + A_n) + 4(A_1 + A_3 + \ldots A_{n-1}) + 2(A_2 + A_4 + \ldots A_{n-2})\right]$$

where

n	= the number of sections
$A_0, A_1, A_2 \ldots$	= the area of successive sections, and
h	= the interval between each section, an appropriate interval being 50 μm.

The areas required may be measured by automatic image analysers, now possessed by most laboratories. They allow the rapid and efficient determination of such parameters as lengths, perimeters and areas of objects in a histological section. The generation of such quantitative information by microprocessor technology now permits the rapid assessment of carotid body volume.

The Forgotten Organ of Morbid Anatomy

As the carotid body and sinus are so intimately concerned with the functions of chemoreception and baroreception it might be anticipated that the morbid anatomist would pay attention to this segment of the systemic circulation at necropsy, particularly if he or she has a special interest in cardiopulmonary disease. In fact the carotid body is the forgotten organ of morbid anatomy. Pathologists have paid little attention to this organ, confining their interest to its tumour, the chemodectoma – one suspects because it has such a characteristic histological appearance (see Chapter 19).

This has led to a peculiar imbalance in our knowledge of the carotid body. Physiologists have been carrying out experiments on laboratory animals for many years, although this has not enabled them to determine precisely the transducer in the carotid body which detects the abnormality of partial pressure of oxygen in the plasma converting it into chemical changes in the cytoplasm of the chief cells and thus into a nerve impulse. In the face of all this activity the morbid anatomist has shown little or no interest in the pathology of chemoreceptor tissue. It is exceptional for examination of the carotid bodies to be included at necropsy, even in cases of heart or lung disease, where it might be expected to show an abnormality. This distortion in our knowledge has led to an unfortunate situation where most of what we know about the histological appearances of the carotid body in health and disease is based on an animal model. This is very unsatisfactory for it cannot be relied on that the animal carotid body will react in the same way to disease as the human (see Chapter 21). Indeed the carotid body of the rat, which is used more frequently than any other animal species in experimental studies in this field, does not react to the hypoxic stimulus like that of man. One can only speculate as to why morbid anatomists so studiously avoid the carotid body at necropsy. In part the answer lies in tradition for, if they have not been trained to include this in dissection, it is most unlikely that they will introduce such an innovation. One suspects too that they may think that examination of the carotid body requires time-consuming preparation and examination of serial sections, which will yield information long after the performance of the necropsy. This is a mistaken view. The carotid body is susceptible to a simple dissection at necropsy and immediate evaluation like the other viscera.

Dissection of the Carotid Body at Necropsy

The following dissection procedure is satisfactory for inclusion in a routine necropsy and we find that it may be completed within a few minutes. A V-shaped incision is made in the neck and the skin is reflected upwards to the lower margin of the mandible. The sternomastoid, stylohyoid and digastric muscles are excised to expose the carotid sheath. After dissection of the jugular vein, the carotid sheath is freed from the vagus nerve and underlying fascia. The common carotid artery is transected 5 cm below its bifurcation and the internal and external carotid arteries at least 3 cm above the bifurcation. Both carotid bifurcations are then removed and pinned to cork boards so that the carotid bodies can be dissected out. This is

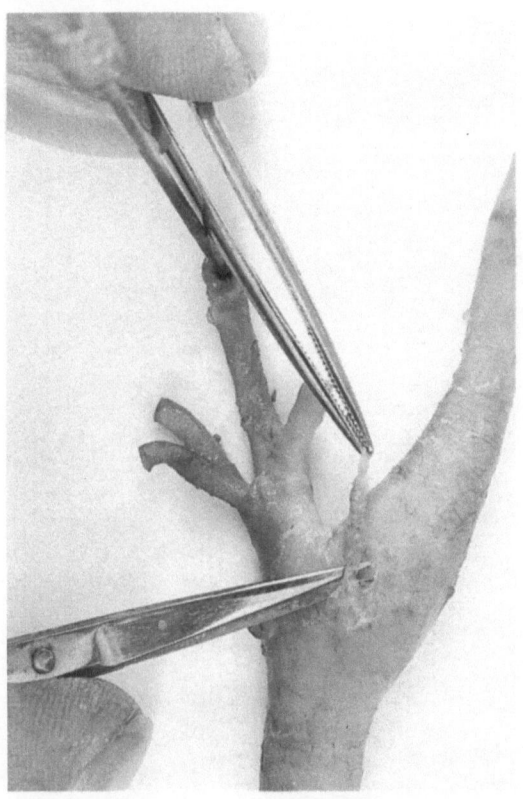

Fig. 2.8. The final stage of dissection of the carotid body. The surrounding fibrous and adipose tissue is held gently with forceps whilst it is cut away with fine scissors from the carotid body. Any remaining connective tissue and fat is then trimmed from the carotid body itself.

achieved by carefully stripping off the adventitial fat and connective tissue with fine dissecting scissors, starting with the common carotid artery and working distally (Fig. 2.8). The tissue is found to strip off readily, leaving the intact carotid body

attached to the surface of its artery of origin by its short glomic artery. Where single carotid bodies are present in the bifurcation, this process takes no longer than 5 minutes, but occasionally, where carotid bodies are situated elsewhere or are multiple, a more extensive dissection is required and this will of course take more time. It is our practice to photograph the carotid body in situ at this point, before removing it to be weighed and then fixed in 10% (v/v) formol saline.

References

Adams WE (1958) The comparative morphology of the carotid body and carotid sinus. Charles C. Thomas, Springfield, Illinois

Dunnill MS (1968) Quantitative methods in histology. In: Dyke SC (ed) Recent advances in clinical pathology. Churchill, London, pp 401–416

Edwards C, Heath D, Harris P (1971) The carotid body in emphysema and left ventricular hypertrophy. J Pathol 104:1–13

Heath D, Edwards C, Harris P (1970) Post-mortem size and structure of the human carotid body. Thorax 25:129–140

Khan Q, Heath D, Smith P (1988) Anatomical variations in human carotid bodies. J Clin Pathol 41:1196–1199

Kohn A (1900) Ueber den Bau und die Entwicklung der sog. Carotisdrüse. Arch Mikr Anat 56:81–148

Luschka H (1862) Ueber die drüsenartige Natur des sogenannten Ganglion intercaroticum. Arch Anat Physiol Lpz 1:405–414

Mönckeberg IG (1905) Die Tumoren der Glandula carotica. Beitr Pathol Anat Allg Pathol 38:1–66

Sato S (1932) Morphologische Untersuchungen über die Carotis-Drüsen bei Wirbeltieren und beim Menschen. Igaku Kenkyuu, Fukuoka 6:707–811 (In Japanese: abstract in Jap J Med Sci I Anat (1934) 5 (ii): 82–83

Schaper A (1892) Beiträge zur Histologie der Glandula carotica. Arch Mikrosk Anat 40:287–320

Smith P, Jago R, Heath D (1982) Anatomical variation and quantitative histology of the normal and enlarged carotid body. J Pathol 137:287–304

3 Normal Histology

In this chapter, the light microscopy of the normal, mature carotid body is presented. Its basic structure and the cytological features of its constituent cells are described. It is essential to have a firm understanding of the histology of the normal carotid body as a basis for comprehension of its physiological functions, and for appreciating the microscopic changes that are associated with ageing, exposure to chronic hypoxia and with neoplasia. Quantitative techniques applied to the histological study of the carotid body are discussed. The microscopy of the glomic vasculature is considered separately in Chapter 16.

General Architecture

The essential glomic and vascular components of the carotid body are embedded within an acellular fibrous stroma which extends to the periphery of the organ but does not form a discrete capsule (Figs. 3.1, 3.2). Instead, it tends to merge with the fibro-fatty connective tissue surrounding the carotid bifurcation. However, this does not interfere with the recognition of the carotid body as a well-defined nodule which can be dissected out cleanly and rapidly (see Chapter 2). In the young the surrounding fibrous tissue is sparse (Fig. 3.1) but in the elderly it becomes more pronounced so that the margins of the carotid body become more blurred (Fig. 3.2). It is rare to find glomic tissue divorced from the main nodule in this surrounding fibrous tissue.

The main components of the carotid body and their distribution are shown diagrammatically in Fig. 3.3. The most conspicuous are lobules consist-

ing of round or oval masses of glomus cells and associated nerve axons. They are the main functional component of the carotid body involved in

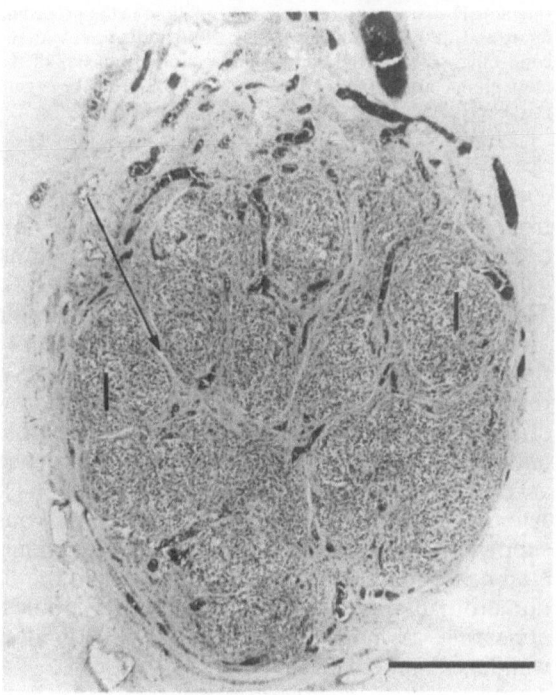

Fig. 3.1. Vertical section through the carotid body of an 11-year-old boy. The organ is cellular, consisting of numerous round lobules (*l*) with narrow septa of stroma between them (*arrow*). Darkly staining venules course through the septa and drain into small veins at the apex (top of the figure). Note that, although there is no capsule, the zone of glomic tissue is well defined. Haematoxylin–eosin (HE) Scale line = 500 μm.

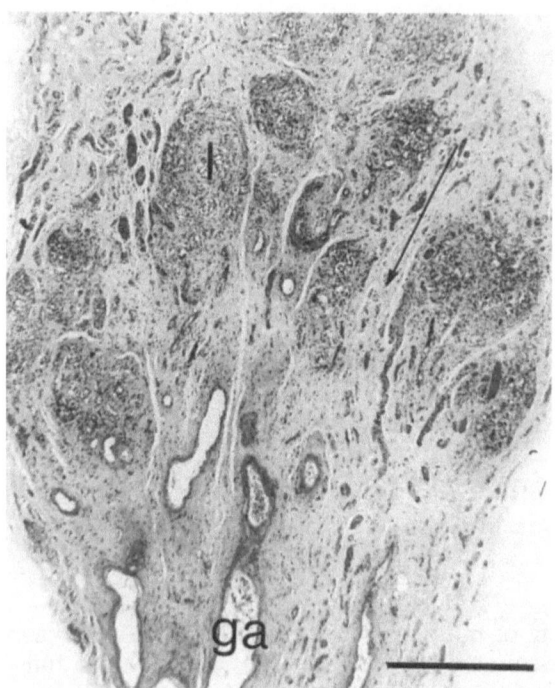

Fig. 3.2. Vertical section through the carotid body of an 80-year-old woman. Lobules (*l*) are of different sizes, irregular in outline and separated by broad bands of fibrous stroma (*arrow*). The stroma is greater in amount than in the preceding figure and, at the outer limits of the carotid body, fuses with the connective tissue of the carotid bifurcation. Branches of the main glomic artery (*ga*) can be seen at the base of the organ. (HE) Scale line = 500 μm.

chemoreception. The main glomic artery, arising from the carotid bifurcation (see Chapter 16), enters the proximal pole of the carotid body and gives rise to several, large branches (Fig. 3.2). These give rise to smaller interlobular glomic arteries which approach the lobules to supply them. Small thin-walled glomic veins run between the lobules to drain into larger veins in the peripheral stroma. They are numerous in the distal pole and add to the histological appearance of intense vascularity at low magnification (Fig. 3.1). Another prominent feature is a prolific nervous supply. There are several large bundles of myelinated nerves in the peripheral stroma. From these, numerous branches extend between the lobules, appearing as many individual bundles as they wander in and out of the plane of section.

Structure of Lobules

Individual glomic lobules are usually round or oval in cross-section, but are occasionally elongated to

form cords of cells. They have a clearly defined border which facilitates measurement of their size. The average diameter of lobules in the adult is about 400 μm, but there is considerable variation so that the range of diameter, calculated from the 95% probability limits, is 250–565 μm (Smith et al. 1982). The configuration of the lobules as they are dispersed throughout the stroma varies according to the age of the subject. In children they are of similar size and are packed closely together, with only narrow bands of stroma between them (Fig. 3.1). With increasing age the lobules become spaced further apart so that in the elderly, who form the majority examined at post mortem, lobules are widely separated by thick bands of acellular collagen (Fig. 3.2). Lobules also become more irregular in outline and show a greater variation in size.

The internal structure of a glomic lobule is represented diagrammatically in Fig. 3.4. The functional units consist of numerous small groups of glomus cells which are referred to as cell

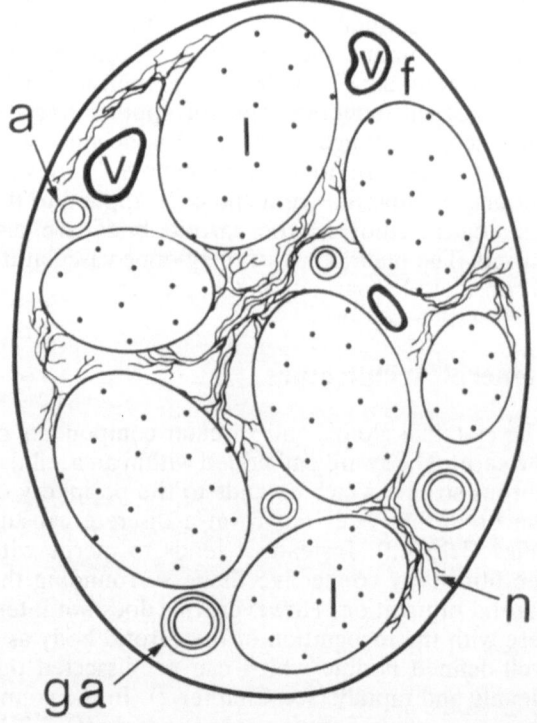

Fig. 3.3. Diagrammatic representation of the carotid body depicting its constituents. Glomic tissue, consisting of chief and sustentacular cells, axons and small blood vessels, forms discrete lobules (*l*). Between the lobules are numerous bundles of myelinated nerves (*n*) and interlobular glomic arteries (*a*). At the base of the carotid body are branches of the main glomic artery (*ga*) and at the apex thin-walled veins (*v*) into which drain interlobular venules. All these histological components are embedded in a largely acellular, fibrous stroma (*f*).

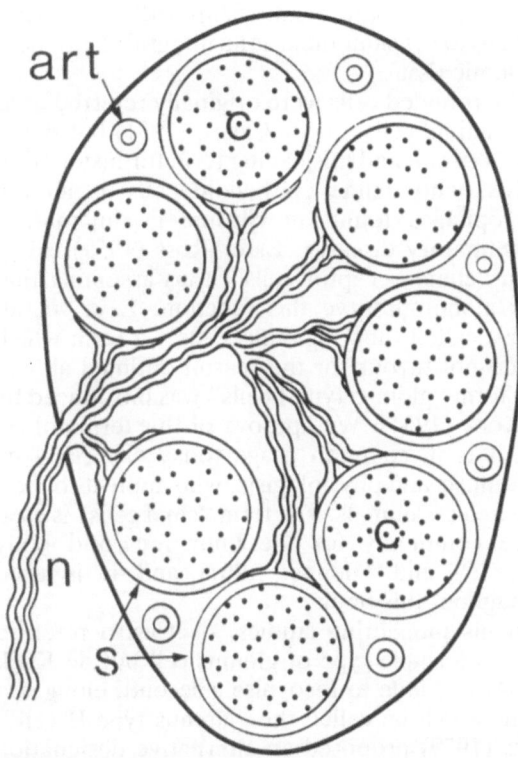

Fig. 3.4. Diagrammatic representation of a glomic lobule. The lobule is composed of round cellular clusters which consist of central cores of chief cells (*c*) surrounded by rims of circularly orientated, elongated, sustentacular cells (*s*). Myelinated nerves (*n*) enter the lobules and branch between the clusters, eventually ending as non-myelinated axons. These enter the clusters, often from one pole, and are taken up by sustentacular cells, which convey them to the central core. The lobule is supplied by minute intra-lobular glomic arterioles (*art*) which terminate as capillaries amongst the chief cells.

clusters. They are typically round in section and consist of central cores of rounded chief cells surrounded by thin rims of elongated sustentacular cells. The nomenclature and cytological features of both types of cell are described in detail below. Cell clusters are separated by thin, indistinct septa of connective tissue in which there are glomic arterioles and venules. Bundles of myelinated nerves can often be found entering the lobules and, in sections which have been cut fortuitously, can be seen to branch amongst the cell clusters (Figs. 3.4, 3.5). In this location the axons are invested by Schwann cells which thus contribute to the population of elongated cells within the lobule and may be mistaken for sustentacular cells.

Interpreting the histological appearance of a section of carotid body may be much more difficult than Fig. 3.4 suggests. Usually each lobule contains several distinct cell clusters in which cores of

chief cells and their surrounding rims of sustentacular cells can be readily made out (Fig. 3.6). In other regions of the same lobule, especially close to the point of entrance of a nerve bundle, the glomus cells may be more diffusely distributed. Sometimes the entire section shows an apparently diffuse organisation of cells due to the plane of section so that adjacent clusters appear to fuse together.

Although we regard the functional histological unit of the carotid body as the cell cluster, Seidl (1975), from studies on animals, preferred to subdivide the glomic tissue even further. She referred to the functional unit as a "glomoid", which consists of a single segment of a capillary with a small group of chief cells closely applied to it. Since each cluster contains several capillaries it follows that there must be a number of glomoids for each cluster. It is rarely possible to distinguish such entities on conventional sections cut from tissue embedded in paraffin-wax, although they can be identified occasionally on thin sections, cut from epoxy resin, which are used for electron

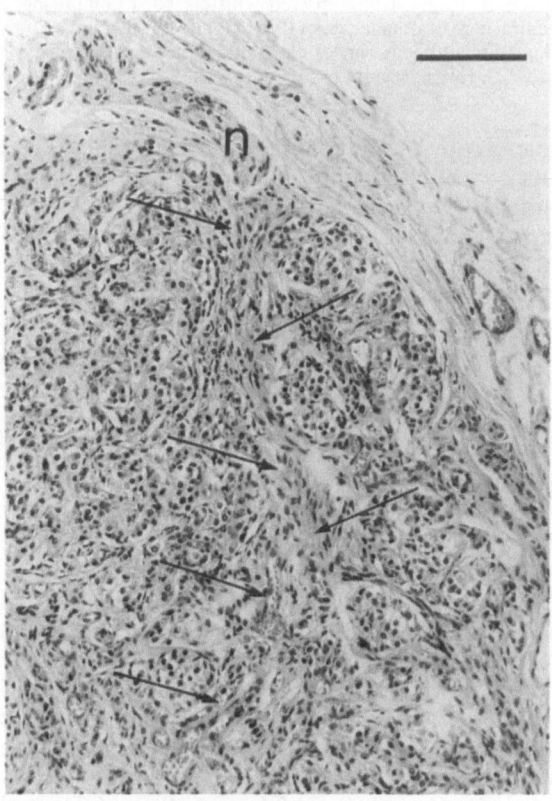

Fig. 3.5. Part of a lobule external to which is a small myelinated nerve (*n*). The nerve enters the lobule and branches amongst the cell clusters. Its course is indicated by arrows. (HE) Scale line = 100 μm.

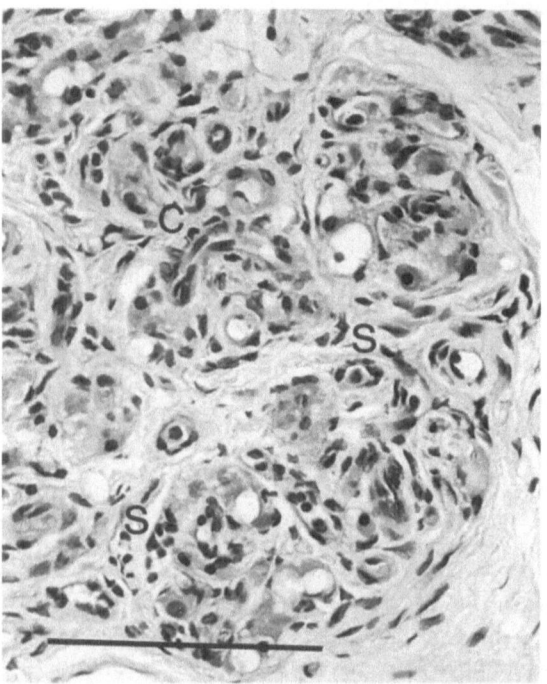

Fig. 3.6. Part of a lobule from a woman of 71 years. The glomic tissue is arranged into discrete clusters each comprising a central core of rounded chief (type I) cells (*c*) surrounded by a thin, discontinuous rim of elongated sustentacular (type II) cells (*s*). (HE) Scale line = 100 μm.

microscopy. The concept of the glomoid was devised by examining scanning electron micrographs of corrosion casts of the vasculature of carotid bodies of animals and by three-dimensional reconstruction from serial sections (Seidl 1975). Although such associations of capillaries and chief cells are of relevance to the physiology of the carotid body, they do not constitute a useful histological feature in man.

Nomenclature of Glomus Cells

In the preceding section we referred to the rounded and elongated cells in the clusters collectively as glomus cells. This is because both form an integral part of the functional units of the carotid "glomus" (see Chapter 1). This is not a universally accepted view and some authors would regard only the rounded cells with their biogenic amines and peptides as glomus cells. They believe that the elongated cells act in merely a supporting, accessory rôle (Grimley and Glenner 1967; McDonald and Mitchell 1975; Verna 1979). However, there is no proof that these elongated cells do not play a functional part in the action of the carotid body

and their intimate relationship with chief cells leads us to consider them as an integral component of glomic tissue.

The rounded cells were originally referred to as "epithelioid cells" by de Castro (1951) but this is an inaccurate and misleading term for histopathologists, who define epithelioid cells as altered macrophages found in chronic granulomatous inflammatory diseases. Later Ross (1959) called them "chemoreceptor cells", thus assuming that they alone subserve this function. Lever et al. (1959) called them "glomus cells", a term which we do not favour for the reason outlined above. The term "glomus type I cells" was introduced by de Kock (1954). We approve of this terminology and it is in common usage today especially by anatomists and physiologists, who apply it to their research on animals. The term "chief cells" is used more commonly for the human carotid body (Grimley and Glenner 1967) and it is used throughout this book.

In his pioneering studies, de Castro referred only to a single type of glomus cell but de Kock (1954) was able to recognise a second, elongated form, which he called the "glomus type II cell". Ross (1959) proposed an alternative designation of "sustentacular cells" because he felt that they closely resembled the supportive cells of the same name in certain sensory organs such as the organ of Corti, taste buds and olfactory mucosa. In these structures the sustentacular cells support both the receptor and the axons that contact it in the same way as they do in the carotid body. The elongated cells are commonly referred to by both of these terms in addition to "sheath" or "supporting cells", but in this book the term "sustentacular cells" is used in preference.

Chief Cells

In humans three distinct forms of chief cell can be recognised. These are the light, dark and progenitor variants and are located in the central cores of the clusters. There are no wholly satisfactory special staining techniques to demonstrate chief cells. The technique of formalin-induced fluorescence is a very sensitive method for demonstrating catechol- and indolamines in their cytoplasm but does not permit the distinction between the three variants. The same criticism can be levelled at argyrophil reactions, in which metallic silver is precipitated onto the cytoplasm where phenolic amines are present. The black or grey deposit of silver, although accurately identifying chief cells, renders their subdivision into three variants impossible. The technique of staining the cells with lead haematoxylin is a crude method for demonstrating polypeptide hormones (Solcia et al. 1969). This stains only a proportion of chief cells, particularly the dark variant. Immunohistochemistry is more precise: the chief cells are labelled with antisera to the neuropeptides leucine- or methionine-enkephalin (see Chapter 8), but this also tends to label only dark cells. We have found that the most information about the three variants of chief cells can be obtained simply by staining them with haematoxylin and eosin, producing appearances described below.

Light Variant

This is by far the commonest variant of chief cells and is so named because of the pale-staining reaction of its nucleus with haematoxylin and eosin (Fig. 3.7). The distinctive nucleus is pale and round or slightly oval in profile. It is larger than that of the other variants and measures, on average, 7 μm in diameter (Heath et al. 1970, 1984). The chromatin pattern consists of fine, branching strands enclosing large, apparently clear, areas which may sometimes resemble vacuoles within the nucleus. These strands frequently radiate from a central nucleolus. A few small clumps of chromatin may be scattered about the nucleus but are most numerous at the periphery. In a minority of

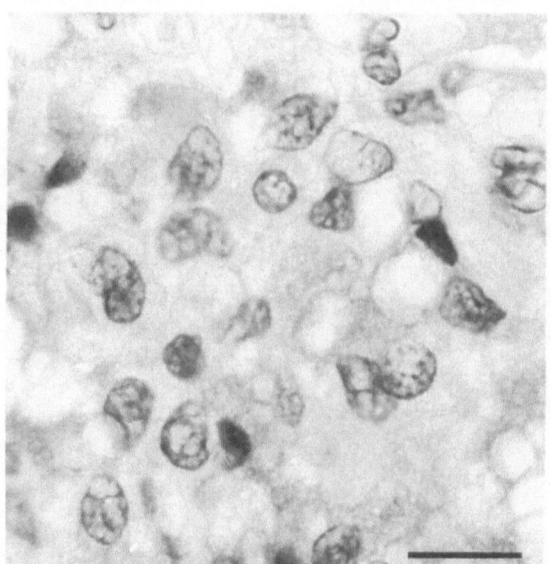

Fig. 3.7. Detail of a group of light chief cells from the core of a cluster. Their nuclei are pale and vesicular, with strands or fine dots of chromatin. The cytoplasm is pale, indistinct and highly vacuolated. (HE) Scale line = 10 μm.

cells the nucleus may lack this vesicular structure but present instead a finely stippled chromatin pattern. The cytoplasm of light cells is abundant and faintly eosinophilic and its borders are usually ill defined so that the appearance of a syncytium is commonly produced (Fig. 3.7). This is a spurious impression brought about by interdigitation of the cell borders and vacuolation of the cytoplasm. Of the three variants, the light cell shows by far the greatest degree of vacuolation, the vacuoles being irregular in shape and sometimes as large as the nucleus. On electron microscopy they appear as large "holes" in the cytoplasm, with irregular borders often associated with disruption of internal and limiting membranes, suggesting autolytic degradation. This variant is thus particularly susceptible to post-mortem autolysis and this should be borne in mind when assessing its histology.

Dark Variant

The nucleus of this cell has an average diameter of 6 μm, which is slightly smaller than that of the light variety. Individual dark cells may occasionally be found in which the nucleus is considerably bigger. The nuclear chromatin pattern consists of numerous small clumps, distributed evenly, so that it has a coarsely granular appearance. Pale vesicular areas may be found in the nucleus but they are small and insignificant in contrast to the light variety. In some dark cells the clumps of chromatin may be packed so closely as to produce an almost homogeneous appearance (Fig. 3.8). The nucleus is conspicuously more haematoxyphilic than that of light cells and this is the main criterion for the designation "dark cell". The cytoplasm is also darker and stains purple with haematoxylin and eosin. It has a more clearly defined border (Fig. 3.8) which is made more conspicuous by its sharp contrast with the surrounding light cells. The cytoplasm commonly has an oval profile but may be elongated, with the formation of streamers. Vacuoles are commonly present but these are small and may be overlooked unless specifically looked for at high magnification. In the normal adult, dark cells are heavily outnumbered by the light variants but, in children and certain pathological states to be described later in this book, dark cells may be more prominent (see Chapters 5 and 9).

The dark cell is not accepted as a distinct entity by McDonald (1981), who regards staining intensity to be an unreliable way of distinguishing variants of chief cell in the rat. We agree that in the rat there is a smooth transition in the intensity of

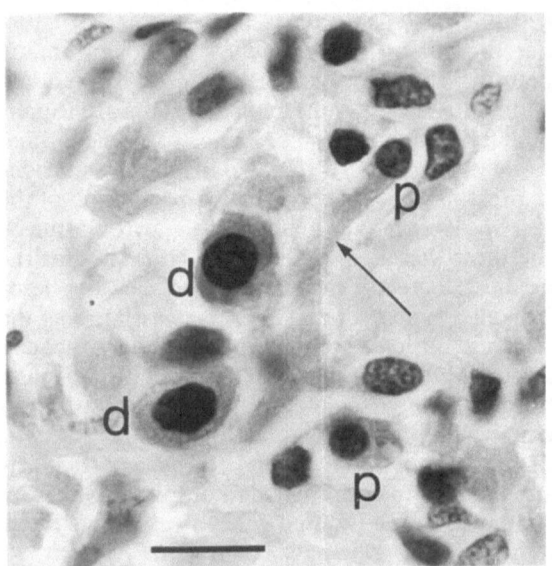

Fig. 3.8. Dark cells (*d*) contain haematoxyphilic nuclei and dark, clearly defined cytoplasm. Progenitor cells (*p*) have small, dark nuclei and less copious cytoplasm of a staining intensity similar to that of dark cells. A cytoplasmic streamer extends from one progenitor cell (*arrow*). (HE) Scale line = 10 *μ*m.

the staining of the chief cells and all appear to be of a single type (Smith et al. 1984). This is in contrast to humans, where a sharply defined distinction can be drawn between the dark and light variants. Most importantly this distinction can also be drawn at the ultrastructural level, implying some functional difference between the two (see Chapter 17).

Progenitor Variant

The nucleus of this variant is small, measuring 4 *μ*m in diameter, and it is intensely, uniformly haematoxyphilic. It is usually situated to one side of the purple-staining cytoplasm, which is less voluminous than in the dark variant. The outline of the cells is usually oval but may occasionally form streamers and is clearly defined (Fig. 3.8). Cytoplasmic vacuoles, when present, are small and inconspicuous.

The application of the term "progenitor cell" to this variant is of recent origin (Heath et al. 1990). This variant was originally designated the pyknotic cell because of its small dark nucleus resembling that of pyknosis (Heath et al. 1970), but the term is misleading. Electron microscopic studies revealed that the cell was far from effete, since it contained a profusion of dense-core vesicles and mitochondria (Jago et al. 1984). Accordingly a more accurate and functional term was required

(Smith 1986). It subsequently transpired from a histological study on the carotid bodies of neonates that this variant is an immature form of chief cell that probably gives rise to the other two forms (Heath et al. 1990). In view of this the alternative designation "progenitor cell" is proposed and is used throughout this book.

Sustentacular and Schwann Cells

Sustentacular cells are elongated, or occasionally triangular, and are restricted in the main to the periphery of the clusters (Figs. 3.6, 3.9), although a few may be located deeper within the core of chief cells. Fine cytoplasmic extensions, invisible on light microscopy, penetrate deep into the core of the cluster, where they encircle the chief cells. The size of their nuclei are on average 13 *μ*m × 4 *μ*m (Heath et al. 1970) but there is considerable variation in these dimensions. The nuclei, although generally dark, show appearances ranging from a stippled chromatin pattern to dense basophilia (Fig. 3.9). The cytoplasm is palely eosinophilic but is difficult to define, since it is so thin and merges with surrounding fibrous tissue. Identification of these calls is thus based on their nuclear characteristics and their location within the lobule.

Sustentacular cells may be confused with other types of elongated cell in the carotid body. Peri-

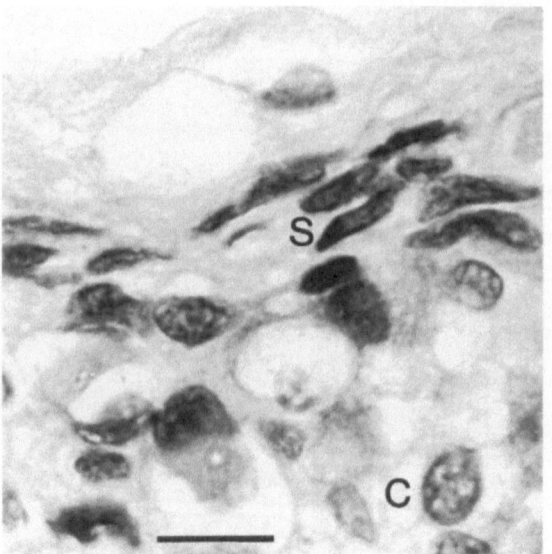

Fig. 3.9. A group of sustentacular cells (*s*) at the periphery of a cell cluster (*c*). Note that, although the elongated nuclei show variation in staining intensity, they are predominantly dark. (HE) Scale line = 10 *μ*m.

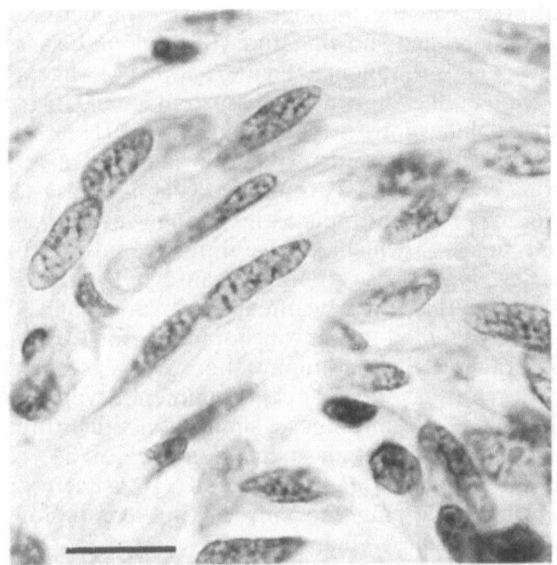

Fig. 3.10. A group of elongated cells in an interlobular septum. These have plump nuclei with a stippled chromatin pattern and are not sustentacular cells but Schwann cells associated with nerve axons. (HE) Scale line = 10 μm.

cytes, which enfold capillaries, also have elongated dark nuclei but here the nucleus is smaller, uniformly basophilic and crescentic in shape, due to curving around its associated blood vessel. Fibrocytes may be difficult to distinguish from the darker, more elongated sustentacular cells. Indeed at the edge of lobules both types of cell merge imperceptibly in the interlobular septa and the boundary between the sustentacular cells and fibrocytes cannot be determined with certainty.

Sustentacular cells may be confused with Schwann cells, which also have elongated nuclei with a stippled chromatin pattern (Fig. 3.10). Although the nuclei of Schwann cells tend to be slightly plumper and paler than those of sustentacular cells, there is some overlap in their histological appearances. Confusion between the two types of cell does not occur where Schwann cells are found in the large myelinated nerves in the stroma, but at the point of entrance of nerves into the lobules. In the latter situation, Schwann cells are arranged as parallel rows in continuity with the nerve bundles in the interlobular septum (Figs. 3.5, 3.10). As the nerves branch within the lobule, their Schwann cells become closely apposed to the sustentacular cells at the periphery of each cluster. There comes a point where it is impossible to determine where Schwann cells end and sustentacular cells begin. From a practical standpoint a precise distinction between the two cells is probably not important, since both support nerve

axons. Schwann cells convey the axons from the nerve bundles to the periphery of cell clusters, where they are taken up by sustentacular cells, which in turn convey them into intimate synaptic contact with the chief cells. Both types of cell are, therefore, analogous (Ross 1959) and, since they are both labelled by antibodies to S-100, an acidic calcium-binding protein found in glial cells, they are also glial in nature (Kondo et al. 1982).

Ganglion Cells and Nerves

Ganglia of the autonomic nervous system are situated close to the carotid body but are excluded from all sections in which the carotid body has been dissected free from adherent connective tissue. Nevertheless, several ganglion cells, either singly or in small groups, are situated within the connective tissue surrounding the carotid body near to its connection with the glossopharyngeal nerve (Pryse-Davies et al. 1964). Hence they are more likely to be encountered in incompletely dissected organs. Ganglion cells in the carotid body are identical in appearance to those in autonomic ganglia. They are several times larger than chief cells, with large, round, vesicular nuclei, and their copious cytoplasm is finely granular and basophilic. Ganglion cells in the carotid body are believed to innervate its blood vessels and hence regulate vascular tone (McDonald and Mitchell 1975).

Large bundles of myelinated nerves are prominent within the connective tissue surrounding the carotid body or within the septa separating the lobules (Fig. 3.3). Soon after entering the lobules they become unmyelinated and the axons form a mesaxonal relationship to Schwann cells. The small nerves which supply the cell clusters are invisible on sections stained with haematoxylin and eosin and can be demonstrated to best advantage by employing an indirect immunofluorescence technique to thick cryostat sections. They can also be demonstrated on conventional, paraffin-wax-embedded sections by impregnation with silver. We have found that the classic silver protargol method of Bodian (1936) gives good results. It demonstrates, at the edges of the clusters, axons which can be seen to form a plexus of wavy black lines which undulate in and out of the plane of section (Fig. 3.11). From these peripheral nerves, several axons weave an irregular course into the centre of the cluster, often from one pole (Fig. 3.11). They terminate on chief cells as a variety of nerve endings such as small boutons, discs, menisci or calyces (Eyzaguirre and Gallego 1975;

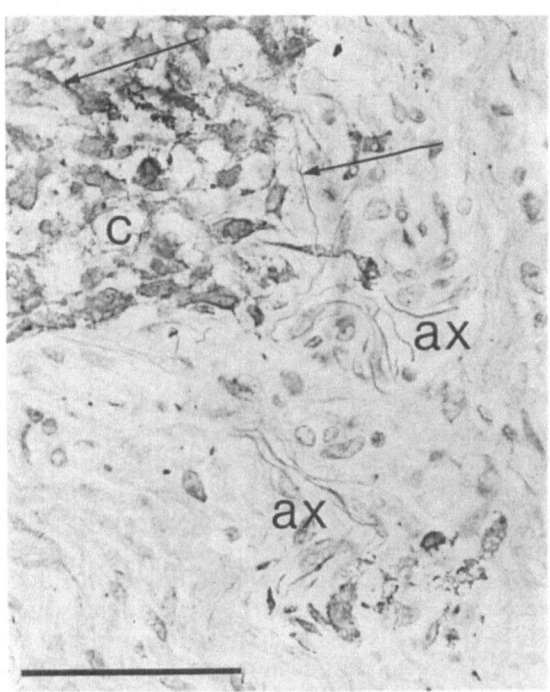

Fig. 3.11. Section of a carotid body impregnated with silver to demonstrate axons, which appear as wavy, black lines (*ax*). They are close to two clusters of chief cells (*c*), and at two points can be seen to enter amongst them (*arrows*). (Bodian's silver protargol method). Scale line = 50 μm.

Eyzaguirre and Fidone 1980). A single nerve may branch several times to supply more than one cell cluster, or a single chief cell may be supplied by more than one nerve ending. Sometimes a lateral branch may arise from a nerve ending to contact an adjacent chief cell. The nature of the nerve axons cannot be inferred from the structure of their endings, since a single axon may branch to produce endings of different types (Eyzaguirre and Fidone 1980).

Mast Cells

Mast cells are common in the carotid body but in the main are confined to the interlobular connective tissue. They are frequently associated with blood vessels. Very few are found in the glomic lobules and these are not associated with the cores of chief cells but with the shells of sustentacular cells. The population of mast cells has been quantified and found to have a broad variation ranging from 18 to 68 cells/mm² as calculated from the 80% confidence limits (Heath et al. 1987).

The number of mast cells bears no relationship to the histology of the carotid body. There is no difference in the number of mast cells between carotid bodies showing the youthful pattern of dark cell prominence (see Chapter 4), ageing carotid bodies showing many sustentacular cells or the acellular fibrosis typical of the aged (Heath et al. 1987). Furthermore, hyperplasia of the carotid body secondary to chronic hypoxaemia or systemic hypertension induces no change in the mast cell density, which is also not related to the underlying disease or to the cause of death. To illustrate this, three women, aged 79, 80 and 93 years, who died from bronchopneumonia with no other underlying pathology, showed a histological pattern of sustentacular cell prominence typical of the aged. Despite these similarities, the three showed disparate mast cell densities of 34, 83 and 8 cells/mm² (Heath et al. 1987). Mast cells do not appear to play a specific rôle in the pathology of the glomic parenchyma.

Tissue Culture

Glomus cells can be grown in culture following dissociation of the tissue by enzymic digestion. In the first attempt to achieve viable cultures, the carotid bodies of young rabbits were dissociated in a mixture of 0.025% trypsin, 0.025% collagenase and 0.05% hyaluronidase in phosphate-buffered saline for 1 hour (Pietruschka 1974). The carotid bodies of cats required twice the concentration of enzymes. Subsequently, the carotid bodies of fetal rats were dissociated in a mixture of 0.1% trypsin, 0.01% DNAse and 0.1% collagenase in calcium- and magnesium-free saline for 30 minutes (Fishman and Schaffner 1984). Although the cells will grow in minimal essential medium enriched with 20% (v/v) fetal calf serum (Pietruschka 1974), Ham's medium F-12 promotes longer life of the cells for up to 7 weeks (Fishman and Schaffner 1984).

Both of the above dissociation techniques do not completely disperse the carotid body into single cells, hence the cultures consist of both isolated cells and cell clusters. Glomus cells spread outwards from the clusters whereupon two distinct forms can be recognised. Chief cells adopt what is termed an "epithelioid configuration", in which the cells are rounded, flattened and send out short cytoplasmic processes. Their nuclei are large, oval and flattened with one to three prominent nucleoli (Pietruschka 1974). The cells maintain close contact with their neighbours and spread out on the substratum to form irregular, flattened sheets. Other cells adopt a "fibroblastoid configuration" and are elongated. Many of these migrate away from the main mass of cells and then send out broad frill-like pseudopodia typical of fibroblasts. Others remain bipolar, do not migrate and may be sustentacular cells (Fishman and Schaffner 1984). Chief cells in culture retain their ability to display formalin-induced fluorescence (Pietruschka 1974) and can be seen on electron microscopy to contain dense-core vesicles (Pietruschka and Schäfer 1976; Fishman and Schaffner 1984). They are thus able to synthesise catecholamines and their adrenergic character can also be demonstrated by an ability to grow in medium devoid of tyrosine. Unlike other adrenergic cells of neural crest origin, such as the small intensely fluorescent cells of the sympathetic ganglia, glomus cells in culture can survive in the absence of nerve growth factor (Fishman and Schaffner 1984). They can also undergo mitosis, as demonstrated by the use of tritiated thymidine to label DNA synthesis (Fishman and Schaffner 1984). In this connection it is of interest that mitotic figures can be demonstrated in chief cells of the carotid bodies in human fetuses, neonates and infants and in chronically hypoxic rats (see Chapter 5).

Quantitative Histology

Since the size of the lobules and population of chief and sustentacular cells changes when the carotid body undergoes hyperplasia, it is useful to have some numerical assessment of them to act as a yardstick against which abnormal carotid bodies can be compared. Glomic lobules have been measured in two ways. In the first, the total area of histological sections occupied by glomic lobules is measured either by point counting or by computerised image analysis (Lack et al. 1985). This produces what has been termed the "area of functional parenchyma" (Arias-Stella and Valcarcel 1973; Lack et al. 1985). The increase in this area has been traced from birth to the age of 36 years and has been found to increase from around 5×10^5 to nearly 25×10^5 μm^2 between those ages (Lack et al. 1985). The greatest change in the functional surface area occurred after the age of 3 years, as did that of the surface area of the whole carotid body.

The problem with such measurements is that they tend to vary depending upon whether the section is cut at right angles or parallel to the long axis of the carotid body and whether or not it includes its widest diameter. This can be overcome by expressing the functional area as percentage of the total area (Lack et al. 1985; Smith et al. 1987). However, since the individual lobules enlarge when the carotid body becomes hyperplastic, a quicker method of assessment is simply to measure their diameter (Smith et al. 1982). This can be estimated by measuring the largest and smallest dimension of each lobule and calculating the average diameter. The mean value of all lobules in the section provides a reasonable estimate of lobule diameter. In 57 normal adult carotid bodies, we found the mean value for lobule diameter to be 411 μm (Smith et al. 1982; see Table 3.1).

Table 3.1. Normal values of the carotid body from 57 adults (after Smith et al. 1982)

Parameter	Mean	Range of 95% probability	
		Lower	Upper
Combined weight (mg)	18	7	30
Diameter of lobules (μm)	411	258	565
Diameter of cell clusters (μm)	82	54	110
Sustentacular cell count (%)	39	31	47
Light cell count (%)	54	—	—
Dark cell count (%)	5	—	—
Progenitor cell count (%)	2	—	—

The diameters of cell clusters can be measured in a similar fashion. This measurement is not so straightforward as that for the lobules, since the elongated cells surrounding one cell cluster may merge with those from a contiguous one. The diameter of the cluster can be measured if its border is taken as the outermost layer of elongated cells that are arranged concentrically to the central core. On this basis the average diameter of clusters is 82 μm (Table 3.1).

The proportions of glomic cells present in the carotid body can be obtained by performing cell counts in which all the components of glomic tissue are counted within randomly selected high power fields. This procedure is facilitated if the eyepiece contains a graticule divided into 40–50 squares to subsample each area. Three high-power fields will provide a total of 600–1000 cells, which is an adequate number for statistical analysis. If the individual count for each type of cell is expressed as a percentage of the total, a differential cell count will be produced. A drawback to this method of expressing the results is that it provides only the *relative* proportions of glomic cells without indicating whether or not they have proliferated. An absolute count can be obtained by dividing the individual counts for each type of cell by the area of section sampled. This provides an index of cell density based upon the number of nuclei per unit area. If this value is then converted into the number of cells per unit volume and then multiplied by the total volume of glomic tissue, calculated by applying Simpson's rule to consecutive sections (see Chapter 2), the total number of each type of glomus cell in the entire carotid body will be arrived at. Clearly such measurements are very tedious and time-consuming and indeed have never been performed on the human carotid body. Although it is the only unequivocal method of demonstrating the presence or absence of hyperplasia, it is likely to be undertaken only by the most devoted research worker and even then there would be time to measure only a few carotid bodies. The differential cell count, although far from ideal, is simple and quick and permits measurement of large series of carotid bodies. It is the method which we would recommend to practising pathologists, provided that they are aware that such counts do not prove strictly whether or not there is hyperplasia but merely indicate a redistribution of cell types. The differential cell counts on 57 subjects with normal carotid bodies are shown in Table 3.1.

Such quantitative data have permitted us to give a numerical definition to the carotid body. By calculating the upper 95% confidence limits we can quote approximate values for the upper limits of normal for various parameters. Thus we define a normal carotid body as having a combined weight of less than 30 mg, a lobule diameter less than 565 μm, cell clusters less than 110 μm in diameter and a sustentacular cell count less than 47% (Table 3.1; Smith et al. 1982). It must be stressed that these values were obtained from adults. The proportions of glomus cells, particularly the dark variant, are quite different in neonates and children (see Chapter 5).

References

Arias-Stella J, Valcarcel J (1973) The human carotid body at high altitudes. Pathol Microbiol 39:292–297

Bodian D (1936) A new method for staining nerve fibres and nerve endings in mounted paraffin sections. Anat Rec 65:89–97

de Castro F (1951) Sur la structure de la synapse dans les chémorécepteurs: leur méchanisme d'excitation et rôle dans la circulation sanguine locale. Acta Physiol Scand 22:14–43

de Kock LL (1954) The intra-glomerular tissues of the carotid body. Acta Anat 21:101–116

Eyzaguirre C, Fidone SJ (1980) Transduction mechanisms in carotid body: glomus cells, putative neurotransmitters, and nerve endings. Am J Physiol 239:C135–C152

Eyzaguirre C, Gallego A (1975) An examination of de Castro's original slides. In: Purves MJ (ed) The peripheral arterial chemoreceptors. Cambridge University Press, London, pp 1–23

Fishman MC, Schaffner AE (1984) Carotid body cell culture

and selective growth of glomus cells. Am J Physiol 246:C106–C113

Grimley PM, Glenner GG (1967) Histology and ultrastructure of carotid body paragangliomas. Comparison with the normal gland. Cancer 20:1473–1488

Heath D, Edwards C, Harris P (1970) Postmortem size and structure of the human carotid body. Thorax 25:129–140

Heath D, Smith P, Jago R (1984) Dark cell proliferation in carotid body hyperplasia. J Pathol 142:39–49

Heath D, Lowe P, Smith P (1987) Mast cells in the human carotid body. J Clin Pathol 40:9–12

Heath D, Khan Q, Smith P (1990) Histopathology of the carotid bodies in neonates and infants. Histopathology 17:511–520

Jago R, Smith P, Heath D (1984) Electron microscopy of carotid body hyperplasia. Arch Pathol Lab Med 108:717–722

Kondo H, Iwanaga T, Nakajima T (1982) Immunocytochemical study on the localization of neuron-specific enolase and S-100 protein in the carotid body of rats. Cell Tissue Res 227:291–295

Lack EE, Perez-Atayde AP, Young JB (1985) Carotid body hyperplasia in cystic fibrosis and cyanotic heart disease. A combined morphometric, ultrastructural and biochemical study. Am J Pathol 119:301–314

Lever JD, Lewis PR, Boyd JD (1959) Observations on the fine structure and histochemistry of the carotid body in the cat and the rabbit. J Anat 93:478–490

McDonald DM (1981) Peripheral chemoreceptors. Structure function relationships of the carotid body. In: Hornbein TF (ed) Regulation of breathing. Marcel Dekker, New York, pp 105–319

McDonald DM, Mitchell RA (1975) The innervation of glomus cells, ganglion cells and blood vessels in the rat carotid body: a quantitative ultrastructural analysis. J Neurocytol 4:177–230

Pietruschka F (1974) Cytochemical demonstration of catecholamines in cells of the carotid body in primary tissue culture. Cell Tissue Res 151:317–321

Pietruschka F, Schäfer D (1976) Fine structure of chemosensitive cells (glomus caroticum) in tissue culture. Cell Tissue Res 168:58–63

Pryse-Davies J, Dawson IMP, Westbury G (1964) Some morphologic, histochemical and chemical observations on chemodectomas and the normal carotid body, including a study of the chromaffin reaction and possible ganglion cell elements. Cancer 17:185–202

Ross LL (1959) Electron microscopic observations of the carotid body of the cat. J Biophys Biochem Cytol 6:253–262

Seidl E (1975) On the morphology of the vascular system of the carotid body of cat and rabbit and its relation to the glomus Type I cells. In: Purves MJ (ed) The peripheral arterial chemoreceptors. Cambridge University Press, London, pp 293–299

Smith P (1986) Electron microscopy of the abnormal carotid body. In: Heath D (ed) Aspects of hypoxia. Liverpool University Press, Liverpool, pp 77–96

Smith P, Jago R, Heath D (1982) Anatomical variation and quantitative histology of the normal and enlarged carotid body. J Pathol 137:287–304

Smith P, Jago R, Heath D (1984) Glomic cells and blood vessels in the hyperplastic carotid bodies of spontaneously hypertensive rats. Cardiovasc Res 18:471–482

Smith P, Greenberg S, Heath D (1987) Effects of glucocorticoids on the rabbit carotid body. Br J Exp Pathol 68:251–258

Solcia E, Capella C, Vassallo G (1969) Lead-haematoxylin as a stain for endocrine cells. Significance of stain and comparison with other selective methods. Histochemie 20:116–126

Verna A (1979) Ultrastructure of the carotid body in the mammals. Int Rev Cytol 60:271–330

4   Histological Changes Associated with Ageing of the Adult Carotid Body

The carotid body does not maintain a uniform histological appearance throughout adult life but shows changes related to ageing. In general terms there is a progressive loss of glomic substance and alteration of the structure, with a change in the proportion of its cytological components.

Area Occupied by Glomic Tissue

An assessment of the amount of glomic tissue in a carotid body may be made by applying methods of tissue morphometry to sections of the organ previously fixed in 10% (v/v) formalin, processed and embedded in paraffin-wax. The areas required may be rapidly measured by automatic image analysers, now possessed by most laboratories. They are the total area in mm^2 of the largest section taken through the mid-plane of the carotid body, and the area occupied by glomic tissue within it. In this way the percentage of the carotid

body occupied by glomic tissue can be assessed. At the same time the area of the largest lobule can be determined. The results of such a study by Hurst et al. (1985) are shown in Table 4.1, where the values expressed are the means of two sets of measurements obtained from both carotid bodies. Hurst et al. studied carotid bodies obtained at necropsy in 47 subjects free from cardiopulmonary disease and systemic hypertension, to ensure that no cases of carotid body hyperplasia were inadvertently included. For the same reason cases with right or left ventricular hypertrophy, as determined by the criteria of Fulton et al. (1952), were excluded. The subjects studied were divided arbitrarily into three groups of "young", "middle-aged" and "old" according to the age ranges shown in Table 4.1.

It will be seen that the relative proportion of the carotid bodies made up by glomic tissue decreases with age. The mean percentage of glomic tissue in the carotid bodies of young persons up to the age

Table 4.1. Means of the total area of carotid body, percentage glomic tissue, and area of the largest lobule in 47 adults grouped according to their age (after Hurst et al. 1985)

Age group	Number of cases	Age (years)		Mean total area of carotid body (mm^2)	Mean % glomic tissue	Mean area largest lobule (mm^2)
		Range	Mean			
Young	11	14–40	26	2.71	45	0.31
Middle-aged	14	41–65	52	3.12	39	0.35
Old	22	66–100	79	4.42	29	0.24

of 40 years is 45%. By the age of 66 years there has been a loss of over a third of the amount of functioning glomic tissue, and its percentage falls to 29% (Table 4.1). At the same time there is a fall in the mean area of the largest lobules of glomic tissue. The mean total area of the carotid bodies increases throughout adult life due mainly to an increase in the connective tissue component.

Histological Appearances and Age

In general terms the carotid bodies of the young are characterised by prominence of dark cells. With middle age, proliferation of sustentacular cells occurs around the clusters of chief cells. In the elderly and aged, fibrosis of the carotid body develops and this is associated with occlusive fibrosis of the glomic arteries. From the age of about 50 years a diffuse infiltration of lymphocytes is found. Each histological appearance may be related to age and to the percentage of glomic tissue in the 47 subjects studied by Hurst et al. (1985) (Table 4.2). The same series of measurements was calculated for each group delineated on the basis of histology and these are shown in Table 4.2. The subjects showing a normal histological appearance of their carotid bodies or one characterised by dark cell prominence have a higher mean percentage of glomic tissue than that found in older subjects with sustentacular cell proliferation and fibrosis, where there is a sharp reduction in percentage of glomic tissue. We may now consider the associations of each histological appearance in turn.

Dark Cell Prominence

Dark cell prominence is a feature of the histology of the carotid bodies of young adults below the age of 40 years. It is often associated with progenitor

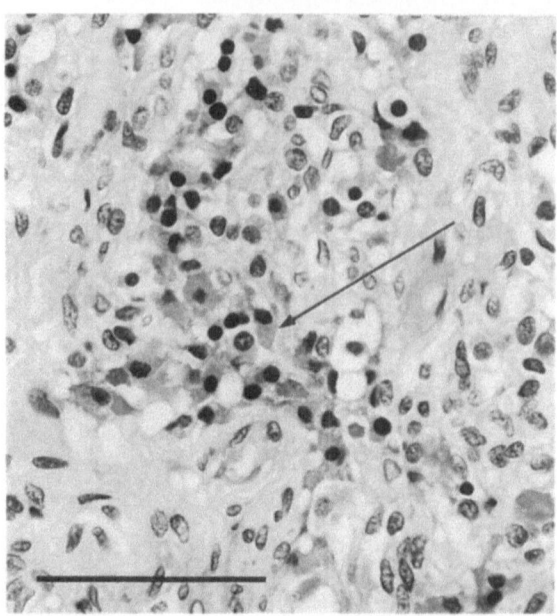

Fig. 4.1. Section of carotid body from a woman of 44 years who died from a coronary thrombosis. There is prominence of dark cells one of which is indicated by an *arrow*. The dark cell shown has a compact nucleus with concentrated heterochromatin situated eccentrically in surrounding ample cytoplasm containing small vesicles. Dark cells are characteristic of the young carotid body. (Haematoxylin–eosin (HE)) Scale line = 50 μm.

cells having dark compact nuclei. Dark cells have characteristic cytological features such as basophilic cytoplasm, often in strap-like cytoplasmic extensions, containing multiple small vesicles (Fig. 4.1; see Chapter 3). On occasion the dark cells spread beyond the confines of the clusters to occur as small groups outside the lobules. Dark cell prominence was found in no less than 8 of the 11 cases constituting the young age group in the study of Hurst et al. (1985). This is presumably indicative of a high functional activity for which the high content of such peptides as methionine- and leucine-enkephalins found in the dark variants of chief cells is required (see Chapter 8). This

Table 4.2. Means of the total area of carotid body, percentage glomic tissue, and area of the largest lobule in 47 subjects grouped according to the histological appearances of their carotid bodies (after Hurst et al. 1985)

Histological group	Number of cases	Age range (years)	Mean total area (mm^2)	Mean % glomic tissue	Mean area largest lobule (mm^2)
Basic	14	16–59	2.92	41	0.34
Dark cell prominence	10	14–96	3.35	40	0.28
Sustentacular cell proliferation	11	68–93	3.94	27	0.22
Fibrosis	12	46–100	4.43	32	0.33

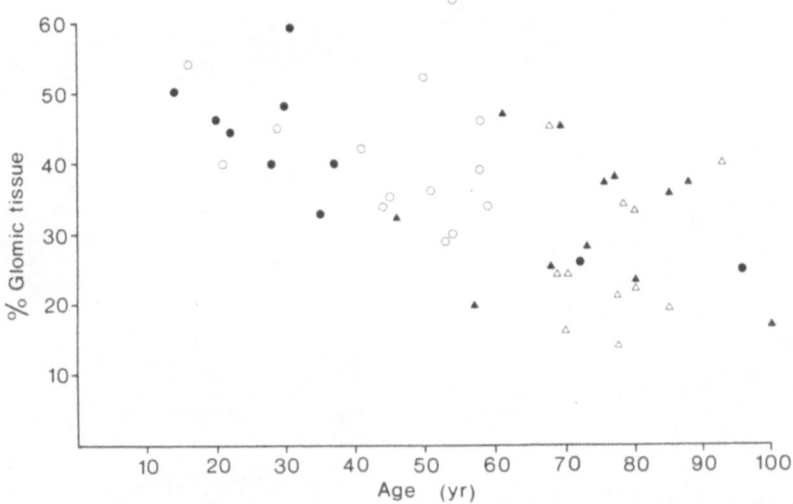

Fig. 4.2. Scatter diagram relating for each of 47 subjects the mean percentage of glomic tissue in a section of carotid body to the age in years. For each case the symbols indicate the histological appearance, as defined in Chapter 3 or in the text of this chapter. Basic histological pattern, ○; dark cell prominence, ●; sustentacular cell proliferation, △; fibrosis, ▲.

is not to say that occasional cases of dark cell prominence are not found in elderly subjects (Fig. 4.2). Under such circumstances it is advisable to investigate the possibility that the person concerned has not been exposed to an acute hypoxic episode which has stimulated into activity some of the dark cell population.

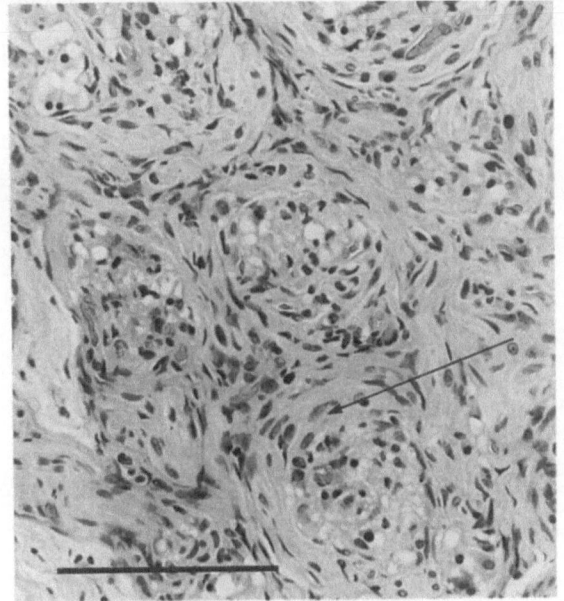

Fig. 4.3. Section of carotid body from a man of 63 years who died from bronchopneumonia. There is prominence of sustentacular cells (*arrow*) a characteristic feature of the carotid body of the elderly. (HE) Scale line = 100 μm.

Sustentacular Cell Proliferation

In the elderly over the age of 62 years there is a progressive proliferation of sustentacular cells around the cell clusters (Fig. 4.3). This progresses almost imperceptibly into fibrosis of the carotid body. In the 22 subjects over this age in the study of Hurst et al. (1985), 11 showed sustentacular cell proliferation and 9 showed fibrosis. When proliferation of sustentacular cells takes place, the normal well-defined lobular structure is lost in many instances, leading to a disorganised appearance of the glomic tissue.

Fibrosis

Commonly sustentacular cell proliferation progresses to unequivocal fibrosis: the plump elongated cells with long nuclei gradually become longer, thinner and more fibrous so that eventually they come to resemble fibrocytes (Fig. 4.4). Fibrous tissue extends inwards from the periphery of the lobules to compress and obliterate the chief cells so that finally only isolated clusters of chief cells, or none at all, remain in the lobules, now converted into whorls of sustentacular cells or fibrous tissue. This leads to a loss of lobular integrity so that the histological appearances become disorganised, with small clusters of glomic tissue separated by larger areas of connective tissue. This is reflected by a decrease in the

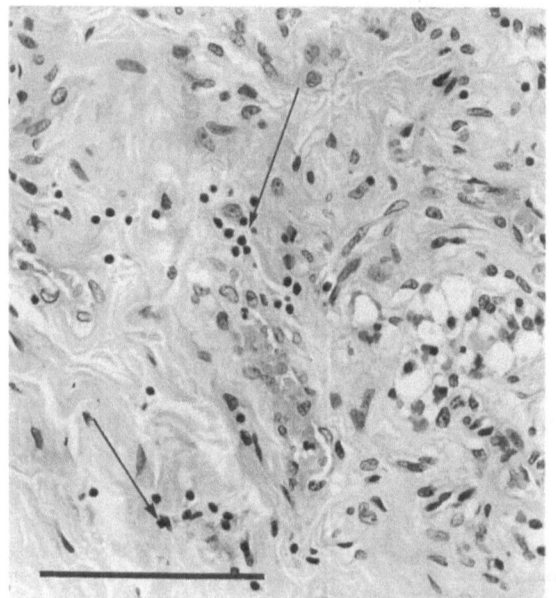

Fig. 4.4. Section of carotid body from an aged man of 87 years who died from carcinoma of the oesophagus. There is extensive loss of glomic tissue with replacement by fibrosis and a diffuse infiltrate of lymphocytes (*arrows*). (HE) Scale line = 80 μm.

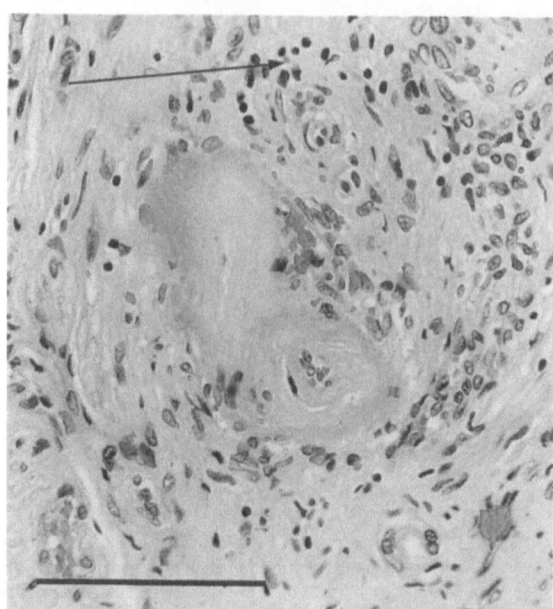

Fig. 4.5. Section of carotid body from an aged man of 73 years who died from hepatocellular carcinoma. In the centre of the figure is a fibrosed small glomic artery. Adjacent to this is a small, diffuse collection of lymphocytes (*arrow*). (HE) Scale line = 80 μm.

maximum size of lobules with increasing age (Table 4.1). In aged subjects the fibrous tissue not infrequently becomes brightly eosinophilic and amorphous or even hyaline in appearance. Areas of eosinophilia reminiscent of focal amyloid degeneration may be found.

Occlusive Changes in Glomic Arteries

With advancing age there is progressive blocking of the branches of the glomic arteries by non-specific intimal fibrosis (Fig. 4.5). In the elderly this may be so severe as to lead to virtual

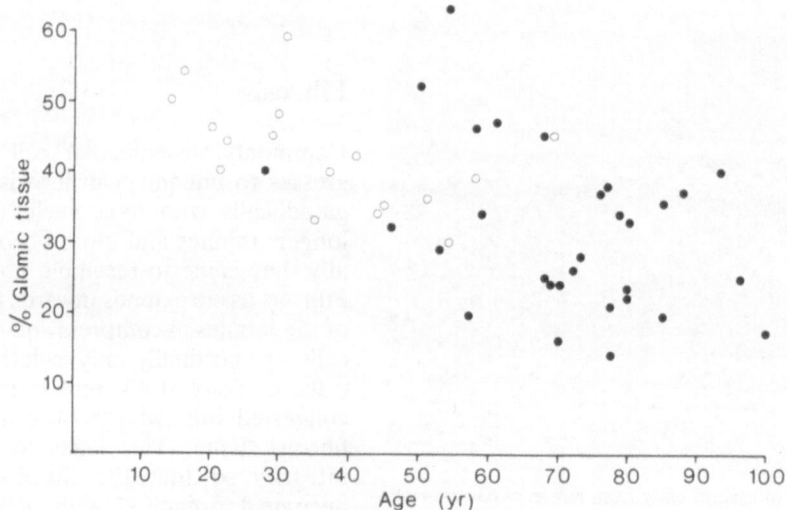

Fig. 4.6. The same diagram as in Fig. 4.2 with the modification that the symbols for each case indicate whether or not lympocytic infiltrates were present in the section. Lymphocytic infiltrate, ●; absence of same, ○.

occlusion. The carotid arterial tree is exceedingly prone to the development of atheroma, especially the lateral wall of the carotid sinus (see Chapter 20). The bifurcation and external carotid artery, from which the glomic artery or arteries arise (see Chapter 16), are much less severely affected, but they are to some extent. This atheroma of the carotid vessels is associated with non-specific intimal fibrosis in the interlobular glomic arteries. It is possible, and even likely, that the histological age-changes in the carotid body are not due to ageing of glomic tissue per se but to progressive ischaemia brought about by increasing blockage of the glomic arteries due to non-specific intimal fibrosis, itself due to age and atheroma of parent carotid arteries.

Lymphocytic Infiltrates

At about the age of 50 years a diffuse infiltration of lymphocytes appears in the carotid bodies and spreads throughout their connective tissues, extending to a varying extent into the glomic tissue (Fig. 4.4). They were found in no fewer than three-quarters of the 47 cases included in the study of Hurst et al. (1985). The relationship to age is very close. Lymphocytes were found in only 2 of the 15 cases below the age of 45 years but in 28 of 32 cases above that age (Fig. 4.6). In the elderly and aged the lymphocytes are found to form large aggregates throughout the glomic substance in a significant minority of subjects. In our opinion this constitutes a distinct disease entity rather than an age change. We have termed it "chronic carotid glomitis" and we consider it in detail in Chapter 6.

References

Fulton RM, Hutchinson EC, Morgan Jones A (1952) Ventricular weight in cardiac hypertrophy. Br Heart J 14:413–420
Hurst G, Heath D, Smith P (1985) Histological changes associated with ageing of the human carotid body. J Pathol 147:181–187

5 The Carotid Bodies in Fetuses, Neonates and Infants

The carotid body is laid down at the site of the third branchial arch very close to its artery, which will come to form the initial part of the internal carotid artery, but its tissues are not thought to be derived directly from those of the branchial arch. In other words the carotid body is branchiomeric rather than branchiogenic. Its primordium has been thought to make its appearance in the human embryo at about 6 weeks of gestation (12.5 mm crown-rump length) as a condensation of undifferentiated mesodermal cells forming a thickening in the adventitia of the branchial arch artery (Boyd 1937). These cells, however, have been considered to give rise to the stroma of the carotid body and not to its functioning glomus cells. That appears to be the rôle of neuroblasts that migrate down the glossopharyngeal nerve from the petrosal ganglion (De Winiwarter 1939). The cytological features of these fetal glomic cells are shown in Fig. 5.1. Using fluorescence microscopy, Korkala and Hervonen (1973) found a second source of neuroblasts in the form of a cord of catecholamine-containing cells that linked the carotid body and the superior cervical ganglion. It is likely that they are concerned with the development of ganglion rather than glomus cells within the carotid body. A different interpretation of the cells in the adventitia of the fetal carotid artery is that they represent the developing carotid sinus (see Chapter 20).

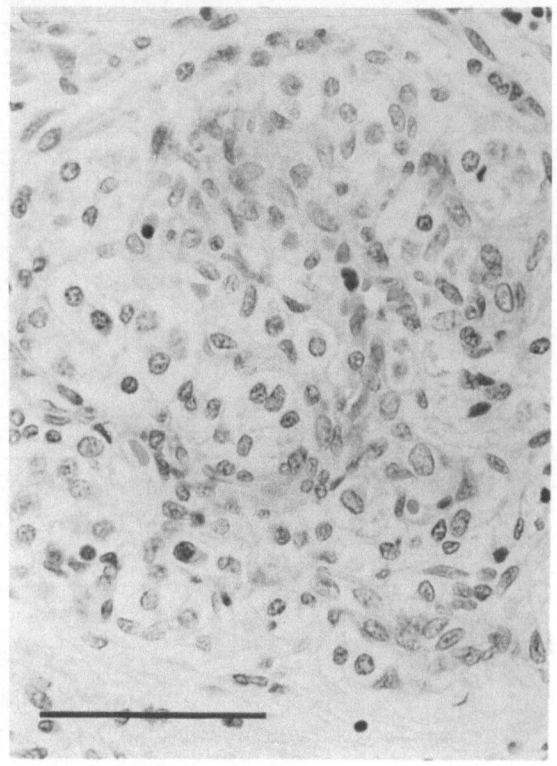

Fig. 5.1. Section of carotid body from a male fetus at 23 weeks of gestation. The lobule of glomic tissue is composed of primitive cells, as defined in the text. Some of the nuclei are elongated and reminiscent of plump sustentacular cells, perhaps consistent with the neurogenic origin of these primitive cells. Note the absence of sustentacular cells from the positions where they would be anticipated in a mature carotid body. (Haematoxylin–eosin (HE)) Scale line = 50 μm.

Table 5.1. Total volume of carotid bodies and glomic tissue in 100 neonates, infants and children including 40 dying as unexpected deaths at home (after Dinsdale et al. 1977)

Age range	No. of cases	Volume carotid body (mm³)			Volume glomic tissue (mm³)		
		Mean	S.D.	Range	Mean	S.D.	Range
29 wk gestation–23 h	17	0.30	0.04	0.07–0.69	0.08	0.05	0.02–0.20
1 day–2 wk	9	0.36	0.02	0.19–0.66	0.12	0.05	0.05–0.23
2–6 wk	9	0.45	0.18	0.25–0.72	0.14	0.05	0.07–0.21
6–12 wk	11	0.60	0.29	0.08–0.91	0.20	0.10	0.02–0.30
12–20 wk	17	0.75	0.48	0.09–2.22	0.27	0.19	0.03–0.81
20–52 wk	17	0.79	0.32	0.29–1.30	0.29	0.14	0.09–0.54
1–10 yr	20	1.26	0.61	0.29–3.23	0.47	0.31	0.09–1.23

Table 5.2. Dimensions of the human carotid body in the fetus, infant and child (after Sato 1932)

A. *Prenatal*

Gestation (months)	Dimensions (mm)		
2	0.45	0.32	0.18
2	0.23	0.28	0.18
3	0.35	0.29	0.13
3	0.57	0.60	0.21
4	0.72	0.51	0.15
5	0.47	0.40	0.36
5	0.62	0.47	0.29
7	0.86	0.92	0.56
8	0.81	0.60	0.48
8	0.98	1.04	0.64
8	0.87	0.75	0.69
9	0.91	0.80	0.52
9	0.42	0.57	0.48
10	1.12	1.22	0.29
10	0.96	1.17	0.70

B. *Infants*

Age	Dimensions (mm)		
1 day	1.05	0.88	0.64
3 days	0.89	1.30	0.59
7 months	0.96	1.02	0.72
10 months	0.97	1.44	0.62
13 months	1.01	1.28	0.88
28 months	1.47	1.28	0.86

C. *Children*

Age (yr)	Dimensions (mm)		
5	1.46	1.68	1.28
6	0.98	1.20	0.86
7	1.44	1.32	0.92
8	1.45	1.65	0.86

Size

The glomera of fetuses are too small to be dissected out for subsequent weighing so that recourse has to be made to serial sections and the application of tissue morphometry to determine their volume. Such methods were used by Dinsdale et al. (1977) to determine the volume of the carotid body as a whole and its component of glomic tissues in 100 neonates, infants and children between 29 weeks of gestation and 9½ years. They employed Simpson's rule (see Chapter 2) to determine the volume of each carotid body from serial sections. Point-counting was then employed to determine the volume of glomic tissue. The results of their study are shown in Table 5.1. There was a progressive rise with age in the mean value of the total volume of the carotid bodies from 0.3 mm³ at birth to 1.2 mm³ at 4½ years. Over the same age the total volume of glomic tissue contained within the carotid bodies rose from 0.08 mm³ to 0.47 mm³. There was a ten-fold variability in glomic tissue at any particular crown-rump length. Nevertheless, carotid body growth is related to the increase in size of the child from birth during infancy, with a glomic tissue volume of 0.02–0.2 mm³ at birth to around 0.1–0.5 mm³ at 10 months. From this age throughout childhood there is a very slow increase.

Sato (1932) provided data on the dimensions of the carotid bodies of fetuses, infants and children in Japan (Table 5.2). His measurements were made from serial sections of paraffin-embedded material. The implication of this is that the dimensions are smaller than those of the natural glomus, since the tissues are shrunken by fixation and processing in the laboratory.

Weight

Lack et al. (1986) reported the mean combined carotid body weight to be 2.0 mg in 19 cases averaging 1 week of age. It had risen to 2.6 mg in 14 cases averaging 2 months of age. In seven infants of average age 5 months the combined weight was 3.5 mg, and in five infants of average age 10 months it was 3.6 mg.

Histological Appearances

Kjaergaard (1973) studied the histological appearances of the carotid bodies in four human fetuses with a fertilisation age of 10–11 weeks and of crown-rump length of between 55 and 62 mm. Each carotid body was in the form of an oval cellular condensation in the mesenchyme, between the internal and external carotid arteries. The maximum transverse and craniocaudal diameters were 250 μm and 390 μm, respectively. Its cells formed strands and whorls around the vascular lumina. The surrounding fibroblasts were oriented concentrically with the surface. In the glomic parenchyma it was possible with some difficulty to distinguish chief cells and sustentacular cells. The chief cells had circular or oval nuclei measuring some 7 μm in diameter. The nucleoplasm was pale and the chromatin particles were delicate, evenly distributed and faintly staining. Eccentric nucleoli were found in contact with the peripheral condensations of chromatin. The cytoplasm was ample and pale and contained deeply staining small granules and a few larger vacuoles. The cytoplasm presented a cloudy appearance and cell borders were difficult to discern.

In the fetus of this age the sustentacular cells do not have the elongated shape characteristic of the adult carotid body but are plumper, with oblong or semilunar nuclei, which cling around chief cells. This accounts for the difficulty in distinguishing between chief and sustentacular cells in the fetal carotid body. The nuclear structure was darker than in the chief cells, the chromatin being assembled in coarser fragments. The cytoplasm was scanty. Pericytes with relatively dark nuclei were identifiable only by virtue of their paravascular situation. Nerve fibres could not be identified on histological examination in the parenchyma, although the sinus nerve and its smaller branches in the periglomeral mesenchyme were distinctly apparent. Lobulation of the parenchyma sets in during the 12th to 16th fetal weeks. The uniformity in histological appearance of these primitive cells is not total, for by 23 weeks of gestation well-defined and recognisable numbers of progenitor, dark and light variants of chief cells can be seen (Table 5.3). In fetuses of up to 28 weeks of gestational age the proportion of cells of fetal type may be as high as 80% but, thereafter, there is a decline in their number so that by a gestational age of 36 weeks the proportion falls to about 6% and after birth these fetal cells cannot be found.

Table 5.3. Representative differential cell counts in carotid bodies of fetuses, neonates and infants (after Heath et al. 1990)

Case no.	Diagnosis	Age[a]	Sex	% of cells				
				Fetal	Progenitor	Dark	Light	Sustentacular
1	Bilateral lung hypoplasia	23W G	M	65	3	<1	3	29
2	Hydrocephalus[b]	37W	F	12	2	5	11	65
3	Cardiopulmonary failure	Full	F	0	2	15	11	71
4	Hypoplasia of left ventricle and atrium	4H	M	0	5	1	14	80
5	Sudden infant death syndrome (SIDS)	9W	M	0	4	1	6	88
				Expressed as % of cell clusters:				
				0	17	7	15	61
				Expressed as % of chief cells in cores:				
				0	44	18	38	
6	Clinically diagnosed as SIDS but associated bronchiolitis	3½M	M	0	4	27	31	38
7	Bronchiolitis. Interstitial pneumonitis	6M	F	0	1	30	32	37
8	Fat embolism to lungs. Multiple fractures	11Y	M	0	3	24	33	40

[a] W, weeks; G, gestation; Full, full term; H, hours, M, months; Y, years.
[b] 5% plasma cells.

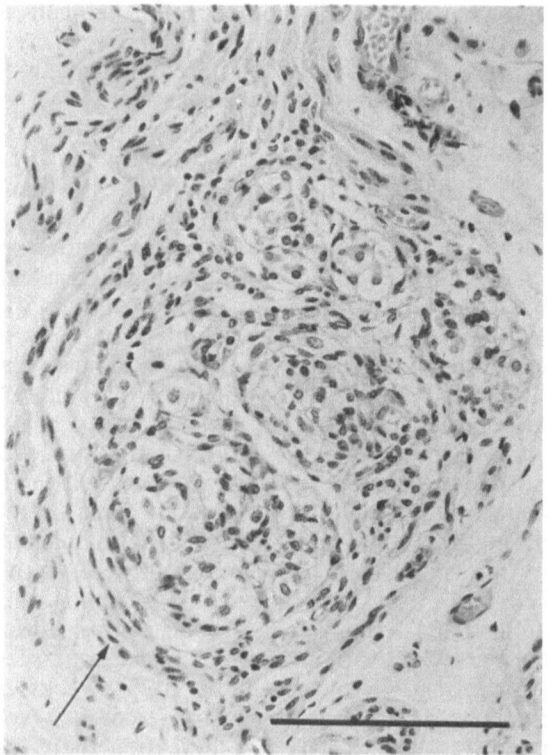

Fig. 5.2. Section of carotid body from a male fetus at 30 weeks of gestation, who had a single atrium and aortic hypoplasia. The glomic tissue is now mature with cellular variants. Sustentacular cells (*arrow*) are now in position, indicating the linking of the chief cells with radicles of the glossopharyngeal nerve. (HE) Scale line = 100 μm.

Ultrastructure of Fetal Carotid Body

In fetuses of 10–11 weeks of fertilisation age (crown-rump length 55–62 mm) the chief cells in three dimensions were rod- or spindle-shaped, with one or more processes (Kjaergaard 1973). The nuclei were oblong, with notches so deep as to render them kidney-shaped. Often the nucleus was situated eccentrically, with a narrow rim of cytoplasm on one aspect and a wider rim with ample organelles on the other side. Nucleoli were seen peripherally in the nucleus, often in relation to the peripheral chromatin. Some distinction could be made between light and dark cells. Granular endoplasmic reticulum was found to occupy only a small part of the cell volume, with the cisterns being situated peripherally parallel to the plasma membrane (Kjaergaard 1973). Shorter cisterns, small ribosome-lined vesicles and free ribosomes were found in the cytoplasm. Small vacuoles due to agranular endoplasmic reticulum or the Golgi complex were seen. Mitochondria were oval, rod- or cigar-shaped. Dense-core vesi-

cles were present in extremely varying numbers in the chief cells. They were unequivocally identifiable showing the centre of electron opaqueness and the surrounding clear halo (see Chapter 17). The mean diameter of dense-core vesicles was found to be 110 nm.

Sustentacular cells were two to three times less common than chief cells. The shape of their nuclei was very irregular. The shape of the sustentacular cells was difficult to analyse, with narrow extensions making their way between the chief cells. There were: small, uncharacteristic, mitochondria; short, flat, granular cisterns; and free ribosomes. Golgi zones, and lysosomes were occasionally observed. Nerve fibres were observed in large numbers between the cells. They had a pale axoplasm which contained longitudinal neurotubules about 20 nm thick.

Our own histological and electron microscopic studies of the carotid bodies in a 16-week fetus confirmed that the fetal cells are, in fact, composed of two distinct cell populations of chief and sustentacular cells. Kondo (1975) also reported similar findings from the light and electron microscopic study of the embryonic development of the carotid body of the rat. As early as the 11 mm embryo he found a cell aggregation of "undifferentiated cells" (corresponding to elongated sustentacular cells) with associated unmyelinated nerve fibres on the anterior wall of the third branchial artery. Cells containing dense-core vesicles appeared in the 12 mm embryo and continued to increase in number as the cellular aggregation increased in size and became separated from the wall of the artery.

The Carotid Body at Birth

After a gestational age of 30 weeks, nerves and bands of classic elongated sustentacular cells are to be found at the periphery of the clusters of chief cells (Fig. 5.2). Before these concentrically arranged type II cells appear, considerable numbers of mature sustentacular cells are to be found beyond the ill-defined lobules of fetal cells, which include developing sustentacular cells. The relatively high percentage of these cells may be explained to some extent by the smaller number of chief cells compared to the adult carotid body. In some instances the high percentage of sustentacular cells may also be partly due to autolysis of some of the chief cells from which eosinophilic masses of vesicular cytoplasm have been extruded. It is not yet established whether nerve axons derived from afferent branches of the glossopharyngeal are to be found enwrapped in either or both of the diffusely arranged sustentacular cells seen in fetuses and the bands of tightly packed sustentacular cells. It is desirable that this investigation be carried out in the future to determine at what stage glomic cells become linked with nerve axons to form functional units. Histological studies (Heath et al. 1990) have shown that the carotid chemoreceptor apparatus appears to be fully mature in most instances at birth, with the appearance of variants of chief cells including the progenitor and dark variants. The maturation of the chief cells and the decline of fetal cells is associated with a steady increase in mature sustentacular cells, reaching a value of about 70% at term (Table 5.3). These high sustentacular cell counts do not signify excessive proliferation of these cells but rather indicate that they are packed closely together around and inside the cell clusters.

As a result of these changes, the chemoreceptor cells now seem to be connected with their nerve supply and the different cytological components are now present to lead to the various histopathological appearances which may be found in the carotid bodies of neonates and infants. Studies have shown that in lambs also the carotid bodies are histologically mature but their function is inhibited up until birth. No carotid nerve activity follows stimulation in the fetal lamb, but immediately after birth normal responses are detectable (Barer et al. 1986). In human infants a reduction in inspired oxygen leads to a biphasic response, increased followed by depressed ventilation; the sleep state may be important in determining the response (Barer et al. 1986). The sustained adult reaction appears 1 to 3 weeks later.

Occasional mitotic figures are seen in chief cells in fetuses, neonates and infants (Fig. 5.3). At first this might seem surprising, for such cells are thought to have a neurogenic origin, as discussed in Chapter 1. Nevertheless, Barer and her colleagues (1986) found unequivocal evidence for cell division in chronically hypoxic rats which were given vincristine to arrest mitosis in metaphase. After a period of hypoxia of 24 to 48 hours, numerous mitoses were found in chief cells, whose identification was confirmed on light microscopy by their site and appearance and, in electron micrographs, by their content of dense-core vesicles.

Dark Cells in States of Hypoxaemia

Virtually no dark cells are found in the carotid bodies for the first 30 weeks of gestation but

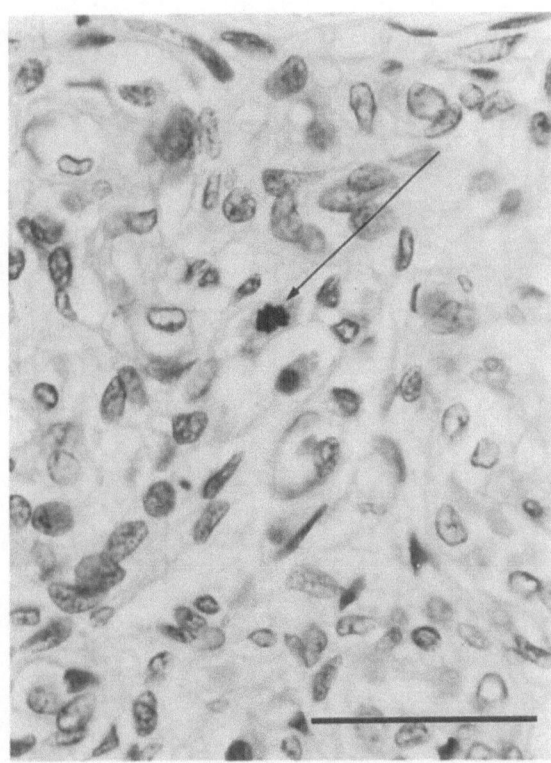

Fig. 5.3. Section of carotid body from a male infant of 7 months with pulmonary arterial hypoplasia. One of the chief cells contains a mitotic figure (*arrow*). (HE) Scale line = 30 μm.

thereafter their number increases rapidly to reach a count of up to 15% at full term (Table 5.3). The dark cell count remains high even up to the age of 1 year. However, in infants suffering from diseases associated with chronic hypoxaemic hypoxia the dark cells show striking prominence. In one case that we studied of a female infant of 6 months with acute bronchiolitis and interstitial pneumonitis the dark cell count was 30% (Table 5.3; Heath et al. 1990). In another case, a boy of 11 years with fat embolism to the lung complicating multiple fractures, the dark cell count was 24% (Table 5.3). In such cases the dark cells may be very striking (Fig. 5.4). The nucleus is dark due to a concentration of heterochromatin and it is commonly eccentric. The cytoplasm is voluminous and basophilic and frequently shows streamers. It contains many vesicles, particularly at its edge. Thus dark cell prominence is characteristic of the young healthy carotid body but it is exaggerated in states of chronic hypoxia. It seems likely that, in infants, dark cells respond to a hypoxic stimulus by forming cytoplasmic streamers packed with methionine- and leucine-enkephalins, although we do not

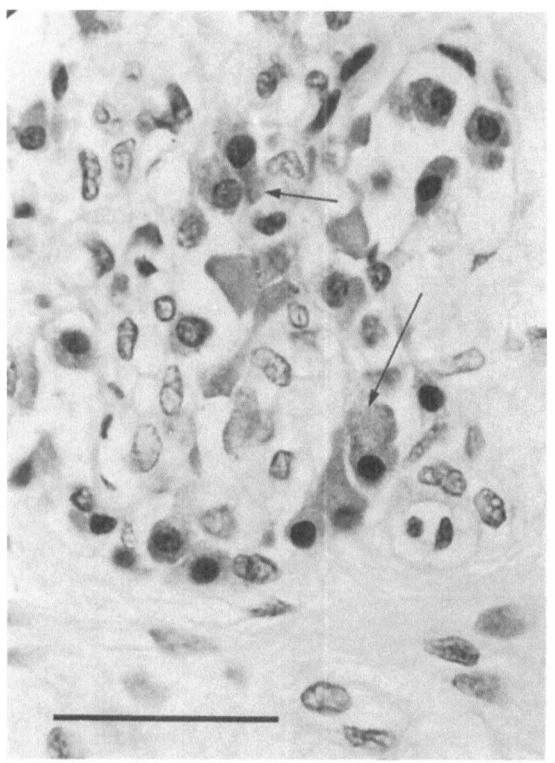

Fig. 5.4. Section of carotid body from a female infant of 6 months who died from acute bronchiolitis and interstitial pneumonitis. The glomic tissue shows prominent dark cells with eccentric nuclei and concentrated heterochromatin. The cytoplasm is voluminous and basophilic and is packed with small cysts. Characteristic streamers of cytoplasm (*arrows*) are present. (HE) Scale line = 30 μm.

know the physiological function of these peptides in states of hypoxia (see Chapter 8).

Enlargement of the carotid body and its lobules has been reported in young adults with cystic fibrosis. Lack and his colleagues (1985) studied the carotid bodies of 30 cases of cystic fibrosis and 17 cases of cyanotic congenital heart disease, occurring mainly in children. A total of 128 age-matched controls was also studied. There was enlargement of the carotid bodies of young adults with cystic fibrosis (Table 5.4). The enlarged carotid bodies were found on histological examination to show an increased number of lobules of greater diameter (Lack et al. 1985). It seems likely that these glomera would have shown dark cell prominence.

The progenitor cell count remains more or less constant at a low level at all stages of development. The light cell count increases from around 10% at birth to about 30% by 3½ months of postnatal life, after which it remains roughly constant (Heath et al. 1990; Table 5.3).

We have seen infiltration of the carotid bodies by plasma cells in a neonate with hydrocephalus who died from cardiac failure after surviving for 16 days. We have observed intense plasma cell activity in carotid bodies before but that was in an aged man of 81 years with chronic glomitis (Heath et al. 1989). In that case the process was considered to be an expression of autoimmunity with local production of antibodies in response to the antigenic stimulus of senescent nerve fibrils. This explanation cannot account for infiltration of the carotid bodies in a neonate at the other extreme of life.

Sudden Infant Death Syndrome

As mortality diminished in the first half of this century, increasing attention was paid to babies dying unexpectedly with no obvious disease to account for this. Much controversy has surrounded these "cot deaths". Unexpected suffocation may even lead to interrogation by the police and social stigma (Emery 1989). Bergman et al. (1970) came to the conclusion that all these cases were natural deaths and recommended that they should be registered as "the sudden infant death syndrome" and this approach became accepted throughout the world. However, 20 years later it is still being questioned by authorities in the field as to whether or not this is "a convenient diagnostic dustbin" (Emery 1989). As such deaths represent a failure of respiration, attention has been given to the possible involvement of the carotid bodies.

Clinical Features

By 1977 Naeye estimated that in the United States there were 8000 to 10 000 deaths each year being

Table 5.4. Combined carotid body weight in young adults with cystic fibrosis and age-matched controls (after Lack et al. 1985)

Age range in years	Controls		Cystic fibrosis	
	Average combined CB wt (mg)	S.D.	Average combined CB wt (mg)	No. of cases
18–21	7	1.8	26	6
21–24	11.3	7.1	32.9	5
24–36	12.7	4.9	22.2	11

CB, carotid body; S.D., standard deviation.

attributed to this syndrome rather confirming Emery's views that many causes of death were being included in this diagnosis. Naeye (1977) regarded sleep apnoea as the central problem. He thought that important epidemiological features were that the great majority of the infants die while they are supposedly asleep and most of these deaths are silent. The deaths start after the neonatal period, reaching peaks at $2\frac{1}{2}$ and 4 months. In some instances there are cardiac arrhythmias.

It is well established that episodes of sleep apnoea are associated with chronic alveolar hypoventilation (Guilleminault and Eldridge 1973). Nocturnal periodic breathing with sleep apnoea is characteristic of lowlanders ascending into high altitude (Heath and Williams 1989). Such incidents, accompanied by cyanosis, have been described in several cases of sudden infant death syndrome (Stevens 1965; Steinschneider 1972). There have been several reported instances of "near misses" in which infants have been resuscitated after an unexpected episode of prolonged apnoea. Such infants have an increased risk of death and should be treated until recurrent apnoea resolves (Kelly and Shannon 1982).

A great variety of clinical and social associations of the syndrome have been reported by many authors and these have been listed and discussed by Kelly and Shannon (1982). The syndrome occurs more commonly in the prematurely born and in the offspring of non-whites and those of low socioeconomic status. Sudden deaths during infancy occur more often in infants whose mothers are young, have poor prenatal care, illness or bleeding during pregnancy, are smokers, or users of methadone. However, it needs to be pointed out that such factors apply to explained as well as unexplained infant deaths. Infants who die from sudden infant death syndrome often have low birth weights and are products of multiple births. Deaths occur more often during the winter months, during which mild upper respiratory infections are common. The syndrome occurs more frequently in families which have already experienced a death so designated, but there is no convincing evidence to determine whether this is genetically transmitted or the result of altered environment prenatally or postnatally.

Morbid Anatomical Associations

These are several features which have been reported repeatedly in infants coming to necropsy and for whom the diagnosis of sudden infant death

syndrome had been arrived at. These include petechiae and muscularisation of the peripheral portions of the pulmonary arterial tree (Naeye 1973), which is characteristic of states of chronic hypoxia (Harris and Heath 1986). Such muscularised pulmonary arterioles lead to increased pulmonary vascular resistance which may lead to right ventricular hypertrophy in some cases of sudden infant death syndrome (Naeye et al. 1976a). Other morbid anatomical markers of hypoxaemia in the syndrome include extramedullary erythropoiesis, an abnormal retention of brown adipose tissue, an abnormal proliferation of astroglial fibres in the brain stem, and a hyperplasia of chromaffin tissue in the adrenal medulla.

Carotid Body Volume and Weight

Controversial as the subject of sudden infant death syndrome is, it would seem that the central factor concerned in mortality is a sudden and unexpected failure of respiration. This suggests that in at least some of the cases diagnosed as falling in the remit of this clinical diagnosis one might expect to find some abnormality of the carotid bodies which initiate and maintain ventilatory drive in the face of hypoxia. There have been several investigations of the chemoreceptors in this syndrome but they have shown a disappointing bias to establishing the volume of the carotid bodies rather than to study the histology of the glomic tissue in an attempt to detect any qualitative abnormality of it. Not surprisingly, this somewhat restricted approach has yielded few worthwhile results.

The volume of the glomic tissue in 40 infants who died unexpectedly at home was measured by Dinsdale et al. (1977). They found no increase in size of the carotid body in the majority of these deaths, although there was a probability that some enlargement occurred in some children dying at the age of 1 year or older. The subjects were included in three groups composed respectively of 21 children dying unexpectedly at home with gross acute disease, 18 showing minor acute disease, and 23 hospital deaths in the same age range. There was no significant difference between these three groups as regards glomic tissue volume against crown-rump length or age. Certainly there was no increase in size of the carotid bodies comparable to those found in chronic obstructive lung disease (see Chapter 10) or the native highlander (see Chapter 11). This is not surprising for one would not anticipate finding the enlargement of the carotid bodies with the associated histological changes of carotid body hyperplasia (see Chapter 10) or dark cell proliferation (see Chapter 9) found after prolonged exposure to hypoxia, in infants with sudden infant death syndrome. Instead one might expect to find a carotid body of normal size manifesting histological components that are unable to come to terms with the stimulus of an acute hypoxic insult.

The total volume of glomic tissue in the carotid bodies of subjects dying with a diagnosis of sudden infant death syndrome was also measured by Naeye et al. (1976b). They studied 17 males and 14 females classified as non-infected sudden infant death syndrome cases; the mean age was 2.4 ($\pm$ 0.2) months. Another 14 males and 11 females were considered to be infected sudden infant death syndrome cases; the mean age in this group was 3.4 ($\pm$ 0.2) months. Fourteen males and 12 females were considered as controls. Naeye et al. found that the ratio of glomic volume to body weight was significantly *smaller* in the non-infected sudden infant death syndrome victims than in the controls. Smaller differences were found between infected sudden infant death syndrome cases and the controls. The controls had 2.6 times as many carotid body glomic cells as did age-matched non-infected sudden infant death syndrome victims. However, it should be noted that once again Naeye and his colleagues (1976b) did not describe the histological appearances of the carotid bodies in the cases studied and did not give any indication as to whether they were normal. Overall, they found that the volume of the carotid body glomic tissue increased with age in the controls; that 63% of cases of sudden unexpected death had a subnormal volume and 23 % of cases had an enlarged volume of glomic tissue in the carotid body.

Lack and his colleagues (1986) studied combined carotid body weight in 38 cases of sudden infant death syndrome and in 51 age-matched controls. They found that there was a gradual increment in combined weight for both groups. Although the average combined weight for subjects with the syndrome was greater than controls in the 1–4 month age interval (mean = 3.4 mg, $p<0.075$) and the 4–8 month interval (mean = 5.0 mg, $p<0.098$) the differences were not statistically significant. Perrin et al. (1984) also found no gross abnormalities in the carotid bodies of ten cases of sudden infant death syndrome as compared to six controls and five cases with congenital heart disease. The percentage of glomic tissue was the same in the sudden infant death syndrome cases (36.4%) compared to the controls (41.0%).

The Histological Appearances of the Carotid Bodies in Sudden Infant Death Syndrome

Most workers have found no appreciable differences in architectural arrangement or cytological features between cases of the syndrome and age-matched controls (Perrin et al. 1984; Lack et al. 1986). However, Cole et al. (1979) studied the carotid bodies in six infants with sudden infant death syndrome and they came to the conclusion that many of what they termed the "chemoreceptor cells" in their cases were small compared to those of control infants. This small size suggested to them that some of the cells in the carotid bodies in cases of sudden infant death syndrome were immature and less likely to be able to cope with hypoxic stress. We shall return to this interesting suggestion below when we consider our own histological findings on this syndrome. On electron microscopic examination, Lack et al. (1986) and Perrin et al. (1984) found no difference in the ultrastructure of the carotid bodies of cases of sudden infant death syndrome from that of age-matched controls. Once again, however, Cole et al. (1979) found a reduction in the size and

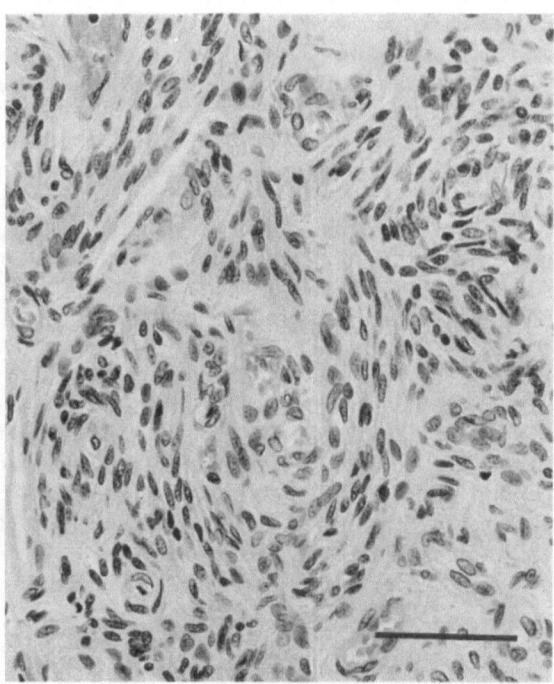

Fig. 5.5. Section of carotid body from a male infant of 9 weeks who died from the sudden infant death syndrome. In some areas there is an overgrowth of sustentacular cells, so that the glomic cells are virtually ablated. (HE) Scale line = 50 μm.

Fig. 5.6. Section of carotid body from the same case. In the left half of the figure is the overgrowth of sustentacular cells illustrated in Fig. 5.5. In the right half of the figure can be seen two surviving lobules of glomic tissue (*arrows g1* and *g2*) in which progenitor cells are prominent. (HE) Scale line = 80 μm.

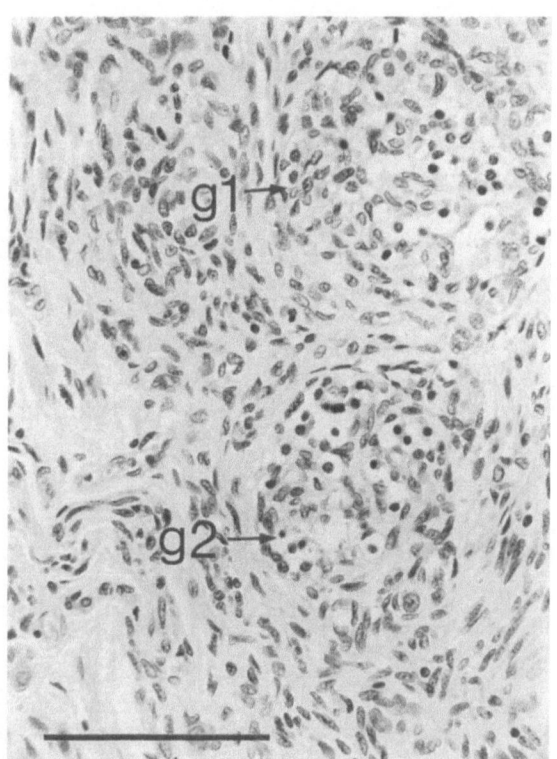

number of chief cells with a reduction in the number or even absence of dense-core vesicles. No statistically significant differences have been reported in the amount of biogenic catecholamines in the chief cells of cases of sudden infant death syndrome compared to those of controls (Lack et al. 1986).

In a study of the histopathology of the carotid bodies in neonates and infants (Heath et al. 1990), three cases were included in the series which had been diagnosed as sudden infant death syndrome. The histological features of the carotid bodies in these three cases differed, lending some support to the view of Emery (1989) that "sudden infant death syndrome" may be a "diagnostic dustbin". In two of them there was prominence of the dark variant of chief cells with a differential count of 27%. The histological appearances were thus comparable to those in infants with diseases leading to sustained hypoxaemia. In one infant (case 6 in Table 5.3) there was a history of hospitalisation for a month for the treatment of bronchiolitis. In a second case with dark cell prominence there was also a history of tracheobronchitis.

In sharp contrast, in the third case (case 5 in Table 5.3), a male infant of 9 weeks with non-infected sudden infant death syndrome, there was no disease predisposing to hypoxaemia. In this case the lobular pattern had been disorganised by an overgrowth of sustentacular cells. In many areas this was gross, with a virtual ablation of chief cells (Fig. 5.5). In other places the proliferated sustentacular cells abutted on lobules of glomic tissue in which the cellular constitution was abnormal, with a predominance of progenitor cells (Fig. 5.6). At higher magnification it was found that there was a virtual absence of dark and light variants of chief cells with predominant immature progenitor cells (Fig. 5.7). In this case (case 5 in Table 5.3) the differential cell count overall showed 88% of sustentacular cells. The prominence of progenitor cells accounted for 17% of cell clusters and 44% of chief cells in cores. It seems to us that these early progenitor cells may correspond to the "immature cells" described by Cole and his colleagues (1979). A carotid body engulfed by sustentacular cells and left with only progenitor cells is much less likely to be capable of responding

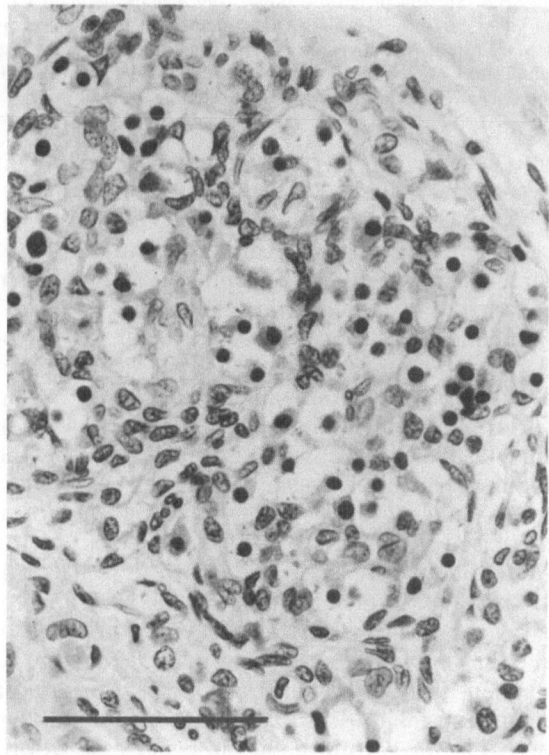

Fig. 5.7. Section of carotid body from the same case showing at higher magnification a lobule of glomic cells in which there is prominence of progenitor cells. (HE) Scale line = 50 μm.

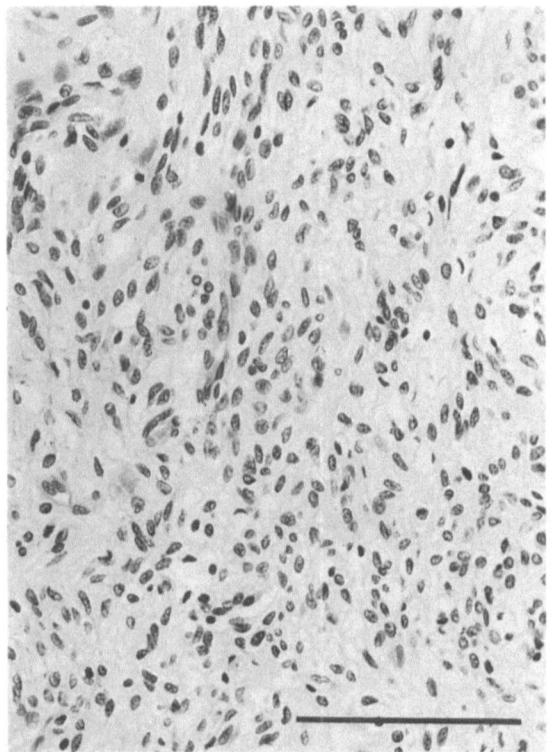

Fig. 5.8. Section of carotid body from a male neonate who survived for only 4 hours. The anticipated glomic tissue is replaced by an overgrowth of sustentacular cells. In this case small surviving lobules of glomic tissue showed a slight predominance of progenitor cells. (HE) Scale line = 80 μm.

to a severe acute hypoxic stimulus such as might occur during a period of sleep apnoea in an infant. This is a very different histological and physiological situation from a carotid body containing prominent dark cells with peptide-rich cytoplasm reacting to sustained hypoxaemia (Fig. 5.4).

In this series of cases (Heath et al. 1990), of particular interest were the histological appearances of the carotid bodies in a neonate who survived for only 4 hours (Table 5.3; Fig. 5.8). In this instance there was disorganisation of the lobular structure by sustentacular cells and only a few islands of glomic tissue, which included an increased proportion of progenitor cells. At the same time hardly any dark cells were to be seen. Heath et al. formed the opinion that such histological appearances were consistent with a case of sudden infant death syndrome "waiting to happen". In this complex and controversial area of sudden infant death syndrome we do not suggest that the carotid bodies in every case will be found to be abnormal histologically. On the other hand it is clear that in some examples of the syndrome these glomera are abnormal. It seems to

us that when a pathologist is called upon to perform a post mortem on a case of sudden infant death syndrome histological examination of the carotid bodies should be regarded as mandatory. Furthermore, a study of their cytological components should be made, and not merely a determination of their volume.

Geerlinger (1976) believed that the basis for the mechanism of sudden infant death syndrome should be sought primarily in the third branchial arch. In his opinion developmental arrest of the third branchial arch and its pharyngeal pouch in the first trimester would probably lead to hypoplasia of the carotid body and hypodevelopment of the glossopharyngeal and superior laryngeal nerves and their mesodermal connections. He thought that hypoplasia of the third parathyroid and its fusion with the thymus might be another anatomical expression of maldevelopment of the third branchial arch.

References

Barer G, Wach R, Pallot D, Bee D (1986) Almitrine, hypoxia, systemic hypertension and the carotid body. In: Heath D (ed) Aspects of hypoxia. Liverpool University Press, Liverpool, pp 113–129

Bergman AB, Beckwith JB, Ray CC (1970) Sudden infant death syndrome. University of Washington Press, Seattle

Boyd JD (1937) The development of the human carotid body. Contr Embryol Carnegie Inst 26:1–31

Cole S, Lindenberg LB, Galioto FM, Howe PE, De Graff AC, Davis JM, Lubka R, Gross EM (1979) Ultrastructural abnormalities of the carotid body in sudden infant death syndrome. Pediatrics 63:13–17

De Winiwarter H (1939) Origine et développement du ganglion carotidien. Appendice: Participation de l'hypoblaste à la constitution des ganglions craniens. Arch Biol (Paris) 50:67–94

Dinsdale F, Emery JL, Gadsdon DR (1977) The carotid body – a quantitative assessment in children. Histopathology 1:179–187

Emery JL (1989) Is sudden infant death syndrome a diagnosis? Or is it just a diagnostic dustbin? Br Med J 299:1240

Geerlinger P (1976) Cot death and the third branchial arch. Lancet ii:716–718

Guilleminault C, Eldridge RL (1973) Insomnia with sleep apnea. A new syndrome. Science 181:856–858

Harris P, Heath D (1986) The human pulmonary circulation, 3rd edn. Churchill Livingstone, Edinburgh

Heath D, Williams DR (1989) High altitude medicine and pathology. Butterworths, London

Heath D, Khan Q, Nash J, Smith P (1989) Carotid body disease and the physician – chronic carotid glomitis. Postgrad Med J 65:353–357

Heath D, Khan Q, Smith P (1990) Histopathology of the carotid bodies in neonates and infants. Histopathology 17:511–520

Kelly DH, Shannon DC (1982) Sudden infant death syndrome and near sudden infant death syndrome: a review of the literature, 1964 to 1982. Ped Clin N Am 29:1241–1261

Kjaergaard J (1973) Anatomy of the carotid glomus and carotid glomus-like bodies (non-chromaffin paraganglia). F.A.D.L.'s Forlag, Copenhagen

Kondo H (1975) A light and electron microscopic study on the embryonic development of the rat carotid body. Am J Anat 144:275–293

Korkala O, Hervonen A (1973) Origin and development of the catecholamine-storing cells of the human fetal carotid body. Histochemie 37:287–297

Lack EE, Perez-Atayde AR, Young JB (1985) Carotid body hyperplasia in cystic fibrosis and cyanotic heart disease. A combined morphometric, ultrastructural, and biochemical study. Am J Pathol 119:301–314

Lack EE, Perez-Atayde AR, Young JB (1986) Carotid bodies in sudden infant death syndrome. A combined light microscopic, ultrastructural, and biochemical study. Pediatr Pathol 6:335–350

Naeye RL (1973) Pulmonary arterial abnormalities in the sudden infant death syndrome. N Engl J Med 289:1167–1170

Naeye RL (1977) The sudden infant death syndrome. Arch Pathol Lab Med 101:165–167

Naeye RL, Whalen P, Ryser M et al. (1976a) Cardiac and other abnormalities in the sudden infant death syndrome. Am J Pathol 82:1–8

Naeye RL, Fisher R, Ryser M, Whalen P (1976b) Carotid body in the sudden infant death syndrome. Science 191:567–569

Perrin DG, Cutz E, Becker LE, Bryan AC (1984) Ulstrastructure of carotid bodies in sudden infant death syndrome. Pediatrics 73:646–651

Sato S (1932) Morphologische Untersuchungen über die Carotis-Drüsen bei Wirbeltieren und beim Menschen. Igaku Kenkyuu, Fukuoka 6:707–811 (In Japanese: abstract in Jap J Med Sci I Anat (1934) 5 (ii):82–83)

Steinschneider A (1972) Prolonged apnea and the sudden infant death syndrome. Clinical and laboratory observations. Pediatrics 50:646–654

Stevens LH (1965) Sudden unexplained death in infancy. Am J Dis Child 110:243–247

6   Chronic Carotid Glomitis

A sparse, diffuse infiltrate of lymphocytes throughout the stroma of the carotid body appears to be an age-related change occurring in subjects over 50 years who show sustentacular cell proliferation around cell clusters or the effects of compression or ablation of glomic tissue by fibrous tissue (Hurst et al. 1985; see Chapter 4). In the elderly or aged, however, discrete, focal aggregations of lymphocytes are to be found. The appearances of these aggregates are so characteristic that they may be regarded as forming a distinct pathological entity which we have designated "chronic carotid glomitis" (Khan et al. 1989). In one study that we carried out, both carotid bodies were obtained at necropsy from each of 75 subjects, 38 male and 37 female, ranging in age from 14 to 90 years. The diffuse infiltration of lymphocytes ascribed to age and referred to above occurred on its own in 34 cases. It was found in two of 18 subjects under 50 years of age (11%) and in 32 of 57 subjects who were older than that (56%). In contrast, in 12 instances there were aggregates of lymphocytes which constitute chronic carotid glomitis.

Histology of Chronic Carotid Glomitis

In chronic carotid glomitis such lymphoid aggregates, measuring up to 430 μm $\times$ 215 μm, are seen in the carotid bodies (Figs. 6.1 to 6.3). Sometimes they are fusiform in shape and situated in the substance of the carotid body, with streamers extending between the components of the glomic tissue (Fig. 6.1). Commonly they are situated in a subcapsular situation (Fig. 6.2). Cuffing of glomic arterioles or small glomic venules running through the capsule or substance of the carotid body by lymphocytes is common. The aggregates lack any follicular or other form of organised structure. The cells consist principally of small lymphocytes of "inactive" (non-transformed) appearance, with a small proportion of larger lymphoid cells, includ-

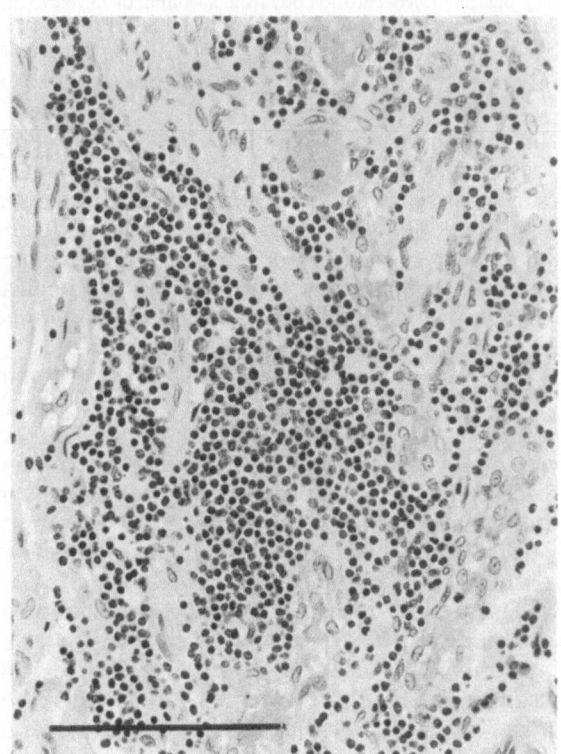

Fig. 6.1. Section of carotid body from a woman of 78 years who died from bronchopneumonia. Chronic carotid glomitis is present, the glomic tissue being infiltrated by small lymphocytes. (Haematoxylin–eosin (HE)) Scale line = 100 μm.

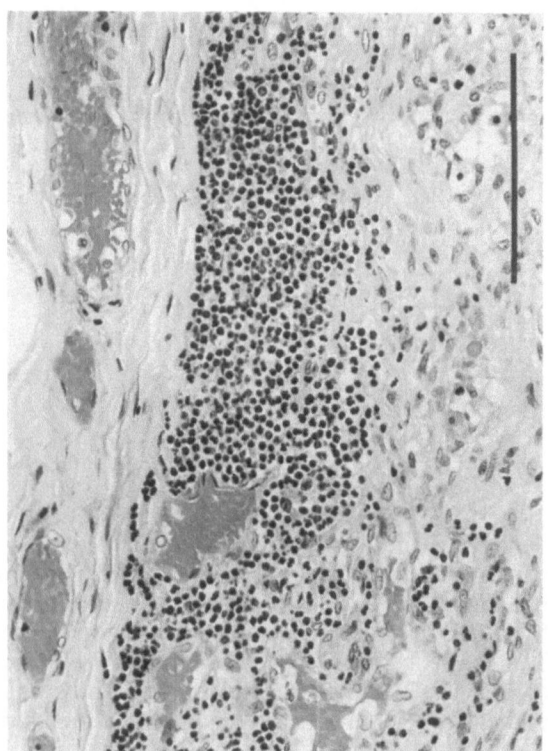

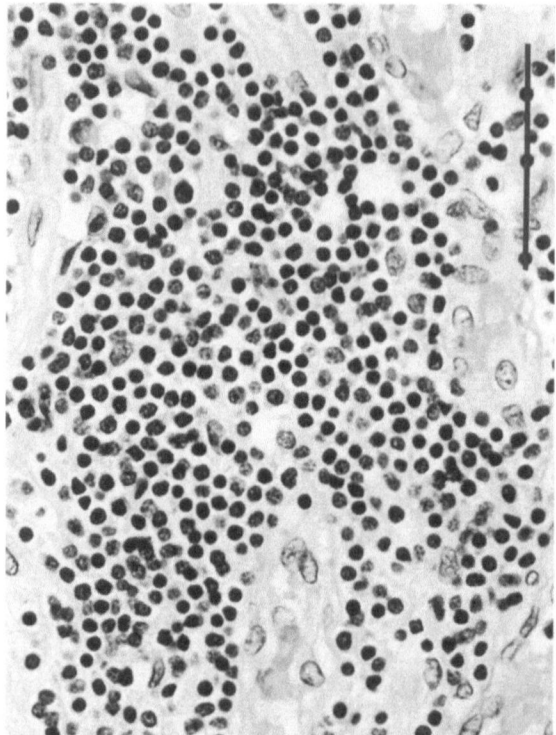

Fig. 6.2. Section of carotid body from a woman of 73 years with myocardial infarction. Chronic carotid glomitis is present but in this instance the aggregate of lymphocytes is subcapsular. (HE) Scale line = 100 μm.

Fig. 6.3. Section of carotid body from the same case as in Fig. 6.1, showing the cytological features of the infiltrating lymphocytes. (HE) Scale line = 50 μm.

ing immunocytes (lymphoplasmacytoid cells) (Fig. 6.3). Diffuse lymphocytic infiltration of the fibrous stroma, or less commonly of the glomic tissue, may be associated with the aggregates of lymphocytes. In areas occupied by focal collections of lymphocytes there are no cytological changes in the surrounding or underlying glomic tissue.

The small lymphocytes diffusely distributed through the stroma with a few within the glomic tissue are virtually all T-cells (Fig. 6.4). Much less commonly there may be diffusely distributed B-cells (Fig. 6.5). Immunohistological analysis of the aggregates shows them to vary in composition, but T-lymphocytes always comprise a half to three-quarters of the cells present.

The 12 cases of chronic carotid glomitis found in our study of 75 cases of pairs of carotid bodies at necropsy all occurred in subjects over the age of 50 years and all but three were found in subjects of 70 years of age or over. Put another way, the overall prevalence of chronic carotid glomitis in the series of 75 cases was 16%, but was 21% in those over the age of 50 years and 29% in those 70 years of age or over (Khan et al. 1989). Six of the cases

were found in women and six in men. In the women the combined carotid body weight equalled or exceeded the upper limit of normal (30 mg) in only one instance. In contrast, in the men the combined carotid body weight equalled or exceeded 30 mg in five of the six cases. Histologically, the glomic tissue showed sustentacular cell proliferation in six cases, fibrosis and atrophy in five cases and normal appearances in only one instance (Khan et al. 1989). In summary, chronic carotid glomitis is a disease of old age showing the development of lymphocytic aggregates on a background of degenerative age-change of the glomic tissue.

Possible Significance of the Lymphoid Aggregates

Infiltrates of similar character occur in chronic inflammation generally, showing a preponderance of T-cells, usually with a normal subset ratio of 2:1 helper to suppressor cells. In such infiltrates B-cells are fewer and invariably polyclonal for light-

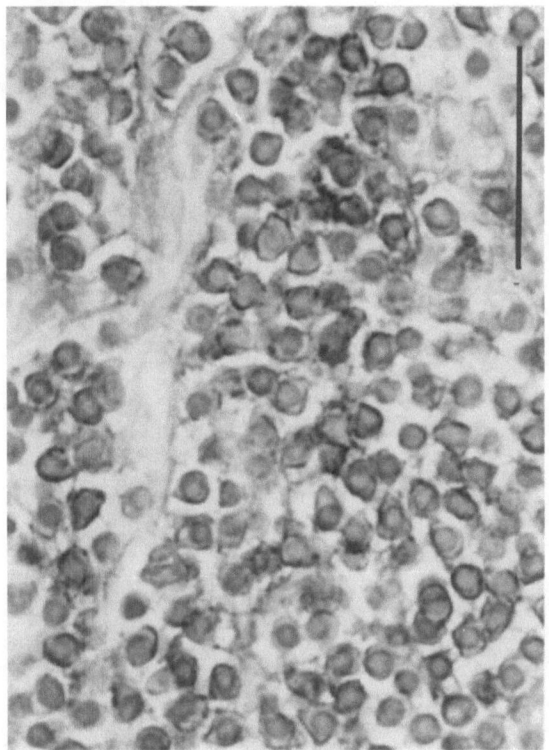

Fig. 6.4. Section of carotid body from a man of 87 years who died from a fracture of the first cervical vertebra. Chronic carotid glomitis is present, the T-lymphocytes showing membrane binding of the antibody UCHL 1. Scale line = 25 μm.

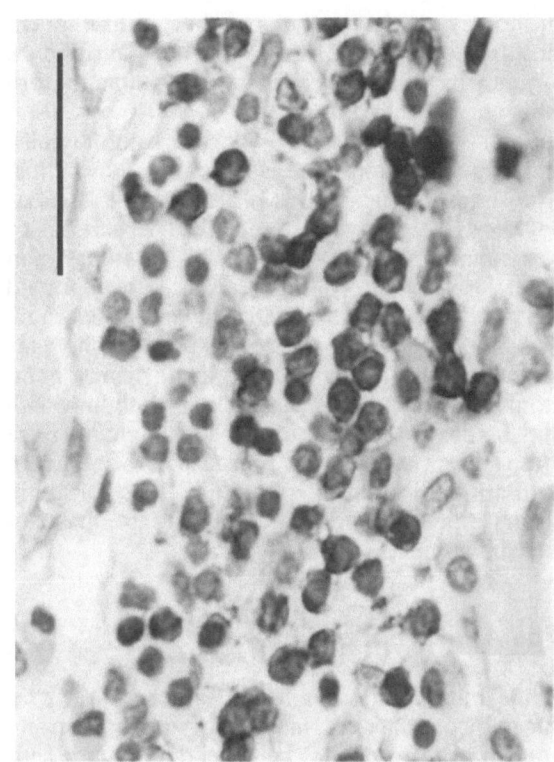

Fig. 6.5. Section of carotid body from a man of 81 years who died from coronary thombosis. Chronic carotid glomitis is present, the B-lymphocytes showing membrane binding of the antibody 4KP5. Scale line = 25 μm.

chain expression, with smaller numbers of immunocytes and plasma cells. As described below, B-cells appear in the infiltrate of chronic carotid glomitis subsequently, just as they do in chronic inflammation, and the plasma cell component may become prominent at particular sites. Similar findings characterise auto-immune disease, but in severe or chronic examples of this, lymphoid follicles may form, with the full range of follicle centre cells in addition to the cell types described above, and again all showing polyclonality. Studies of the salivary glands in Sjögren's syndrome have shown predominantly B-cells (Chused et al. 1974), or fewer B-cells and a predominant population of T-cells, mostly of CD4 (helper) phenotype (Lindahl et al. 1985). In thyroid glands affected by Hashimoto's or Graves' disease, the infiltrating population of lymphocytes between the thyroid follicles is composed predominantly of T-cells, of which the majority are of CD8 (suppressor) phenotype. The germinal centres and their mantle zones contain the expected B-cells (Misaki et al. 1985). In chronic carotid glomitis the carotid bodies do not show germinal centres, but contain lymphoid aggregates of mixed B- and T-cell content, as well as a diffuse T-cell infiltrate.

A Possible Auto-immune Disease

The histological picture of chronic carotid glomitis is reminiscent of that found in other organs such as the thyroid and salivary glands, where it has been shown, or suspected, to be a manifestation of auto-immunity. Focal collections of lymphocytes were found in the thyroid gland in half the women and a quarter of the men patients in a prospective study of 197 thyroid glands obtained at necropsy (Mitchell et al. 1984). In this condition large lymphoid aggregates are found to be associated with microsomal auto-antibodies. Focal chronic thyroiditis is usually present in patients with Graves' disease. Such aggregates of lymphocytes indicate a form of auto-immune thyroiditis in which there are auto-antibodies against thyroglobulin and against a lipoprotein of the membrane of the endoplasmic reticulum of thyroid epithelial cells, although the frequency and titres of such antibodies are lower in focal thyroiditis than in Hashimoto's disease. Most of the auto-antibody is formed by plasma cells in the inflammatory infiltrate in the gland. It is well known that patients with this disease of the thyroid may also manifest auto-immune gastritis or adrenalitis. Auto-immune sialadenitis is also known to occur but it is associated with connective tissue diseases, particularly rheumatoid arthritis, rather than other forms of auto-immune disease (MacDonald 1985, p. 19.9). In affected salivary or lacrimal glands dense focal aggregates of lymphocytes centre around intraglandular ducts which commonly show hyperplasia of their epithelium which may be destroyed. These histological changes are often accompanied by the presence in the serum of auto-antibodies to the epithelium. This condition occurs especially in middle and old age, particularly in women.

The Appearance of Plasma Cells

While most antibodies found in serum are produced by plasma cells, in lymph nodes, spleen and bone marrow some may be formed by plasma cells at the site of a chronic, inflammatory reaction around antigenic material. In the case of auto-immune disease the site of focal accumulations of plasma cells in a complex tissue containing several components may offer some insight into which component is degenerating and forming an antigenic stimulus. In a man aged 81 years with chronic carotid glomitis we found intense plasma cell activity throughout the carotid bodies (Figs. 6.6 and 6.7). Some glomic venules were surrounded by a cuff of lymphocytes and many more by prominent collections of large, mature plasma cells (Fig. 6.6). There were large collections of lymphocytes closely approximated to lengths of myelinated nerves coursing through the stroma of the carotid body. Small nerves, commonly seen in transverse section, had around them an infiltrate of large, mature plasma cells (Fig. 6.7). A severe, diffuse infiltrate of lymphocytes and plasma cells within the glomic parenchyma was situated mainly around the clusters in close proximity to the encircling sustentacular cells, which ensheathe axons. The small lymphocytes and plasma cells around glomic blood vessels and nerve fibrils proved to be, or were derived from, B-cells. Hence it seems likely that the antigenic stimulus for the apparent auto-immune reaction is related to degenerating nerve tissue in the ageing glomus. The fibrils involved are likely to be branches of the glossopharyngeal and unmyelinated axons

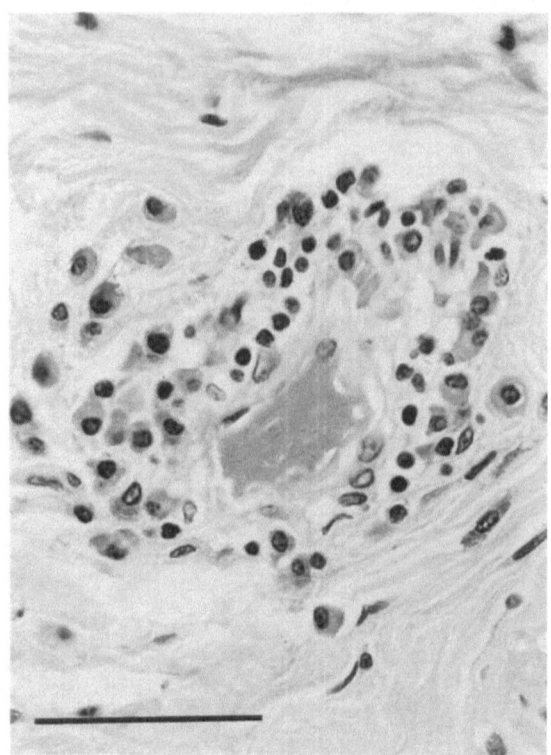

Fig. 6.6. Section of carotid body from a man of 81 years who died from a myocardial infarction secondary to coronary artery disease. He had chronic carotid glomitis characterised in his case by a prominent exudate of plasma cells. In this section the plasma cells are congregated around a small glomic venule. (HE) Scale line = 50 μm.

ensheathed by the sustentacular cells surrounding clusters of chief cells in the glomic parenchyma. It is possible that with involvement of the nerves in the carotid body chemoreceptor activity is compromised.

Focal Chronic Thyroiditis and Chronic Carotid Glomitis

In view of the fact that the most frequent manifestation of auto-immune disease in man occurs in the thyroid gland, as focal chronic thyroiditis characterised by prominent aggregates of lymphocytes similar to those which occur in chronic carotid glomitis, we thought it would be of interest to study the relation between the two diseases. Accordingly we studied the incidence of lymphoid aggregates in the thyroid gland and the carotid bodies in 50 subjects over the age of 50 years coming to necropsy (Heath and Khan 1989). We found that the two conditions appear to be distinct, affecting different age groups. Focal chronic thyroiditis occurred as early as the sixth decade

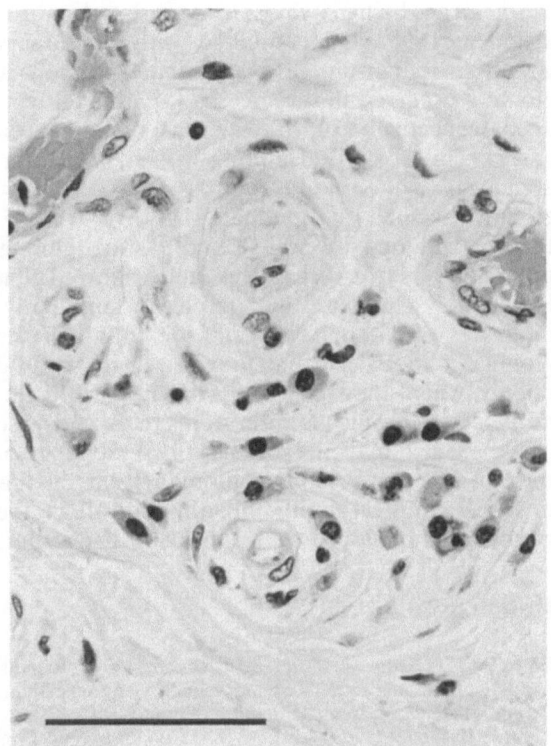

Fig. 6.7. Section of carotid body from the same case as in Fig. 6.6. In this instance the plasma cells have accumulated around bundles of nerve fibrils. (HE) Scale line = 50 μm.

and was found on its own in 34% of subjects. In contrast, chronic carotid glomitis was characteristic of the eighth and ninth decades and occurred on its own in 14% of subjects. In a further 12%, thyroiditis and glomitis coexisted.

Chronic Carotid Glomitis and the Non-chromaffin Paraganglionic System

When we considered the nature of the carotid body in Chapter 1 it was pointed out that it is part of a generalised non-chromaffin paraganglionic system. All the components of this system share the basic histological unit of clusters of chief cells enveloped in a shell of sustentacular cells. The distinctive nature of the carotid body lies in the fact that, like the aortic bodies, it is a chemoreceptor. In other words the distinction is functional rather than structural. Hence the question immediately arises as to whether the lymphoid aggregates in the carotid bodies described in this chapter are to be found in other components of the non-chromaffin paraganglionic system. We have investigated this problem and selected the glomus pulmonale as a readily accessible representative of the system anatomically distant from the carotid body and not a chemoreceptor.

The Glomus Pulmonale

Many glomera are present around the heart and great vessels and one of them is present on the dorsal aspect of the bifurcation of the pulmonary trunk. Krahl (1960) found it to be situated there so constantly that he designated it "the glomus pulmonale", believing it to be the homologue of the carotid body for the sixth branchial arch from which the pulmonary circulation is derived. He found it in the cat, dog, rat, cow and chimpanzee, as well as in humans (Krahl 1962). The glomus is readily demonstrated by staining the surface of the pulmonary trunk with 1% methylene blue solution (Edwards and Heath 1969). A small plexus of nerves just posterior and caudal to the bifurcation is revealed and a block of tissue from this area of the wall of the pulmonary trunk will contain in its adventitia the glomus pulmonale. In passing we may note that the mere situation of a glomus in the wall of the pulmonary trunk does not prove that it is the "glomus pulmonale" of derivatives of the sixth branchial arch. The crucial point is its source of blood supply. Krahl (1962) believed that this glomus receives its blood supply from the pulmonary trunk, but Comroe (1962) found no physiological evidence of a chemoreceptor attached to the pulmonary arteries and responsive to low oxygen tension. He was unable to produce immediate respiratory stimulation by injection into the pulmonary arteries of cyanide, which is a known stimulator of chemoreceptor tissue. Comroe believed that to prove conclusively the existence of a glomus pulmonale which is quite distinct from the aortic body it is essential to demonstrate both glomera in the same subject. One should be able to demonstrate a pulmonary glomus with a blood supply from the pulmonary trunk, and an aortic glomus with a blood supply from the aorta or one of its branches. We confirmed from histological studies in a male infant and from corrosion cast studies on rats that the glomus pulmonale, like the other intertruncal glomera, has a solely systemic arterial blood supply through branches of the coronary arteries and aorta. There is no supply from the pulmonary circulation to support the concept of a "glomus pulmonale". This was the readily accessible nodule of tissue that we chose in order to determine whether focal chronic glomitis is confined to the carotid body or involves other components of the non-chromaffin paraganglionic system.

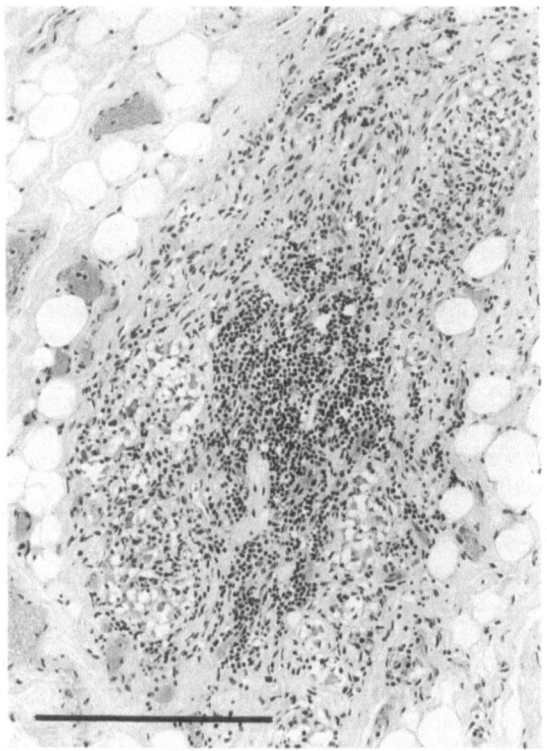

Fig. 6.8. Section of glomus pulmonale from a woman of 73 years who died from carcinoma of the gall bladder with metastases. Chronic pulmonary glomitis is present, with an aggregate of small lymphocytes in the centre of the glomus. (HE) Scale line = 200 μm.

Pulmonary Glomitis

A histological study was made by Khan and Heath (1990) of the carotid bodies and glomus pulmonale in 50 consecutive subjects over the age of 50 years (who came to necropsy), to determine whether chronic glomitis is confined to the carotid bodies or whether it also occurs in other glomera. Lymphocytes formed aggregates in the glomic substance and fibrous capsule of the glomus pulmonale just as in cases of chronic carotid glomitis (Fig. 6.8). The aggregates were composed of small lymphocytes and were commonly found around small, thin-walled blood vessels and around branches of nerve trunks as they entered the glomus pulmonale. In some instances the structure of the glomus was disturbed so that the cellular clusters had been replaced by fibrous tissue in some areas and distorted by a proliferation of sustentacular cells in others. Such abnormal glomera frequently showed infiltration by plasma cells.

Chronic carotid glomitis was found in seven subjects (14%) ranging in age from 67 to 90 years,

the mean age being 79 years. It was isolated in three instances (6%) and combined with pulmonary glomitis in four instances (8%). Isolated pulmonary glomitis occurred in three cases (6%) in subjects aged, respectively, 76, 76 and 77 years. In the seven cases of chronic carotid glomitis women predominated at six to one. In the seven cases of pulmonary glomitis women predominated at five to two. Plasma cell activity was found in the glomus pulmonale in five cases. This study showed that lymphoid aggregates and, less commonly, infiltrates of plasma cells may be found in the glomus pulmonale just as they are in the carotid bodies, where they constitute chronic carotid glomitis. Thus chronic glomitis seems to be a disease process which may affect at least two members of the non-chromaffin paraganglionic system, but it is not clear why focal chronic inflammation affects the glomus pulmonale of one person but not of another.

References

Chused TM, Hardin JA, Frank MM, Green I (1974) Identification of cells infiltrating the minor salivary glands in patients with Sjögren's syndrome. J Immunol 112:641–648

Comroe JH (1962) In: de Reuck AVS, O'Connor M (eds) Ciba Foundation Symposium on pulmonary structure and function. Churchill, London, pp 70–73 (discussion)

Edwards C, Heath D (1969) Microanatomy of glomic tissue of the pulmonary trunk. Thorax 24:209–217

Heath D, Khan Q (1989) Focal chronic thyroiditis and chronic carotid glomitis. J Pathol 159:29–34

Heath D, Khan Q, Nash J, Smith P (1989) Carotid body disease and the physician – chronic carotid glomitis. Postgrad Med J 65:353–357

Hurst G, Heath D, Smith P (1985) Histological changes associated with ageing of the human carotid body. J Pathol 147:181–187

Khan Q, Heath D (1990) Chronic carotid glomitis and the glomus pulmonale. J Clin Pathol 43:39–42

Khan Q, Heath D, Nash J, Smith P (1989) Chronic carotid glomitis. Histopathology 14:471–481

Krahl VE (1960) The glomus pulmonale. A preliminary report. Bull Sch Med Univ Maryland 45:36–38

Krahl VE (1962) The glomus pulmonale: its location and microscopic anatomy. In: de Reuck AVS, O'Connor M (eds) Ciba Foundation Symposium on pulmonary structure and function. Churchill, London, pp 53–69

Lindahl G, Hedfors, E, Klareskog L, Forsum U (1985) Epithelial HLA-DR expression and T lymphocyte subsets in salivary glands in Sjögren's syndrome. Clin Exp Immunol 61:475–482

MacDonald DG (1985) Alimentary tract. I: The oral cavity, salivary glands and oropharynx. In: Anderson JR (ed) Muir's Textbook of Pathology 12th edit. Edward Arnold, London, pp 19.1–19.12

Misaki T, Konishi J, Nakashima T, Iida Y, Kasagi K, Endo K, Uchiyama T, Kuma K, Torizuka K (1985) Immunohistological phenotyping of thyroid infiltrating lymphocytes in Graves' disease and Hashimoto's thyroiditis. Clin Exp Immunol 60:104–110

Mitchell JD, Kirkham N, Machin D (1984) Focal lymphocytic thyroiditis in Southampton. J Pathol 144:269–273

7 Physiology

Although well over half a century has elapsed since the classic studies of de Castro (1928) and Heymans et al. (1930) revealed the carotid body to be the main peripheral arterial chemoreceptor, neither the mechanism of the process nor the identification of the transducer has yet been established. Throughout this book we give an account of the histological and ultrastructural features of the various tissue components of the glomus in health, at different ages throughout life, and in disease. In this chapter we consider the physiology of chemoreception and its relation to the cells, nerves and blood vessels of the carotid body. This affords us an opportunity to refer to part at least of the prodigious volume of research by many distinguished physiologists working in this field. It will become apparent that, in spite of all this intense activity, the precise mechanism of chemoreception is still far from being understood. It must also always be kept in mind that the carotid body may subserve other functions, perhaps of an endocrinological nature, in addition to that of chemoreception.

Stimuli

The carotid body is unusual amongst sensory organs in that it is responsive to several stimuli. These include hypoxia, hypercarbia and acidosis as well as temperature, blood pressure and osmolarity (Eyzaguirre and Zapata 1984).

Oxygen

The classical studies of Heymans et al. (1930) demonstrated that the carotid body responds to a reduction in the partial pressure of oxygen, an elevation in the partial pressure of carbon dioxide, and lowered pH in systemic arterial blood. This response has been confirmed subsequently many times by using carotid sinus nerve preparations, a technique which involves dissecting out the carotid bifurcations of experimental animals to expose the carotid bodies and their nervous supply. The carotid sinus nerve is then sectioned and its peripheral stump teased out to expose individual fibres, which are placed on electrodes to measure the nervous impulses generated within the glomic tissue. Chemosensory fibres can be distinguished from those subserving baroreception by their random discharge frequency, as contrasted with the periodic firing of the latter synchronously with the heart beat. Such preparations demonstrate that reducing the oxygen tension of blood in the carotid artery causes an increase in the frequency of action potentials in the carotid sinus nerve. The relationship between discharge frequency and oxygen tension is exponential, so that as hypoxaemia increases there is a disproportionate rise in neural activity (Torrance 1968; Eyzaguirre and Zapata 1984; O'Regan and Majcherczyk 1982). Severe levels of hypoxaemia, with an arterial oxygen tension less than 20 mmHg, or hyperoxia may inhibit the discharge.

Carbon Dioxide

When the arterial oxygen tension is held constant, the discharge frequency increases as the partial pressure of carbon dioxide is raised, but in this case the relationship is linear and not exponential (O'Regan and Majcherczyk 1982; Eyzaguirre and Zapata 1984). This difference in response between oxygen and carbon dioxide raises the possibility that the two gases are detected in the carotid body by different mechanisms. When the two stimuli of hypoxia and hypercarbia are applied simultaneously, there is a dramatic increase in frequency of discharge which is greater than simple addition of the two individual responses (Torrance 1968; O'Regan and Majcherczyk 1982; Eyzaguirre and Zapata 1984). This potentiating effect means that the response of the carotid body to hypoxia is greatly enhanced where there is superimposed hypercarbia. Indeed, when arterial carbon dioxide tension is abnormally low – in the range 10–15 mmHg – the chemosensory response to hypoxia is virtually abolished (O'Regan and Majcherczyk 1982).

Acidosis

There is a linear increase in discharge from the carotid sinus nerve as the hydrogen ion concentration is increased. Clearly, a fall in pH will reflect increases in carbon dioxide tension and so the two factors may simply be separate manifestations of the same stimulus. There is evidence, however, that carbon dioxide can stimulate the carotid body even when the pH is held constant (Eyzaguirre and Zapata 1984). Fluctuations in pH due to changes in carbon dioxide tension during the ventilatory cycle have been shown to induce oscillations in chemosensory discharges in the carotid sinus nerve. It may be that dynamic changes in pH are more important in potentiating the chemosensory response to hypoxia than the mean value (O'Regan and Majcherczyk 1982).

Temperature

Experiments with superfused carotid bodies have shown that glomic tissue is sensitive to changes in temperature. The technique of superfusion is used extensively for in vitro studies on the physiology of the carotid body and involves its removal, with the attached carotid sinus nerve, from the experimental animal. These are then mounted within a transparent superfusion chamber through which Ringer's solution is pumped at constant flow over the organ. The nature of the perfusate can then be altered and its influence upon the carotid body detected by attaching electrodes to the carotid sinus nerve. This in vitro technique has the advantage that the glomus is divorced from the influences of the central nervous system and autonomic input. It has the disadvantage that any substances under investigation are applied to the surface of the carotid body and do not pass through its vasculature as in life. By increasing the temperature of the perfusate, an increase in frequency of discharges can be detected (Eyzaguirre and Zapata 1984). The altered frequency is related more to the change in temperature than to its steady level. Thus, when the temperature is increased in a step-wise manner, there is a rapid increase in frequency as the temperature rises, followed by a gradual decline to a steady level for the duration of the temperature change. The effect of temperature is not merely an idiosyncracy of the in vitro preparation, since the carotid body in situ responds in a similar fashion to graded changes in the temperature of circulating blood (Eyzaguirre and Zapata 1984).

Osmolarity

The technique of superfusion has also been used to show that the carotid body is responsive to changes in osmolarity. Thus, when the superfusing solution is hyperosmotic, there is an increase in the rate of discharge, whereas hyposmotic solutions have an opposite effect. Thus the carotid body may be involved in some way with regulating the osmolarity of the environment (Eyzaguirre and Zapata 1984).

The concentrations of individual ions in the blood also influence the activity of the carotid body. For example, an increase in the level of potassium provokes an enhanced steady-state discharge in the carotid sinus nerve (Band and Linton 1983). With superimposed hypoxia the discharge frequency is augmented by a degree greater than the additive effects of the two stimuli alone (Band and Linton 1987). It is suggested that this reflex may enhance the hyperventilation associated with exercise during which an increase in plasma potassium often occurs (Band and Linton 1983).

Blood Pressure

Transient changes in the frequency of discharge in chemosensory fibres of the carotid sinus nerve occur during abrupt alterations of blood pressure in the carotid artery (Biscoe et al. 1970a). Chemosensory discharges do not change within a wide range of constant blood pressure except when it falls to 60 mmHg or below; then there is an increase in discharge frequency in the carotid sinus nerve (Biscoe et al. 1970a). This is likely to be the result of reflex vasoconstriction in the glomic vasculature leading to hypoxia of the glomic tissue (Eyzaguirre and Zapata 1984). This relative insensitivity to changes in blood pressure is in marked contrast to the way in which baroreceptors in the carotid sinus accurately reflect changes in intravascular pressure.

Chemical Agents

The carotid body is stimulated by a wide range of chemicals either in vivo or in vitro. Many of these substances are neurotransmitters or their antagonists, as described below, but another group, comprising poisons of oxidative metabolism, is also of interest. These poisons are used widely in physiological studies of the carotid bodies of animals and include sodium cyanide, dinitrophenol, antimycin A and oligomycin. Cyanide in particular causes chemosensory stimulation which is increased during hypoxia but not hypercarbia. When the carotid sinus nerves are intact, cyanide causes a reflex increase in ventilation. This response is so predictable that it is used as an experimental tool to determine the presence or absence of neural pathways from the carotid bodies to the brain or to distinguish between baroreceptor and chemoreceptor fibres in sinus nerve preparations.

Response to Stimulation

The carotid bodies, and to a lesser extent the aortic bodies, mediate major reflex effects on the lungs, heart and blood vessels. These effects have been analysed in experimental animals as well as in men who have had both carotid bodies removed for the treatment of bronchial asthma.

Hyperventilation

It is well known that acute episodes of hypoxia result in a rapid increase in minute ventilation. The relationship between reduced arterial oxygen tension and minute ventilation is hyperbolic and thus similar to the response curve for the rate of discharge in the carotid sinus nerve (Torrance 1968). That this effect is mediated by the carotid bodies can be demonstrated in animals following bilateral glomectomy in which the ventilatory response to hypoxia is virtually abolished (O'Regan and Majchercyzk 1982; Olson et al. 1988). In humans, bilateral glomectomy causes no

change in ventilation when air is breathed, even during exercise (Lugliani et al. 1971), but acute hypoxia fails to induce the hyperventilation typical of subjects with intact carotid bodies (Holton and Wood 1965; Lugliani et al. 1971; Honda 1985). Honda (1985) was able to detect a small residual hypoxic response amounting to 10% of normal in subjects 30 years after operation. He attributed this to the activity of the aortic bodies or central nervous system.

Hypercarbia also stimulates ventilation but a major contribution to this reflex is mediated by the central nervous system. A technique for determining the sensitivity of the carotid bodies to hypercarbia involves introducing carbon dioxide into the respiratory gases for two breaths only and noting the immediate change in ventilation before the gas reaches the brain (Dejours 1963). In the dog this "double-breath carbon dioxide test" leads to a sharp rise in ventilation followed by a gradual decline over a period of approximately 30 seconds. When the same test is applied to chemodenervated dogs the change in ventilation is delayed, smaller and develops gradually (Dejours 1963), indicating that the hyperventilatory response to hypercarbia is mediated in part by the carotid bodies. The calculated response curve relating the partial pressure of carbon dioxide to minute volume is linear (Torrance 1968), but its gradient is small in contrast to that for oxygen. However, in the presence of superimposed hypoxia, the gradient becomes much steeper, beyond a threshold value which varies according to the level of oxygen tension. Thus, hypercarbia and hypoxia both increase ventilation through the activity of the carotid bodies, and their combined effects are slightly greater than additive (Torrance 1968). In patients with bilateral glomectomy the ventilatory response to carbon dioxide is reduced to about 70% of normal, indicating that in humans the carotid bodies account for 30% of this reflex, the remainder being mediated centrally (Lugliani et al. 1971; Honda 1985).

Effects on Vascular Tone

Stimulation of the carotid bodies by introducing hypoxic blood or saline into the carotid arteries of dogs causes an elevation of systemic blood pressure associated with a 12% increase in peripheral vascular resistance (Parker et al. 1975; Olson et al. 1982; O'Regan and Majcherczyk 1982). Vasoconstriction is particularly evident in the renal vessels but occurs also in the muscles of the limbs and in the mesenteric blood vessels (Parker et al. 1975; Marshall 1987). It is apparently enhanced following vagotomy (Parker et al. 1975; Olson et al. 1982). The carotid bodies also tend to induce pulmonary vasoconstriction but this is normally antagonised by the vagus nerve. Thus, local hypoxic stimulation of the carotid body of vagotomised dogs causes a 21% increase in pulmonary arterial pressure and resistance but does not do so when the vagus nerve is intact (Olson et al. 1982).

In contrast to the above vasoconstrictor effects, the coronary circulation undergoes dilatation when the carotid bodies are stimulated (O'Regan and Majcherczyk 1982; Murray et al. 1984). This reflex is partly dependent upon vagal vasodilator fibres but in some cases may be followed by vasoconstriction mediated by cardiac sympathetic nerves and adrenal catecholamines (Murray et al. 1984).

Effects on Heart Rate

Hypoxic stimulation of the carotid bodies commonly leads to hyperventilation and tachycardia. If, however, hyperventilation is prevented by the control of breathing, bradycardia ensues (O'Regan and Majcherczyk 1982; Marshall 1987). Furthermore, if brief pulses of hypoxia are applied to the carotid bodies during expiration there is a prompt reduction in heart rate but when the stimulus occurs during inspiration it has little effect. This suggests that activation of pulmonary stretch receptors during inspiration antagonises the cardiac reflex from the carotid bodies (Haymet and McCloskey 1975). Presumably the increased depth of inspiration during hyperventilation has a similar effect. The carotid bodies are also antagonised by the aortic bodies, stimulation of which leads to tachycardia (O'Regan and Majcherczyk 1982). Thus the cardiovascular effects of chemoreceptor stimulation can be antagonised by vagal afferent activity and pulmonary stretch receptors, which may reduce or even reverse the peripheral vasoconstriction and bradycardia. On the other hand, episodes of apnoea may enhance the cardiovascular reflexes (Haymet and McCloskey 1975; O'Regan and Majcherczyk 1982).

The Carotid Body as a Sensory Organ

The basic functional unit of the carotid body, sometimes called a glomoid, consists of a capillary close to a small group of chief cells to which are applied the endings of nerves that are ensheathed by sustentacular cells (see Chapter 3). By analogy with other sensory organs the chief cell should be responsive to changes in blood gases within the capillary, whereupon it transmits signals across synapses to the nerve endings, depolarising them and initiating chemosensory discharges in the carotid sinus nerve. This signal is then conveyed along the glossopharyngeal nerve to the petrosal ganglion from which dendrites interreact with the central respiratory centres thereby effecting the chemosensory responses. This simple, and apparently logical, scheme was proposed by de Castro (1928) but has evaded precise characterisation despite intensive efforts over the ensuing 63 years. Almost every stage in this theoretical sequence is controversial, with evidence to support or refute it, and with supporters or antagonists in the scientific community.

In order for the scheme to work at all the impulses must travel from the carotid body to the petrosal ganglion in sensory afferent nerve fibres. The pioneering work of de Castro (1926) involved sectioning the glossopharyngeal nerve at its exit from the cranium, which induced degeneration of all nerves within the glomoids. Removing the superior cervical ganglion resulted in degeneration of nerves supplying only the glomic vasculature, from which he concluded that glomoids are supplied by axons derived from the ninth cranial nerve. He then went on to section the roots of the glossopharyngeal nerve central to the petrosal ganglion and 5–12 days later found no degeneration of nerve endings. de Castro (1928) concluded that the cell bodies of sinus nerve fibres are in the petrosal ganglion and the nerve endings are, therefore, preganglionic sensory afferents. Electron microscopic studies of the carotid body revealed that nerve endings adjacent to chief cells contained numerous clear-cored synaptic vesicles resembling those of motor neurons (see Chapter 17). These findings suggested that the direction of nervous impulses may be from nerve ending to chief cell, the latter being regarded as a postsynaptic secretory cell similar to that in the adrenal medulla. In further experiments Biscoe et al. (1970b) sectioned the intracranial roots of the glossopharyngeal nerve and observed the subsequent changes with the electron microscope. Degeneration of nerve fibres occurred in the glomic parenchyma over a prolonged period of several months, adding further support to the idea that glomic nerves are efferent. The experiments of Biscoe et al. (1970b) were repeated by other workers and this time it was found that section of the glossopharyngeal nerve distal to its ganglion caused rapid degeneration of nerves in the glomic parenchyma within 3 days but section central to the ganglion resulted in normal chemosensory ultrastructure and function (Eyzaguirre and Zapata 1984). The discrepancy between these new studies and the original ones of Biscoe et al. (1970b) was explained on the grounds that the latter group sectioned the nerves too close to the ganglion, thus damaging it and inducing slow, retrograde degeneration of peripheral processes. Pallot (1987) was cautious in accepting this explanation too readily and pointed out that the new experiments extended for only 7–87 days after operation whereas those of Biscoe et al. (1970b) involved waiting up to 378 days after section of the nerve roots.

The nature of the nerve endings apposed to chief cells has been investigated using labelling techniques. Radiolabelled amino acids have been injected into the petrosal ganglion and traced to the nerve endings, indicating that their neurons are situated within the sensory ganglion (Fidone et al. 1977). Conversely, horseradish peroxidase has been traced from the carotid sinus nerve centrally to neurons in the petrosal ganglion (Eyzaguirre and Zapata 1984). Thus, the balance of evidence strongly suggests that many, but not all, of the nerve endings on chief cells are sensory afferents establishing the carotid body as a sensory organ responding to the various stimuli described above. The identity of the structure which detects these stimuli is controversial.

A prime candidate must be the chief cell which contains numerous chemical transmitters including acetylcholine, dopamine, noradrenaline, adrenaline, serotonin, enkephalins, substance P and vasoactive intestinal peptide. The possible rôle which each of these play as transmitters of the chemosensory signal is discussed below.

Acetylcholine

Acetylcholine has been identified in the carotid bodies of animals and does not appear to be contained within nerve endings since, when these are destroyed by cutting the carotid sinus nerve, the concentration of the amine is unchanged (Eyzaguirre and Fidone 1980). Furthermore, choline, the precursor molecule of acetylcholine, is actively taken up by the glomus even after its nervous supply is cut. Localising the site of this uptake with tritiated choline followed by autoradiography demonstrates that it is concentrated in glomus cells and chief cells in particular (Eyzaguirre and Fidone 1980). Acetylcholine is thus in a position to act as a neurotransmitter from chief cell to nerve ending, a supposition supported by the identification of cholinesterase and the fact

that exogenously applied acetylcholine produces a dose-dependent excitation of the carotid bodies in some species. This excitatory influence is enhanced by inhibiting cholinesterase (Eyzaguirre and Zapata 1984). There is a number of observations which mitigate against acetylcholine being involved in chemosensory transmission. In the rabbit it has an inhibitory influence through muscarinic actions whereas in the cat its response is excitatory through nicotinic mechanisms (Monti-Bloch and Eyzaguirre 1980; Ponte and Sadler 1989a). When the effect of acetylcholine is blocked by curarising agents the carotid body continues to respond to natural stimuli. Furthermore, for acetylcholine to be a neurotransmitter, there should be cholinoceptor sites on the nerve endings and yet labelling them with α-bungarotoxin demonstrates that the majority of such sites are located on the membranes of chief cells (Eyzaguirre and Zapata 1984).

Catecholamines

Catecholamines can be demonstrated in the carotid body by the technique of formaldehyde-induced fluorescence and their nature has been analysed by chromatography. They comprise dopamine, noradrenaline and adrenaline, the last being present in low concentration and in one study being undetectable in several species except for the cat (Mir et al. 1982). The other two are present in significant amounts, although there is considerable variation in their reported concentrations. Mir et al. (1982) noted from a survey of the literature that the concentration of noradrenaline in the carotid body of the cat was reported to range from 479 to 957 pmol/carotid body and that of dopamine from 162 to 1331 pmol/carotid body. These differences are probably a reflection of the various analytical techniques employed with earlier, fluorimetric methods of assay tending to overestimate the concentrations of amines (Pallot 1987). The situation is complicated by the fact that the relative proportions of noradrenaline and dopamine differ widely between species. Mir et al. (1982) conducted a systematic study of catecholamines in several species using the highly sensitive techniques of radioenzymic assay and high-pressure liquid chromatography. They found that the highest ratio of noradrenaline to dopamine is in the guinea pig at nearly 5:1 followed by the cat at 4:1, the Sprague–Dawley rat at 9:5, the rabbit at 1:3 and the lowest ratio in the ferret at 1:5. Furthermore, different strains of the same species may have disparate ratios (Mir et al. 1982). In most species removal of the superior cervical ganglion reduces the level of noradrenaline by 80%–90% indicating that this amine is confined predominantly to sympathetic nerves, whereas dopamine is localised to the chief cells (Fidone and Gonzalez 1982; Mir et al. 1982). An exception is the cat in which both amines are apparently contained within the chief cells (Pallot 1987). Catecholamines can also be demonstrated in the human carotid body by formalin-induced fluorescence (Hamburger et al. 1966), where their relative concentrations are dopamine 64%, noradrenaline 14.8% and adrenaline 2.5%, the remaining 18.8% consisting of 5-hydroxytryptamine (Steel and Hinterberger 1972). Ultrastructural studies demonstrate that catecholamines are contained within the dense-core vesicles (see Chapter 17).

Effects of Hypoxia

If catecholamines are involved in transmitting the chemosensory signal, there should be evidence that they are released during stimulation of the carotid body by hypoxia. Indirect evidence for release has been obtained by measuring the concentrations of amines following brief episodes of hypoxia. Thus, exposure of rats to 5% oxygen for 15 minutes reduced the level of dopamine by 30% while 10% oxygen took up to 1 hour for a similar level of depletion to occur (Hanbauer and Hellström 1978). Release of dopamine still occurred when the carotid sinus nerve was severed, indicating that it is triggered directly by hypoxia and not by efferent nerves. Noradrenaline on the other hand was not affected by hypoxia. A similar reduction in the content of dopamine in response to acute hypoxia has been found in the cat (Fitzgerald et al. 1983; Starlinger and Acker 1986; Pallot 1987). In this species, however, there was also a small (Fitzgerald et al. 1983) or large (Starlinger and Acker 1986) reduction in the content of noradrenaline consistent with the observation above that this amine is located predominantly in chief cells of the cat.

Depletion of catecholamines by hypoxia is associated with their increased synthesis by chief cells. This was shown by subjecting rabbits to hypoxia for 3 hours and then removing the carotid bodies and incubating them with tritiated tyrosine, a catecholamine precursor (Fidone et al. 1982a). Dopamine synthesis, as estimated by the incorporation of labelled tyrosine, was increased by an amount dependent upon the level of hypoxia. With 14% oxygen, synthesis was increased by 53% and with 10% oxygen a 72% elevation was observed; this was independent of whether or not

the nervous supply was intact during exposure to hypoxia. The production of noradrenaline was not affected, which might be anticipated from the small amount of this amine in chief cells of the rabbit. Direct evidence that the synthesised dopamine is secreted by the chief cells was obtained by studying the effects of hypoxia on the carotid body of rabbits in a superfusion chamber (Fidone et al. 1982b). It was found that there was release of labelled dopamine into the perfusate which increased progressively with declining oxygen concentration. The ability to synthesise and release dopamine in response to hypoxia is apparently an intrinsic property of chief cells, since this process occurs even when they are dissociated in tissue culture (Fishman et al. 1985).

The increased synthesis of dopamine in response to hypoxia appears to exceed its release over a prolonged time, leading to its accumulation within the carotid body. Thus after exposure of rats to diminished barometric pressure of 450 mmHg for 7 days there was a four-fold increase in the level of dopamine, which rose by a further six-fold on recovery from hypoxia (Olson et al. 1983). Exposure of rats to 10% oxygen for 21 days caused the dopamine levels to rise from 6.9 to 85.8 pmol/carotid body (Barer et al. 1986). There was also a substantial rise in the concentration of noradrenaline from 8.3 to 147.0 pmol/carotid body. Hypercarbia on the other hand had only a small influence on the metabolism of catecholamines (Barer et al. 1986).

Catecholamines as Chemosensory Transmitters

Experiments to investigate the possibility that catecholamines transmit the chemosensory signal involve injecting them into the carotid artery. The results of such studies are often contradictory. For example, in the cat exogenous noradrenaline increases sensory discharges after a brief period of inhibition (Eyzaguirre and Zapata 1984; Pallot 1987) but the time course of this response closely parallels that of systemic vasoconstriction and it is abolished by agents which prevent a pressor response. Furthermore, chemoexcitation is not induced by noradrenaline in superfusion chambers where the glomic vasculature is by-passed (Eyzaguirre and Zapata 1984). It is probable, therefore, that noradrenaline is normally released from sympathetic nerves adjacent to the glomic arteries, causing them to constrict. The subsequent hypoxia of the glomic tissue then initiates a chemosensory response. Noradrenaline thus affects the neural discharges indirectly and is unlikely to be a primary chemosensory transmitter (Eyzaguirre and Zapata 1984).

Dopamine has a predominantly inhibitory effect on the discharges in the sinus nerve of the cat (Monti-Bloch and Eyzaguirre 1980; Cárdenas and Zapata 1980; Eyzaguirre and Zapata 1984; McQueen 1984). Other species may respond differently, such as the rabbit in which it causes excitation in vitro but inhibition in vivo (Monti-Bloch and Eyzaguirre 1980) or the dog where low doses induce inhibition of the chemosensory discharge and high doses increase it (Eyzaguirre and Zapata 1984). The level of stimulation of the carotid body also influences the effect of exogenous dopamine. Thus, when the carotid body of the cat is stimulated by low doses of cyanide, dopamine has an inhibitory effect but, when the dose of cyanide is high, the same concentration of dopamine is excitatory (Cárdenas and Zapata 1980). This suggests that there may be two types of dopaminoceptor; inhibitory ones responding to small amounts of dopamine and excitatory ones responding to a larger release of dopamine during powerful stimulation (Cárdenas and Zapata 1980). Indeed, different proportions of these receptors in different species might explain the variable effects of dopamine in experimental animals (Eyzaguirre and Zapata 1984). In humans the effect of exogenous dopamine is inhibitory, since it depresses ventilation both in normoxia and hypoxia (Welsh et al. 1978).

A common criticism of experiments involving application of exogenous transmitters is that it is difficult to determine what constitutes a physiological dose and to what extent the responses are due to their secondary effects. As a consequence, specific blockers of the activity of endogenous transmitters are employed. One of these, haloperidol, produces a variable effect on the activity of the carotid body, sometimes inducing chemoexcitation, sometimes inhibition, depending upon the species and the dose (Chow et al. 1986). In the cat low doses of haloperidol appear to oppose the inhibitory action of endogenous dopamine. Since the drug can cross the blood–brain barrier, it is possible that higher doses influence the central respiratory centres (Chow et al. 1986). Domperidone is a dopamine antagonist which does not affect the central nervous system and when it is infused into the carotid artery of the cat or rabbit it causes an increase in the rate of discharge in the sinus nerve and opposes the depressant action of exogenously applied dopamine (Mir et al. 1984; McQueen 1984; Zapata and Torrealba 1984; Ponte and Sadler 1989a). It has little effect on ventilation in these animals (Zapata and Tor-

realba 1984) or in humans (Delpierre et al. 1987). Domperidone acts by specifically blocking dopamine D2 receptors and agonists of these receptors cause chemosensory inhibition whereas D1 receptor agonists do not (Mir et al. 1984; McQueen 1984). This suggests that dopamine stimulates D2 receptors but their precise location is not known.

The difficulty in accepting dopamine as a transmitter of the chemosensory response lies in the experimental evidence that its effects are largely inhibitory rather than excitatory and also that dopamine antagonists do not block the response of the carotid body to hypoxia (Eyzaguirre and Zapata 1984; Mir et al. 1984; Ponte and Sadler 1989a). In fact the depressant action of dopamine is greatest in normoxia, becoming progressively less effective as the oxygen tension falls (Ponte and Sadler 1989a). For these reasons it has been suggested that dopamine has merely a modulatory rôle in chemoreception, possibly being released through the action of efferent nerves (Eyzaguirre and Zapata 1984). In this context Ponte and Sadler (1989a) observed that direct application of domperidone to the carotid body of hypoxic rabbits or cats for a prolonged period of time induced a gradual decline in discharge frequency in fibres of the carotid sinus nerve. This suggested to them that there is normally a tendency for the nerve endings to adapt to the hypoxic stimulus and that dopamine opposes this. They concluded that dopamine may not be involved in the acute response to hypoxia but is important in the long-term maintenance of the chemoreceptor reflex. This would explain why prolonged hypoxia is associated with increased synthesis and release of dopamine.

5-Hydroxytryptamine

This indolamine is present in significant amounts in the adult human carotid body (Hamburger et al. 1966) accounting for 18.8% of the total amine content (Steele and Hinterberger 1972). Its concentration is higher in patients with systemic hypertension (Steele and Hinterberger 1972) and in the infant accounts for 72% of all amines (Perrin et al. 1986). Dearnaley et al. (1968) could find no fluorescence in the carotid body of the rabbit corresponding to the wavelength associated with 5-hydroxytryptamine (5-HT). However, in this species there is evidence that glomic tissue may possess receptors for serotonin (Mir et al. 1984). There is little evidence to suggest that 5-HT is a neurotransmitter. In the dog it causes increased chemosensory discharge in vivo when applied exogenously but will do so only rarely in the superfusion chamber (Torrance 1968). Furthermore, although cultured glomic cells secrete 5-HT into the medium, this is not influenced by hypoxia as is the release of dopamine (Fishman et al. 1985). For such reasons this indolamine has been largely ignored by physiologists, but since there are strong grounds for rejecting both acetylcholine and catecholamines as chemosensory transmitters its rôle should be reassessed.

Peptide Neurotransmitters

The carotid bodies of humans and animals contain significant concentrations of peptides which elsewhere in the body may act as neurotransmitters (see Chapter 8). Foremost among these are methionine- and leucine-enkephalins, substance P and vasoactive intestinal peptide (VIP). The effects of these peptides or their antagonists on chemoreception in experimental animals are as contradictory as are those for dopamine and they are reviewed briefly in Chapter 8. The overall impression that one gains from these studies is that methionine-enkephalin inhibits the chemosensory reflex, substance P augments it and vasoactive intestinal peptide has a variable effect dependent upon the dose. There are, however, many exceptions to this generalisation, so that a rôle for peptides acting as neurotransmitters of the chemosensory response is far from assured as yet. It is considered by some workers that peptides may be involved in an endocrine function of the carotid body (Chapters 8 and 15).

The Chief Cell as Transducer

The component of the carotid body responsible for detecting changes in arterial blood gases and converting this information into an electrical signal has so far evaded precise indentification. Such a transducer is often considered to be in the chief cell, largely because of its complement of neurotransmitters, its synaptic contact with nerve endings and its proximity to capillaries. Some experiments to establish it in this rôle have sought to destroy chief cells selectively whilst leaving the nerve endings intact. Verna et al. (1975) applied a cold probe to the rabbit carotid body to kill the chief cells, with loss of formaldehyde-induced fluorescence. Between 3 and 5 months later chemoreceptor activity in the carotid sinus nerve was much reduced or absent, despite the fact that

the nerve endings appeared to be intact. In similar studies Monti-Bloch et al. (1983) ligated the glomic artery of the rat and the resulting ischaemia caused necrosis of chief and sustentacular cells followed by regeneration of nerve axons. The carotid bodies then showed little or no response to stimulation by hypoxia, from which it was concluded that chief cells are essential for providing the chemosensory response.

Another approach involved transplanting foreign nerves into the carotid body and testing whether they conducted chemosensory impulses. de Castro (1940 cited by Pallot 1987) was the first to perform such a procedure, using vagal nerve fibres. Perfusion of the carotid artery with hypercarbic solutions elicited an increase in ventilation. However, such a response could have been due to chemoreceptor fibres from the aortic bodies travelling in the vagus nerve. To circumvent such criticisms, the carotid and lingual branches of the ninth cranial nerve of cats were bilaterally crossed over and left to regenerate (Dinger et al. 1984). Electron microscopy showed that the nerves grew into the carotid body and became apposed to chief cells, although their numbers were reduced to 10% of normal. Stimulation of the carotid body by asphyxia provoked discharges in the transplanted nerve which were also reduced to 10% of normal. Interestingly, stimulation of the taste buds with acid caused short bursts of hyperventilation, indicating that the carotid sinus nerves were still active but now responded to a different sensory organ. This type of experiment was taken a step further by transplanting the carotid bodies of cats into the tenuissimus muscle of the hind leg. The nerve supplying the muscle was sutured close to the carotid body and the nerves allowed to regenerate for 95–174 days (Monti-Bloch et al. 1984). The cats then breathed hypoxic or hypercarbic gas mixtures and recordings from the muscle nerve showed that 40% of its fibres responded to these stimuli as well as to stretching of the muscle. The inference from these experiments is that the nerve endings which regenerate from the transplanted nerves are responding passively to stimulation by their adopted sensory organ. Since they grow into close juxtaposition with chief cells, the conclusion can be drawn that the chief cell is essential for chemoreception and is thus the transducer. Against this argument must be balanced the observations made in the preceding sections in which there are strong grounds for rejecting each of the potential neurotransmitters in chief cells as effectors of the chemosensory signal.

Irrespective of whether or not the chief cell is the primary transducer of chemoreception it is

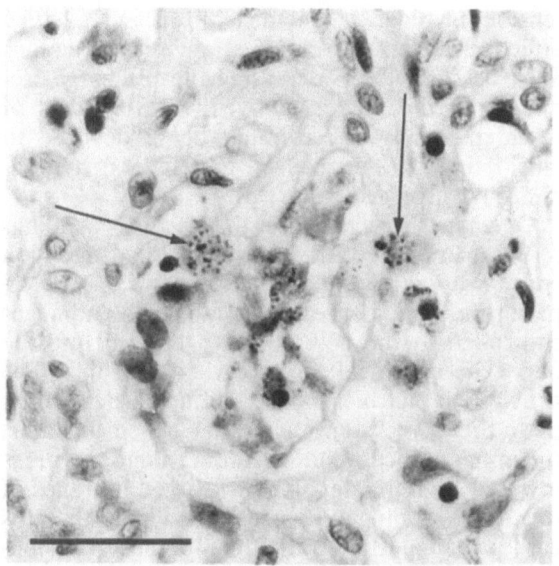

Fig. 7.1. Lobule from the carotid body of a man of 47 years with haemosiderosis due to repeated blood transfusions. Several chief cells contain granular deposits of haemosiderin (*arrows*) which stain with Perls's Prussian blue reaction. (Perls's) Scale line = 25 μm.

sensitive to changes in the composition of the blood. Histological evidence supporting this came from our examination of the carotid body at necropsy from a man of 47 years who had had repeated blood transfusions in the treatment of massive haematemeses from ruptured oesophageal varices. The chief cells contained deposits of haemosiderin appearing as brown granules within their cytoplasm (Fig. 7.1) which stained positively with Perls's reagents. Iron-overload due to repeated blood transfusions commonly leads to deposition of haemosiderin in hepatocytes and Küpffer cells but our observation makes it clear that the same deposition of ferric iron occurs in the chief cells of the carotid body, indicating that they too are sensitive to biochemical changes in the plasma.

The Nerve Endings as Transducer

An alternative possibility is that nerve endings in the carotid body have an intrinsic chemoreceptor property. Most experiments to investigate this involve cutting the carotid sinus nerve close to the carotid body and suturing its central stump into the adventitia of the external carotid artery. After several months, nerve fibres from the stump regenerate and grow into the surrounding tissue to form a neuroma. These fibres can be attached to

electrodes and their response to various stimuli assessed. Several workers have found that such neuromas respond to hypoxia and cyanide. However, they did not remove the carotid bodies and so it is possible that some nerve fibres may have re-established contact with glomic tissue (Eyzaguirre and Fidone 1980). Bingmann and Kienecker (1984) studied neuromas after removal of the carotid bodies. Spontaneous discharges could be recorded from the regenerated nerves which increased during hypoxia, although to a lesser degree than in the intact carotid body. The neuromas did not respond to changes in carbon dioxide tension or pH, providing further evidence that these stimuli are detected differently from hypoxia. These authors also grew spinal ganglia in culture and found that their axons did not respond to hypoxia by changes in membrane potential. They concluded that a response to hypoxic stimulation is not a universal property of nerves in general but is peculiar to those supplying the carotid body.

Ponte and Sadler (1989b) were more circumspect in their interpretations. They teased out the fibres from neuromas of the cut carotid sinus nerve in rabbits and measured the electrical responses in each. From a total of 700 fibres tested, 98.3% were mechanosensitive and hence presumably baroreceptor. In animals with both carotid bodies removed, only 1.7% of fibres responded to hypoxia but this proportion increased more than two-fold if the glomera were left intact. Although sectioning the sinus nerves is reported to cause degeneration of rabbit carotid bodies, Ponte and Sadler (1989b) were able to find small groups of surviving chief cells by electron microscopy. Nerve endings, derived from the neuromas, that responded to cyanide were found close to these cells. Nerves confined to the neuromas could not be stimulated by cyanide. Ponte and Sadler (1989b) concluded that nerve endings forming in neuromas are not chemosensitive on their own and that for chemoreception to occur they must become closely associated with chief cells. This does not necessarily preclude nerve endings as the transducers of chemoreception but it adds the caveat that they must be near to chief cells in order to acquire that function. Such considerations are supported by the experiments of Zapata et al. (1976) in which the carotid sinus nerve was crushed at varying distances from the carotid body. Chemosensory activity was re-established after intervals of time which were longer the further the regenerating nerves had to travel to reach the carotid body. Ultrastructural examination established that regenerated nerve endings apposed to chief cells appeared at the same time as chemosensory activity recovered. No responses were found before such contact was made, indicating that nerve endings on their own cannot respond to chemical stimuli but require the close presence of chief cells. Interestingly, this apposition does not require synaptic contact, suggesting that neurotransmitters are not involved (Ponte and Sadler 1989b). Thus, one can speculate that chief cells exert a trophic influence over nerve endings inducing in them the property to respond to changes in oxygen tension. The chief cells themselves may serve as modulators of this response.

The Sustentacular Cell as Transducer

Paintal (1968) suggested that the sustentacular cell might be the oxygen sensor and transducer. He argued that the chief cell is highly metabolically active, utilising large quantities of oxygen and tending to lower the oxygen tension in its vicinity, which is sensed by the adjacent sustentacular cell. This responds by contraction of the filaments in its cytoplasm, thereby deforming the axons contained within it, which respond by discharging impulses. During hypoxaemia the oxygen tension near to sustentacular cells falls dramatically so that the mechanical stimulation of nerves is greatly accentuated. According to this hypothesis the glomic nerves are mechanosensory, yet poking the carotid body has never been shown to influence the rate of discharge in the carotid sinus nerve. Mills and Jöbsis (1972) provided evidence for the existence in the carotid body of a unique form of cytochrome which has a low affinity for oxygen as described below. They suggested that it might be in the sustentacular cells, which are as a consequence unusually sensitive to changes in oxygen tension and can thus act as oxygen sensors. These authors did not indicate how this information could be translated into signals to the chief cells or nerve endings. Such considerations have largely fallen into disfavour, but it should be borne in mind that the sustentacular cell is well placed to act as an oxygen sensor, since it is closer to the blood than either chief cells or nerves and is in intimate contact with both.

Biochemistry of Transduction

Irrespective of where in the carotid body the process of transduction occurs, it is highly likely that the initial detection of changes in oxygen tension involves biochemical processes. To be effective in transduction these chemical reactions must be capable of increasing the discharge frequency in the sensory nerves. A great deal of research activity is currently being devoted to elucidating this mechanism and several hypotheses have emerged so far.

One hypothesis proposes that the nerve endings respond to changes in extracellular hydrogen ion concentration (Eyzaguirre and Zapata 1984). The pH in the small space surrounding the endings is controlled by the flow of hydrogen ions into the nerve in exchange for bicarbonate ions. The activity of this bicarbonate pump is dependent upon the oxygen tension so that, when this falls, the efflux of bicarbonate is reduced and the pH is lowered, thereby depolarising the membranes of the nerve endings. It is assumed that there is carbonic anhydrase on the membranes of sustentacular and chief cells but, although this enzyme has been identified in both, its precise location is uncertain (Eyzaguirre and Zapata 1984). This theory has the advantage of explaining why hypoxia, hypercarbia and low pH all stimulate the carotid body and also explains why inhibitors of carbonic anhydrase, such as acetazolamide, reduce the response of the carotid body to carbon dioxide.

A different approach originated from the observation that certain poisons of oxidative phosphorylation such as cyanide and dinitrophenol cause excitation of the carotid body, suggesting that hypoxia acts at the level of oxidative phosphorylation in the mitochondria. The process of oxidative phosphorylation, in which oxygen is employed to generate ATP, occurs in three stages. The first is the electron transport chain, involving cytochromes and flavoproteins. This is inhibited by cyanide and antimycin A. The second is referred to as "initial energy conservation" and represents an intermediate high-energy state before ADP is phosphorylated. It is blocked by uncouplers of oxidative phosphorylation, such as dinitrophenol, which dissipates the energy conserved. The third stage is that of phosphorylation in which ADP is converted into the high-energy phosphate ATP. This stage is inhibited by oligomycin (Mulligan et al. 1981). There is dispute as to precisely at what stage in this process transduction of the chemosensory stimulus might occur. One school of thought believes that it is at the level of initial energy conservation. An important function of this intermediate, high-energy state is to drive the pump which transports calcium ions into the mitochondria. Any reduction in initial energy conservation will lead to release of calcium ions from the mitochondria into the cytosol of the cytoplasm. This in turn will cause release of potassium ions into the extracellular space, depolarise the cell membrane and initiate a nervous discharge (Mulligan et al. 1981; Acker 1987; Biscoe and Duchen 1990). This scheme is referred to as "the calcium hypothesis".

Another school of thought considers the important step to be at the level of phosphorylation and that reduced synthesis of ATP causes chemoexcitation by producing instability of membranes in the nerve endings (Acker 1987). This is known as "the metabolic hypothesis". Mulligan et al. (1981) sought to distinguish between these hypotheses by employing a battery of blockers and uncouplers of oxidative phosphorylation on cats. Oligomycin caused a brisk increase in discharge in the carotid sinus nerve which subsided to a lower constant frequency after about 25 seconds. If the cats were then rendered hypoxic there was little or no increase in discharge frequency. Subsequent application of other blockers such as cyanide, antimycin A and dinitrophenol failed to induce their usual chemoexcitation. Mulligan et al. (1981) argued that oligomycin blocks the process of phosphorylation and deprives the cell of ATP. It does not influence the earlier stage of initial energy conservation and should not, therefore, have any effect upon calcium transport; yet it is still capable of causing vigorous stimulation of the carotid body. Furthermore, in the presence of oligomycin, uncouplers of initial energy conservation and electron transport should still be capable of releasing calcium from the mitochondria and stimulating a chemosensory response. The fact that they do not mitigates against the calcium hypothesis of transduction and suggests that reduced synthesis of ATP is responsible. As an interesting aside to their study, Mulligan et al. (1981) noted that oligomycin did not block the response of the carotid body to hypercarbia, indicating that carbon dioxide causes excitation through a different pathway. In order for either of the above hypotheses to work, the electron transport chain in the carotid body must be unusually sensitive to small reductions in oxygen tension.

Normal cytochromes are not influenced by hypoxia until the oxygen tension falls below 10 mmHg, which is well below that at which chemoreception occurs. Mills and Jöbsis (1972) provided evidence that the carotid body contains two types of cytochrome a_3, one being normal with a high affinity for oxygen and the other unique with a very low affinity for oxygen. This low-affinity cytochrome a_3 would begin to become reduced at an oxygen tension of about 100 mmHg

(Eyzaguirre and Zapata 1984). It would thus limit the rate of oxidative phosphorylation and ATP production within a range of oxygen tension similar to that to which the carotid body responds.

Central Regulation of the Carotid Body

The afferent sensory impulses travelling in the carotid sinus nerve to the petrosal ganglion appear to be under some form of efferent control but there is disagreement and controversy over the pathways of this control and its effects. The most obvious form of efferent control is a sympathetic supply from the superior cervical ganglion to the ganglioglomerular nerve, some fibres of which innervate the carotid body. Electrical stimulation of the ganglioglomerular nerve produces excitatory responses in the carotid body which are of two types (O'Regan 1981). The first (type I) is not α-adrenergic in nature but the second (type II) is. Sympathetic stimulation causes a reduction in blood flow in the carotid body which is attenuated by α-adrenergic blockers, indicating that the type II response is vasoconstrictor. Removal of the superior cervical ganglion leads to an increased blood flow without materially affecting the rate of chemosensory impulses (Eyzaguirre and Zapata 1984). As discussed previously, sympathectomy leads to degeneration of noradrenaline-containing nerve endings around the glomic vasculature which may be effectors of the type II response. The type I response probably arises from activation of sympathetic fibres in the glomic tissue (O'Regan 1981). Verna et al. (1984) used 6-hydroxydopamine to create selective degeneration of sympathetic nerves in rabbits. Not only were degenerating nerves found around glomic arteries but some were also seen close to cell clusters. He suggested that sympathetic activity may also act to reset the chemoreceptor response of the carotid body during exercise, stress or acclimatisation to high altitude.

Efferent signals travelling in the severed carotid sinus nerve can be detected from electrodes applied to its central stump. Electrical stimulation of the distal stump, sending impulses towards the carotid body, inhibits sensory discharges in unstimulated fibres (Eyzaguirre and Zapata 1984). Such observations suggest that the carotid body normally receives centrifugal impulses from fibres in the carotid sinus nerve and that these have an inhibitory influence. This efferent inhibitory activity is also operative during prolonged exposure to hypoxia (Lahiri et al. 1983). The origin and nature of these neural pathways is uncertain. They may be efferent, motor nerves which synapse with chief cells or they might be parasympathetic fibres which synapse with ganglion cells near to the carotid body from which postganglionic fibres innervate glomic arteries. It has also been suggested that retrograde (antidromic) stimulation of sensory afferents may collide with centripetal impulses thereby blocking the chemosensory signal (Eyzaguirre and Zapata 1984). Whatever their nature, it is possible that these nerves provide the pathway by which the central respiratory centres can modify the response of the carotid body to hypoxia and hypercarbia (O'Regan and Majcherczyk 1982). One can speculate that they stimulate chief cells to release inhibitory neurotransmitters such as dopamine and methionine-enkephalin which reduce the discharge frequency of sensory nerve endings.

References

Acker H (1987) The involvement of nerve terminals in the paraganglionic chemoreceptor system. Ann NY Acad Sci 519:369–384

Band DM, Linton RAF (1983) Plasma potassium and the carotid body chemoreceptor in the cat. J Physiol (Lond) 345:33P

Band DM, Linton RAF (1987) The interaction of hypoxia and K^+ on the carotid chemoreceptor in anaesthetized cats. J Physiol (Lond) 394:65P

Barer G, Wach R, Pallot D, Bee D (1986) Almitrine, hypoxia, systemic hypertension and the carotid body. In: Heath D (ed) Aspects of hypoxia. Liverpool University Press, Liverpool, pp 113–129

Bingman D, Kienecker EW (1984) Effects of hypoxia on regenerated sinus nerve fibres in vivo and on neurons in vitro. In: Pallot DJ (ed) The peripheral arterial chemoreceptors. Croom Helm, Beckenham, Kent, pp 243–252

Biscoe TJ, Duchen MR (1990) Cellular basis of transduction in carotid chemoreceptors. Am J Physiol 258: L271–L278

Biscoe TJ, Bradley GW, Purves MJ (1970a) The relation between carotid body chemoreceptor discharge, carotid sinus pressure and carotid body venous flow. J Physiol (Lond) 208:99–120

Biscoe TJ, Lall A, Sampson SR (1970b) Electron microscopic and electrophysiological studies on the carotid body following intracranial section of the glossopharyngeal nerve. J Physiol (Lond) 208:133–152

Cárdenas H, Zapata P (1980) Dual effects of dopamine upon chemosensory responses to cyanide. Neurosci Lett 18:317–322

Chow CM, Winder C, Read DJC (1986) Influences of endogenous dopamine on carotid body discharge and ventilation. J Appl Physiol 60:370–375

Dearnaley DP, Fillenz M, Woods RI (1968) The identification of dopamine in the rabbit's carotid body. Proc R Soc Lond [Biol] 170:195–203

de Castro F (1926) Sur la structure et l'innervation de la glande intercarotidienne (glomus caroticum) de l'homme et des mammifères, et sur un nouveau système d'innervation autonome du nerf glossopharyngien. Trav Lab Rech Biol 24:365–432

de Castro F (1928) Sur la structure et l'innervation du sinus carotidien de l'homme et des mammifères. Nouveaux faits sur l'innervation et la fonction du glomus caroticum. Études anatomiques et physiologiques. Trav Lab Rech Biol 25:331–380

Dejours P (1963) Control of respiration by arterial chemoreceptors. Ann NY Acad Sci 109:682–693

Delpierre S, Fornais M, Guillot C, Grimaud C (1987) Increased ventilatory chemosensitivity induced by domperidone, a dopamine antagonist, in healthy humans. Bull Eur Physiopathol Respir 23: 31–35

Dinger BG, Stensaas LJ, Fidone SJ (1984) Chemosensory end-organs reinervated by normal and foreign nerves. In: Pallot DJ (ed) The peripheral arterial chemoreceptors. Croom Helm, Beckenham, Kent, pp 225–234

Eyzaguirre C, Fidone SJ (1980) Transduction mechanisms in carotid body: glomus cells, putative neurotransmitters, and nerve endings. Am J Physiol 239:C135–C152

Eyzaguirre C, Zapata P (1984) Perspectives in carotid body research. J Appl Physiol 57:931–957

Fidone S, Gonzalez C (1982) Catecholamine synthesis in rabbit carotid body in vitro. J Physiol (Lond) 333:69–79

Fidone SJ, Zapata P, Stensaas LJ (1977) Axonal transport of labeled material into sensory nerve endings of cat carotid body. Brain Res 124:9–28

Fidone S, Gonzalez C, Yoshizaki K (1982a) Effects of hypoxia on catecholamine synthesis in rabbit carotid body in vitro. J Physiol (Lond) 333:81–91

Fidone S, Gonzalez C, Yoshizaki K (1982b) Effects of low oxygen on the release of dopamine from the rabbit carotid body in vitro. J Physiol (Lond) 333:93–110

Fishman MC, Greene WL, Platika D (1985) Oxygen chemoreception by carotid body cells in culture. Proc Natl Acad Sci USA 82:1448–1450

Fitzgerald RS, Garger P, Hauer MC, Raff H, Fechter L (1983) Effect of hypoxia and hypercapnia on catecholamine content in cat carotid body. J Appl Physiol 54:1408–1413

Hamburger B, Ritzén M, Wersall J (1966) Demonstration of catecholamines and 5-hydroxytryptamine in the human carotid body. J Pharmacol Exp Ther 152:197–201

Hanbauer I, Hellström S (1978) The regulation of dopamine and noradrenaline in the rat carotid body and its modification by denervation and hypoxia. J Physiol (Lond) 282:21–34

Haymet BT, McCloskey DI (1975) Baroreceptor and chemoreceptor influences on heart rate during the respiratory cycle in the dog. J Physiol (Lond) 245:699–712

Heymans C, Bouckhaert JJ, Dautrebande L (1930) Sinus carotidien et réflexes respiratoires. II. Influences respiratoires réflexes de l'acidôse de l'alkalôse, de l'anhydride carbonique, de l'ion hydrogène et de l'anoxémie. Sinus carotidiens et échanges respiratoires dans les poumons et au delà des poumons. Arch Int Pharmacodyn Ther 39:400–448

Holton P, Wood JB (1965) The effects of bilateral removal of the carotid bodies and denervation of the carotid sinuses in two human subjects. J Physiol (Lond) 181:365–378

Honda Y (1985) Role of carotid chemoreceptors in control of breathing at rest and in exercise: studies on human subjects with bilateral carotid body resection. Japn J Physiol 35:535–544

Lahiri S, Smatresk N, Pokorski M, Barnard P, Mokashi A (1983) Efferent inhibition of carotid body chemoreception in chronically hypoxic cats. Am J Physiol 245:R678–R683

Lugliani R, Whipp BJ, Seard C, Wasserman K (1971) Effect of bilateral carotid body resection on ventilatory control at rest and during exercise in man. New Engl J Med 285:1105–1111

Marshall JM (1987) Analysis of cardiovascular responses

evoked following changes in peripheral chemoreceptor activity in the rat. J Physiol (Lond) 394:393–414

McQueen DS (1984) Effects of selective dopamine receptor agonists and antagonists on carotid body chemoreceptor activity. In: Pallot DJ (ed) The peripheral arterial chemoreceptors. Croom Helm, Beckenham, Kent, pp 325–333

Mills E, Jöbsis FF (1972) Mitochondrial respiratory chain of carotid body and chemoreceptor response to changes in oxygen tension. J Neurophysiol 35:405–428

Mir AK, Al-Neamy K, Pallot DJ, Nahorski SR (1982) Catecholamines in the carotid body of several mammalian species: effects of surgical and chemical sympathectomy. Brain Res 252:335–342

Mir AK, McQueen DS, Pallot DJ, Nahorski SR (1984) Direct biochemical and neuropharmacological indentification of dopamine D_2-receptors in the rabbit carotid body. Brain Res 291:273–283

Monti-Bloch L, Eyzaguirre C (1980) A comparative physiological and pharmacological study of cat and rabbit carotid body chemoreceptors. Brain Res 193:449–464

Monti-Bloch L, Stensaas LJ, Eyzaguirre C (1983) Effects of ischaemia on the function and structure of the cat carotid body. Brain Res 270:63–76

Monti-Bloch L, Stensaas LJ, Eyzaguirre C (1984) Induction of chemosensitivity in muscle nerve fibres by carotid body transplantation. In: Pallot DJ (ed) The peripheral arterial chemoreceptors. Croom Helm, Beckenham, Kent, pp 235–242

Mulligan E, Lahiri S, Storey BT (1981) Carotid body O_2 chemoreception and mitochondrial oxidative phosphorylation. J Appl Physiol 51:438–446

Murray PA, Lavallee M, Vatner SF (1984) α-Adrenergic-mediated reduction in coronary blood flow secondary to carotid chemoreceptor reflex activation in conscious dogs. Circ Res 54:96–106

Olson EB, Vidruk EH, McGimmon DR, Dempsey JA (1983) Monoamine neurotransmitter metabolism during acclimatization to hypoxia in rats. Respir Physiol 54:79–96

Olson EB, Vidruk EH, Dempsey JA (1988) Carotid body excision significantly changes ventilatory control in awake cats. J Appl Physiol 64:666–671

Olson NC, Robinson NE, Anderson DL, Scott JB (1982) Effect of carotid body hypoxia and/or hypercapnia on pulmonary vascular resistance. Proc Soc Exp Biol Med 170:188–193

O'Regan RG (1981) Responses of carotid body chemosensory activity and blood flow to stimulation of sympathetic nerves in the cat. J Physiol (Lond) 315:81–98

O'Regan RG, Majcherczyk S (1982) Role of peripheral chemoreceptors and central chemosensitivity in the regulation of respiration and circulation. J Exp Biol 100:23–40

Paintal AS (1968) Some considerations relating to studies on chemoreceptor responses. In: Torrance RW (ed) Proceedings of the Wates Foundation Symposium on arterial chemoreceptors. Blackwell, Edinburgh and Oxford, pp 253–261

Pallot DJ (1987) The mammalian carotid body. Adv Anat Embryol Cell Biol 102:1–91

Parker PE, Dabney JM, Scott JB, Haddy FJ (1975) Reflex vascular responses in kidney, ileum and forelimb to carotid body stimulation. Am J Physiol 228:46–51

Perrin DG, Chan W, Cutz E, Madapallimattam A, Cole MJ (1986) Serotonin in the human infant carotid body. Experientia 42:562–564

Ponte J, Sadler CL (1989a) Interactions between hypoxia, acetylcholine and dopamine in the carotid body of the rabbit and cat. J Physiol (Lond) 410:395–410

Ponte J, Sadler CL (1989b) Studies on the regenerated carotid sinus nerve of the rabbit. J Physiol (Lond) 410:411–424

Starlinger H, Acker H (1986) The norepinephrine and dopamine content of the cat carotid body in vivo under normoxic and hypoxic conditions. Neurosci Lett 64:65–68

Steele RH, Hinterberger J (1972) Catecholamines and 5-hydroxytryptamine in the carotid body in vascular, respiratory, and other diseases. J Lab Clin Med 80:63–70

Torrance RW (1968) Prolegomena. In: Torrance RW (ed) Proceedings of the Wates Foundation Symposium on arterial chemoreceptors. Blackwell, Edinburgh and Oxford, pp 1–40

Verna A, Roumy M, Leitner LM (1975) Loss of chemoreceptive properties of the rabbit carotid body after destruction of the glomus cells. Brain Res 100:13–23

Verna A, Barets A, Salat C (1984) Distribution of sympathetic nerve endings within the rabbit carotid body: a histochemical and ultrastructural study. J Neurocytol 13:849–865

Welsh MJ, Heistad DD, Abboud FM (1978) Depression of ventilation by dopamine in man: evidence for an effect on the chemoreceptor reflex. J Clin Invest 61:708–713

Zapata P, Torrealba F (1984) Blockade of dopamine-induced chemosensory inhibition by domperidone. Neurosci Lett 51:359–364

Zapata P, Stensaas LJ, Eyzaguirre C (1976) Axon regeneration following a lesion of the carotid nerve: electrophysiological and ultrastructural observations. Brain Res 113:235–253

8 Peptides

Biologically active peptides have been identified in the carotid bodies of animals over the last decade (Table 8.1). In contrast there are few papers concerned with peptides in the human carotid body, which, however, contains at least six. More than 20 years ago Pearse (1969) predicted that a hypothetical peptide hormone, which he called "glomin", was likely to be present in the glomic chief cells for structurally these are typical APUD cells, which are known to secrete polypeptide hormones. In this chapter the concentration and location of six peptides in the human carotid body are considered and their possible functions discussed.

Concentrations of Peptides

The human carotid body contains methionine (Met)- and leucine (Leu)- enkephalins, bombesin, neurotensin, substance P and vasoactive intestinal polypeptide (VIP). Their concentrations are shown in Table 8.2. In the study from which these data are derived (Heath et al. 1988) carotid bodies were removed from 15 necropsies carried out less than 24 hours after death. The subjects ranged in age from 60 to 87 years. Nine were male and six female. None had died from conditions which were likely to be associated with hyperplasia of the glomic tissue (see Chapter 10). After the tissues were extracted with boiling acetic acid for 15 minutes, they were stored in solid carbon

Table 8.1. Peptides identified in the carotid bodies of animals

Peptide	Species	Reference
Met-enkephalin	Cat	Hansen et al. 1982; Lundberg et al. 1979; Varndell et al. 1982; Wharton et al. 1980
	Dog	Kobayashi et al. 1983
	Rabbit	Hanson et al. 1986
	Pig	Fried et al. 1989; Varndell et al. 1982
Leu-enkephalin	Cat	Hansen et al. 1982; Lundberg et al. 1979; Wharton et al. 1980
Substance P	Cat	Cuello and McQueen 1980; Lundberg et al. 1979; Prabhakar et al. 1989; Scheibner et al. 1988; Wharton et al. 1980
	Rabbit	Gallagher et al. 1985; Hanson et al. 1986
	Chicken	Kameda 1989
VIP	Cat	Lundberg et al. 1979; Wharton et al. 1980
	Chicken	Kameda 1989
Cholecystokinin	Cat	Kummer et al. 1985; McQueen and Ribeiro 1981
Neurokinin A	Cat	Prabhakar et al. 1989
Galanin, neuropeptide Y	Pig	Fried et al. 1989
Calcitonin gene-related peptide, somatostatin	Chicken	Kameda 1989
Neuron-specific enolase, S-100 protein	Rat	Kondo et al. 1982

VIP, vasoactive intestinal peptide.

Table 8.2. Concentrations of six peptides in the human carotid body as determined by radioimmunoassay (after Heath et al. 1988)

Peptide	No. carotid bodies assayed	Concentration in pmol/g	
		Mean	Range
Met-enkephalin	10	612	219–1267
Leu-enkephalin	10	162	53–353
Bombesin	13	73	30–124
Neurotensin	13	67	36–114
Substance P	8	16	4–25
VIP	13	9	4–16

VIP, vasoactive intestinal peptide.

dioxide prior to radioimmunoassay using commercial kits employing antisera to the various peptides.

Two of the above peptides in the human carotid body, Met- and Leu-enkephalins, were first discovered in the brain. Both consist of five amino acid residues and act on the central nervous system in a fashion analogous to that of morphine. Other larger but chemically related peptides have also been identified in the brain and included with the enkephalins in a group referred to as endorphins (Polak and Bloom 1980). Because of their pharmacological similarities to morphine they are also collectively called "opiate peptides". In the periphery, enkephalins are localised to the adrenal medulla and autonomic nerves.

The enkephalins in the human carotid body are present to a much greater concentration than are the other four peptides listed in Table 8.2. Although there is a wide variation in concentration between individuals, there is consistently four times as much Met- as Leu-enkephalin. Indeed there is a close statistical correlation between the levels of the two peptides. A similar four-fold preponderance of Met-enkephalin has been reported in the carotid body of the cat (Wharton et al. 1980). Also present in significantly high concentrations are bombesin and neurotensin, although there is a poor correlation between them. The neuropeptides substance P and VIP are found at very low concentrations, close to the limits of detection by radioimmunoassay. The levels of VIP are closely correlated with those of neurotensin and to a lesser extent with those of bombesin. Alpha-atrial natriuretic factor is not present or at least is undetectable by radioimmunoassay of tissues obtained post mortem. Radioimmunoassay gives no indication as to where these peptides are located. For this purpose histological sections labelled by an immunohistochemical technique are required.

Immunohistochemistry of the Normal Carotid Body

Many studies on the peptides within the carotid bodies of animals have employed an indirect immunofluorescence technique. This requires the use of thick, frozen sections and is a very sensitive means of demonstrating nerve axons. Although major components of the organ such as lobules and stroma can be distinguished in this way, fine distinctions, such as the identity of the three variants of chief cell, are obscured. For a detailed study of the precise location of peptides, thin sections cut from tissues embedded in paraffin-wax are to be preferred. The commonest immunohistochemical technique applied to such sections involves a peroxidase enzyme conjugated to a secondary antibody or else the more reliable peroxidase–antiperoxidase (PAP) technique in which a sandwich of three antibodies is used. The location of the peroxidase is then rendered visible by adding hydrogen peroxide and an oxygen-sensitive chemical called a chromogen. The oxygen which is liberated is captured by the chromogen, which is converted into a dye that accurately pin-points the location of the primary antibody and hence the peptide. In our experience this method produces very weak labelling in many human carotid bodies and we have employed instead the more sensitive, although technically more exacting, method of immunogold–silver labelling. In this technique the secondary antibody is conjugated with colloidal gold particles which become bound to the site of the primary antibody and hence the peptides. These gold particles are too small to resolve under the light microscope and so a silver solution such as the Janssen Intense M kit is added. This causes metallic silver to be deposited onto the gold particles making them bigger and hence visible. As a result labelled cells appear black or brown depending upon the density of silver granules in their cytoplasm.

We employed the immunogold–silver technique on 26 carotid bodies obtained at necropsy (Smith et al. 1990). The interval between death and fixation of the tissues in all cases was less than 48 hours. Cases were excluded from study if they showed evidence of systemic hypertension or pulmonary emphysema, since both conditions may lead to hyperplasia of the carotid body (see Chapter 10). Histological sections were treated with primary antisera directed against Met- and Leu-enkephalin, bombesin, neurotensin, substance P and VIP. Dilutions of these antisera were chosen such that a visible deposit of silver granules was present but not so dense that it obscured the nuclei, the recognition of which was necessary to identify the variants of chief cell. The results of this study on normal carotid bodies are described below.

Enkephalins

Immunolabelling for both Met- and Leu-enkephalins is predominantly within glomic chief cells, with no labelling at all in the sustentacular cells (Fig. 8.1). There is considerable variability in the intensity of the labelling between cases, with some showing a dense deposit of silver rendering the cells black whilst others have a weaker reaction in which the chief cells appear pale brown. There is also considerable variation in the proportion of chief cells that is labelled and this wide variability between cases is in agreement with the findings at radioimmunoassay discussed above. Myelinated nerves and blood vessels are also labelled for the enkephalins in a minority of cases. Nerves are labelled only for Met-enkephalin, whereas blood vessels are diffusely labelled with antisera to both peptides. Vascular labelling is, however, probably

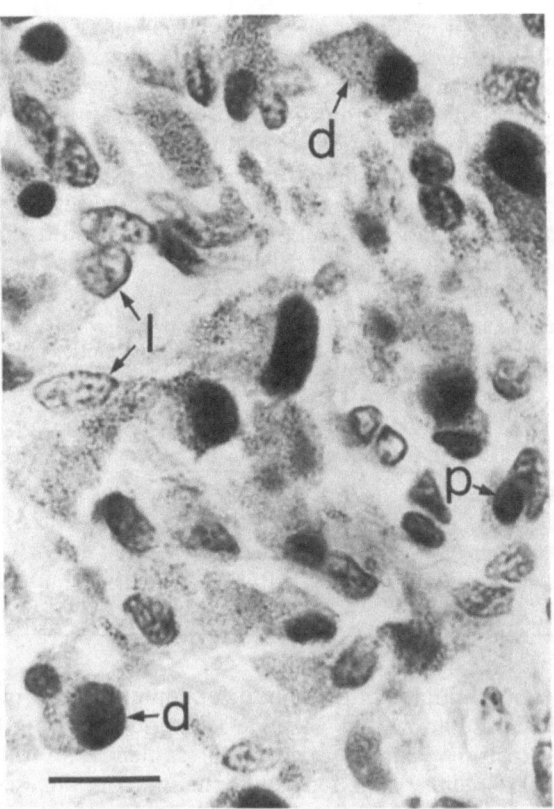

Fig. 8.2. Detail of a cell cluster immunolabelled for Leu-enkephalin. The fine granules of silver can be seen in the cytoplasm of dark cells (*d*) and progenitor cells (*p*). The nuclei of light cells (*l*) can be seen but there is no immunolabelling of their cytoplasm. Scale line = 10 μm.

non-specific binding of silver to plasma proteins which have diffused into the vessel wall after death.

Not only is immunolabelling for both enkephalins predominantly on glomic chief cells, but there is a marked tendency for the dark variant to be preferentially selected (Fig. 8.2). Virtually all of the dark chief cells show a heavy deposit of silver, whilst most of the light cells have little reaction, if at all. Progenitor cells also tend to be heavily labelled, like the dark cells, but, since these are uncommon variants, they are inconspicuous.

The extent of this immunoreactivity may be quantified by counting the numbers of labelled cells per unit area (Table 8.3). Each labelled cell was assigned to a category of light, dark or progenitor variants (Smith et al. 1990). A further category of unidentifiable cells was necessary because some did not include the nucleus in the section, and hence could not be identified, whereas in others the nucleus was obscured by a dense deposit of silver. These measurements showed a

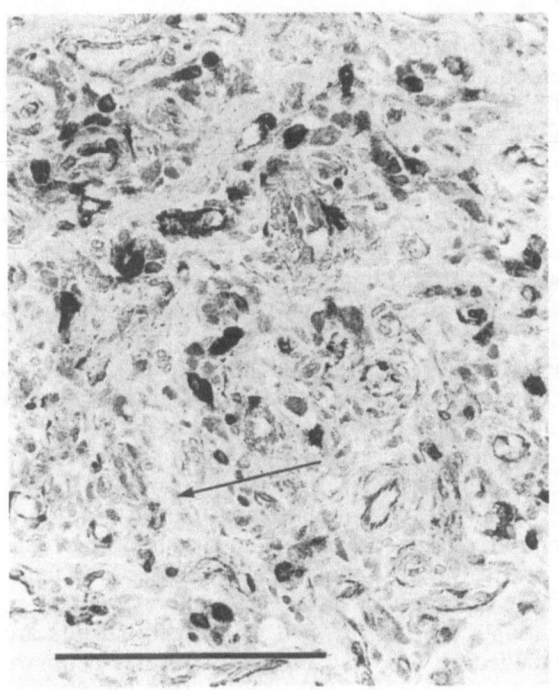

Fig. 8.1. Part of a lobule of carotid body immunolabelled for Met-enkephalin. Many of the chief cells show a black deposit of silver granules, whereas others contain a sparser deposit and are grey. Most of the labelled cells are dark chief cells. Sustentacular cells (*arrow*) are unlabelled. (Immunogold–silver technique). Scale line = 100 μm.

striking preponderance of immunolabelling on the dark chief cells. There was a broad range of values amongst the ten cases (Table 8.3), which may have indicated an inherent variability within the population or varying degrees of autolytic degradation of the peptides. The latter possibility is supported by the fact that three-quarters of those cases with numerous labelled cells were dissected and fixed less than 24 hours after death.

Table 8.3. Number of chief cells labelled for Met- and Leu-enkephalin per mm^2 of tissue from ten pairs of carotid bodies (after Smith et al. 1990)

Cell type	Met-enkephalin		Leu-enkephalin	
	Mean	Range	Mean	Range
Light	31	6–88	38	6–88
Dark	100	19–206	125	19–244
Progenitor	13	0–31	19	0–50
Unidentifiable	44	13–88	75	13–181
Total	188	44–363	256	38–556

Immunolabelling for both enkephalins can be localised to within a single cell and this is not surprising, since the precursor molecule of the enkephalins, called proenkephalin, contains six Met- and one Leu-enkephalin molecule in its structure. Immunoreactivity to both may, therefore, be expected if both constituent molecules are immunologically expressed (Van Noorden and Varndell 1987). Immunohistochemistry at the electron microscopical level using carotid bodies of cats has demonstrated that Met-enkephalin is stored within the dense-core vesicles, especially at their periphery (Varndell et al. 1982). Similar studies have been carried out on dogs in which antisera to both Met-enkephalin and noradrenaline were conjugated to gold particles of different diameter, permitting recognition of both on the same section. Enkephalin and noradrenaline were demonstrated not only within the same cell but also within the same granule (Kobayashi et al. 1983). Thus biogenic amines and enkephalins are located at the same site in the chief cells of animals. Comparable studies have not been performed in man but it is reasonable to assume that enkephalins are located within the dense-core vesicles in the human carotid body.

Substance P

Substance P is an 11 amino acid residue peptide which was first isolated from the gut and brain. Pharmacological studies suggest that it is a sensory neurotransmitter and constrictor of smooth

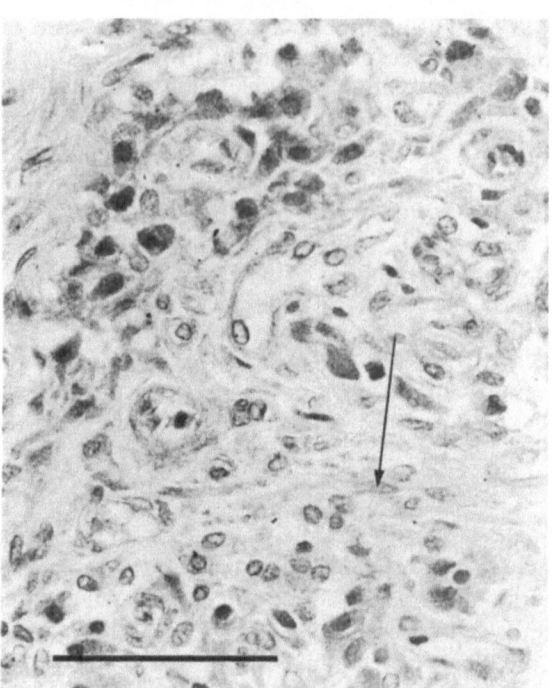

Fig. 8.3. Glomic lobule immunolabelled for substance P. All chief cells show a faint, greyish deposit of silver but elongated sustentacular cells (*arrow*) are unlabelled. Scale line = 50 μm.

muscle, hence its predominant localisation to autonomic nerves in several organs (Polak and Bloom 1980). Early immunohistochemical studies on the carotid bodies of cats indicated that substance P was located exclusively within fine nerve fibres throughout the organ but especially those adjacent to clusters of chief cells (Lundberg et al. 1979; Wharton et al. 1980). However, it was reported that 20% of glomic chief cells were also immunoreactive for this peptide (Cuello and McQueen 1980). This was confirmed by Prabhakar et al. (1989). Although substance P is confined to nerves in the fetal cat, it is prevalent in chief cells postnatally (Scheibner et al. 1988).

We found that substance P was present in the chief cells of man in 16 of 24 cases studied (Smith et al. 1990). Immunolabelling for this peptide is not selective for any particular variant of chief cell, as with the enkephalins, but consists of a faint, diffuse deposit of silver granules over all three variants (Fig. 8.3). Most of the cell clusters are labelled in this way but sustentacular cells are immuno-negative. Sometimes the bundles of myelinated nerves between the lobules may show a faint deposit of silver, but in no instances can the fine axons which penetrate the cell clusters be discerned. This absence of axonal labelling is, at first sight, anomalous, since in all the animal studies substance P is localised to fine varicose nerves irrespective of whether or not it is located in chief cells. This is probably a shortcoming of the technique, since thin sectioning of paraffin-wax-embedded tissue is not a recommended way of detecting nerves and the technique of immuno-fluorescence of thick cryostat sections is the preferred method of choice, as outlined above (Van Noorden and Varndell 1987). However, thin sections do permit the ready indentification of the variants of chief cell. Further studies on human material by the immunofluorescence technique applied to frozen sections are desirable.

Vasoactive Intestinal Peptide

The distribution of VIP within the human carotid body is similar to that of substance P in that it occurs in all variants of chief cell but is absent from sustentacular cells and nerves. In most carotid bodies the immunoreactivity is faint and patchy in distribution, with many cell clusters showing no reaction at all. This weak reaction for both substance P and VIP is to be expected in view of the small quantities of these peptides detected by radioimmunoassay (Table 8.2). VIP has not been demonstrated within the chief cells of animals but is confined to varicose nerve fibres, especially those around glomic blood vessels (Lundberg et al. 1979; Wharton et al. 1980). An explanation for the lack of immunoreactivity of nerves in human material is the same as that suggested for substance P above.

Neurotensin

Radioimmunoassay reveals a significant concentration of neurotensin in the human carotid body (Table 8.2), but we found it very difficult to demonstrate by the immunogold–silver technique. We identified it in only 8 of 23 pairs of carotid bodies and even then the labelling was weak and involved only a few chief cells. No other cells or tissues were labelled. Autolytic hydrolysis of the peptide is unlikely to explain this discrepancy in a demonstration of neurotensin by radioimmuno-assay and immunohistochemistry. For both studies (Heath et al. 1988; Smith et al. 1990) the carotid bodies were obtained from the same post-mortem room by the same method of dissection. Subsequently the tissues were studied in two countries using primary antisera from different manufacturers. It may be that the preparation of tissues for immunohistochemistry degraded or blocked the activity of neurotensin or else the primary antiserum that Smith et al. (1990)

employed was more specific, and restricted in its binding, than that used for radioimmunoassay.

Bombesin

Immunostaining for bombesin is entirely different from that of the other five peptides in that it is confined almost entirely to the glomic vasculature. Chief cells are occasionally faintly labelled and in some cases the myelinated nerves in the stroma may also contain a dense deposit of silver. However, the most striking labelling is seen in the glomic arteries. Thus, from a total of 26 carotid bodies which we examined, 13 showed dense deposits of silver in their glomic arteries but not in the glomic veins (Smith et al. 1990). The labelling for bombesin is well circumscribed, clearly delineated and confined to cellular structures, unlike the diffuse, non-specific deposit of silver on plasma proteins which sometimes occurs with the other antisera. In transverse sections of the inter-lobular glomic arteries immunolabelling is seen as dark, fusiform deposits scattered throughout an otherwise clear media (Fig. 8.4). These labelled cells are presumably smooth muscle, since no other type of cell has been identified in the media of glomic arteries by electron microscopy (Jago et al. 1982).The position of silver granules within these smooth muscle cells is predominantly confined to the periphery of the cytoplasm, leaving a clear central area in which the nucleus can be seen. In longitudinal sections of glomic arteries, where the smooth muscle cells have been cut transversely, the silver deposits form a series of rings with a clear, central nucleus in each cell (Fig. 8.5). The immunoreactivity for bombesin is present in radicles of the glomic arteries of all dimensions and extends to the most terminal branches of the intralobular glomic arterioles (Fig. 8.6). Even these minute vessels, embedded within the glomic lobules, show labelling of elongated smooth muscle in their medias.

Bombesin is a 14 amino acid residue peptide originally extracted from amphibian skin. It is probable that it does not occur in mammals, but antisera to it react with a similar mammalian peptide, containing 27 amino acid residues, called gastrin-releasing peptide (GRP). Thus, bombesin-like and GRP-like immunoreactivity are frequently regarded as the same thing. This is strictly speaking inaccurate, since the two molecules are not the same; they merely share some of the same amino acid sequences. In this chapter the word "bombesin" has been used, since that was the peptide against which antisera were raised both for radioimmunoassay and immunohistochemistry. Bombesin is typically located within the neuroendocrine cells of the lung (Stahlman et al. 1985) and GRP has been detected within nerves surrounding the bronchi (Barnes 1987). Since pulmonary neuroendocrine cells and glomic chief cells are ultrastructurally similar, one might anticipate the presence of bombesin in chief cells. It is surprising, therefore, that its main location appears to be in the glomic arteries. It is not clear whether it is secreted by the smooth muscle cells of these vessels or bound to receptor sites on their surface. The peripheral localisation of immuno-reactivity favours the latter.

This finding is of interest in view of the unusual structure of glomic arteries, which is quite different from that of other systemic arteries of comparable diameter (see Chapter 16). They contain much more elastic tissue than do other systemic arteries of the same size and these elastic fibres are arranged in a loose anastomotic pattern. In places

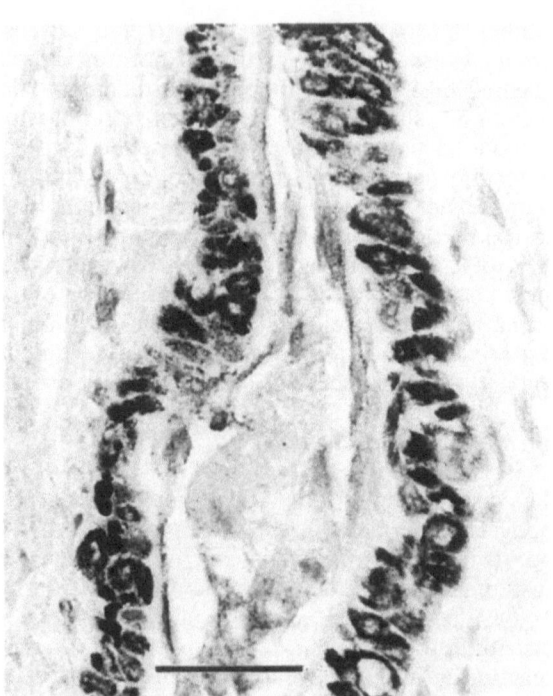

Fig. 8.5. Longitudinal section of an interlobular glomic artery immunolabelled for bombesin. Smooth muscle cells of the media are sectioned transversely and show a peripheral deposit of silver surrounding a clear centre. Scale line = 25 μm.

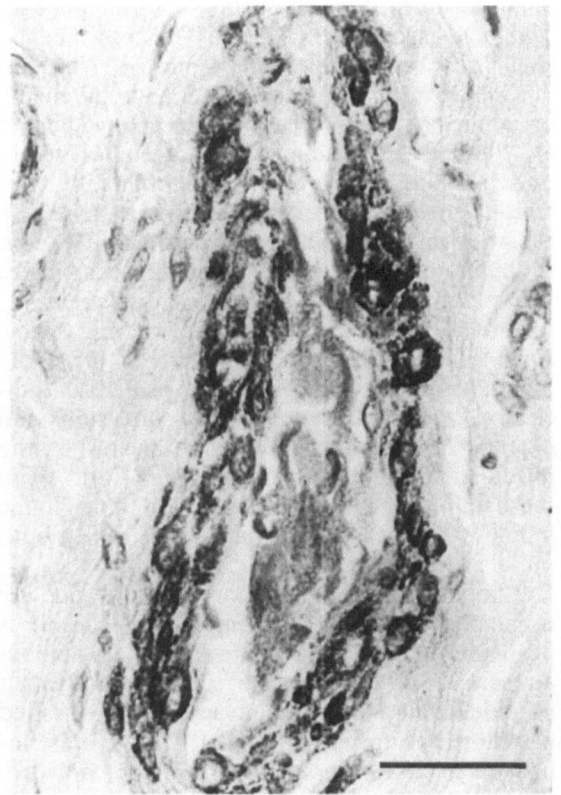

Fig. 8.4. Slightly oblique section of an interlobular glomic artery immunolabelled for bombesin. The fusiform smooth muscle cells of the media show a heavy deposit of silver which is largely peripheral in distribution. A central clear area occupied by the nucleus can be seen. The plasma in the lumen of the vessel shows faint, non-specific labelling with silver. Scale line = 25 μm.

the media becomes thin and dilated, reminiscent of the carotid sinus, and for this reason a baro-receptor function has been tentatively ascribed to them. Bombesin attached to receptor sites on their smooth muscle cells might, therefore, be involved in some way with this postulated baroreceptor rôle or alternatively with regulating vascular tone. Exogenously applied bombesin causes contraction of smooth muscle cells in several organs (Kulik et al. 1983).

Putative Function of Carotid Body Peptides

There are numerous physiological and pharmacological studies which attempt to elucidate the rôle which peptides play in the function of the carotid body. These all involve experiments on animals, usually the cat, and most employ either the enkephalins or substance P, since both are located within the chief cells. The experimental technique in most of these experiments involves exposing the carotid sinus nerve and recording nervous discharges when peptides or their antagonists are

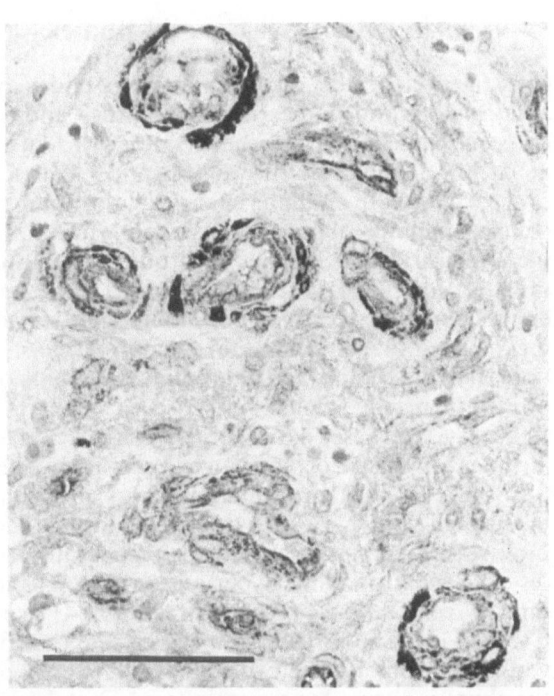

Fig. 8.6. A group of intralobular glomic arterioles embedded deep within the glomic tissue. The vessels show immunolabelling for bombesin with dense deposits of silver within fusiform smooth muscle cells in their medias. Scale line = 50 μm.

injected into the carotid artery. Using this type of preparation McQueen and Ribeiro (1980) found that Met-enkephalin depressed the discharge from the carotid sinus nerve. Furthermore, when the carotid body was stimulated by the infusion of cyanide or acetylcholine, Met-enkephalin had a significant depressor effect on the increased discharge caused by the former but actually potentiated the effect of the latter. Other opiate peptides such as morphine and β-endorphin behaved similarly but were much less potent (McQueen and Ribeiro 1980, 1981). Experiments in which the isolated carotid body was superfused with fluid in a chamber confirmed the inhibitory action of Met-enkephalin but also showed a striking rebound effect after perfusion of the peptide was discontinued (Monti-Bloch and Eyzaguirre 1985). Small doses actually resulted in an increased discharge. However, the application of chemosensory stimulants produced results which were at odds with those of McQueen and Ribeiro. Thus Met-enkephalin had little effect upon stimulation by cyanide but produced a striking inhibition of stimulation by acetylcholine (Monti-Bloch and Eyzaguirre 1985). The two studies used different experimental techniques, which may explain the discrepan-

cies between them. It is also doubtful whether perfusing the carotid body with exogenous peptides can be regarded as representative of its normal function. Probably a more realistic way to study their effects is to observe the influence of specific antagonists, since these reveal the activity of endogenous peptide. An antagonist to opiate peptides such as Met-enkephalin is naloxone. In all published studies this substance causes excitation of the carotid body as measured by the rate of discharge from the sinus nerve (McQueen and Ribeiro 1980; Pokorski and Lahiri 1981; Monti-Bloch and Eyzaguirre 1985). Furthermore, naloxone blocks the inhibitory influence of exogenous Met-enkephalin and stimulates ventilation in the absence of hypoxia (Pokorski and Lahiri 1981). Thus, these experiments suggest that chemosensitivity and ventilation are normally suppressed by endogenous opiate peptides.

Injection of substance P into the carotid artery produces an effect on the rate of nervous discharge opposite to that of Met-enkephalin. Thus there is a transient decrease followed by a sustained increase in chemoreceptor discharge (McQueen 1980; Prabhakar et al. 1984). The effect of chemosensory-stimulating agents on this response is, however, contradictory. Thus, whereas cyanide potentiated the response, acetylcholine actually inhibited it (McQueen 1980). In the superfused carotid body, cyanide depressed the response of substance P in some cases and potentiated it in others. Acetylcholine on the other hand increased the response to substance P (Monti-Bloch and Eyzaguirre 1985). Similar conflicting results were obtained for the interaction between dopamine and substance P. The customary inhibitory effect of this amine on chemosensory discharge was reduced by substance P in one experiment (McQueen 1980) and increased in another (Monti-Bloch and Eyzaguirre 1985). When a specific antagonist to substance P was injected intra-arterially into cats it caused the firing rate in the sinus nerve to decline, suggesting that endogenous substance P normally exerts a stimulatory influence (Prabhakar et al. 1984). The antagonist also blocked the effect of exogenous peptide and almost completely abolished the response of the carotid body to inhalation of a hypoxic gas mixture. However, the antagonist had no effect on the response of the carotid body to hyperoxic hypercarbia, suggesting that substance P plays a rôle in the response to the level of oxygen in the blood but not to that of carbon dioxide (Prabhakar et al. 1987).

The effects of other peptides have been studied in less detail. Neurokinin A immunoreactivity has been found in the carotid body of cats. This substance belongs to a group of structurally related peptides called tachykinins to which substance P belongs. Neurokinin A appears to have a stimulating effect on the carotid body (Prabhakar et al. 1989) as does cholecystokinin (McQueen and Ribeiro 1981). VIP, on the other hand, has a variable effect dependent upon the dose administered. Thus low doses decrease the chemoreceptor discharge whereas high doses stimulate it (McQueen and Ribeiro 1981).

No clear picture emerges from all these studies of the mechanism by which peptides govern or modulate the activity of the carotid body. Some appear to stimulate it, others to depress it. Some influence stimulation of the organ in a predictable fashion but with others the results are contradictory. It may be that their overall effect on chemoreception depends upon a balance between their respective concentrations or rates of release. Alternatively it has been suggested that the carotid body may have both excitatory and inhibitory reactive sites for peptides and that the effect of a particular one depends upon which of these sites are active (Monti-Bloch and Eyzaguirre 1985). However, the complexity of the situation raises the question as to whether peptides may not be primarily involved in chemoreception at all. All the above experiments were designed on the assumption that the carotid body functions only as a chemoreceptor. However, it may influence sodium absorption by the kidney (see Chapter 15). Perhaps one or more of these peptides, plus others yet to be discovered, may constitute the hormone "glomin" proposed by Pearse (1969).

Peptides in Abnormal Carotid Bodies

Investigation of the distribution of peptides in abnormal carotid bodies is in its infancy and to date only nine cases have been studied (Khan et al. 1990). Results on this small series of cases suggest that there might be changes in the number of cells immunoreactive for enkephalins in certain pathological states. Sections of carotid body in this investigation were labelled for Met- and Leu-enkephalins, using the same immunogold–silver technique as that employed for the normal carotid bodies discussed above. Differential cell counts of positively labelled cells were also made in the same way. The cases in the series were selected to include a variety of histological abnormalities (Table 8.4). Thus there were four cases of sustentacular cell hyperplasia. Two of these (cases 1 and 2) were secondary to chronic obstructive airways disease, one to systemic hypertension (case 3) and one to coarctation of the aorta (case 4) (see Chapter 14). The series also included one case of ventricular septal defect with subacute reversal of the shunt inducing hypoxaemia (case 5) and four cases of chronic carotid glomitis (cases 6–9) (see Chapter 6).

The distribution of immunolabelling in the various components of the carotid body in all nine cases was much the same as in normal organs. Thus labelling was confined almost entirely to chief cells, with a strong tendency to favour the dark variant. There was considerable variation in the total number of chief cells labelled for either enkephalin between the nine cases (Table 8.4). In all but one (case 7) there was a similar number of chief cells labelled for either enkephalin in each

Table 8.4. Total number of chief cells labelled for Met- and Leu-enkephalin per mm^2 of tissue from nine pairs of abnormal carotid bodies (after Khan et al. 1990)

Case number	Age (yr)	Necropsy diagnosis	Cellular abnormality	No. of cells immunoreactive for	
				Met-enkephalin	Leu-enkephalin
1	52	COAD	S	131	75
2	55	COAD	S	31	44
3	61	SH	S	13	19
4	61	CA	S*	106	100
5	62	VSD	DCP	231	281
6	78	B	CCG	75	119
7	87	RTA	CCG	19	189
8	90	PT	CCG	56	38
9	87	LC	CCG	175	119

COAD, chronic obstructive airways disease; SH, systemic hypertension; CA, coarctation of the aorta; VSD, ventricular septal defect; B, bronchopneumonia; RTA, road traffic accident; PT, pulmonary thromboembolism; LC, liver cirrhosis; S, sustentacular cell hyperplasia; *, sheets of sustentacular cells and atrophy of chief cells; DCP, dark cell prominence; CCG, chronic carotid glomitis.

case. The counts for the four cases of sustentacular cell hyperplasia were low when compared with the mean values from the ten normal carotid bodies in which there were 188 and 256 cells per mm^2, respectively, for Met- and Leu-enkephalin. In all but one case, the counts were within the normal range, albeit at the lower end (Table 8.3). In case 3 the total number of labelled cells fell below the range of the ten normal cases.

It is tempting to postulate that these low numbers of cells containing enkephalins have physiological significance. It has been shown, for example, that many patients with chronic bronchitis and emphysema show a reduced ventilatory response to hypoxia and hence by implication a reduced chemosensory response (Flenley et al. 1970). Animal experiments have demonstrated that acute exposure to hypoxia results in depletion of both Met-enkephalin and substance P from the carotid body (Hanson et al. 1986). However, it is difficult to equate such data with the apparent chemosensory inhibition which Met-enkephalin induces. On this basis one would anticipate increased levels of enkephalin to be associated with a reduced ventilatory drive to hypoxia. It is possible that the low counts of labelled cells in cases of sustentacular cell hyperplasia were simply a reflection of their abnormal histology. Thus, in this condition, cores of chief cells become spaced further apart by thick bands of sustentacular cells. This would leave fewer chief cells per unit area available for labelling than in a normal carotid body. Allowing even for this inequality in measurement, it is surprising that the number of enkephalin-positive cells was not increased.

It is equally difficult to interpret the results from the case of ventricular septal defect. The histology of this case showed a striking prominence of the dark variant of chief cell and, by immunohistochemistry, most of these dark cells were labelled. Consequently there was a high count for this case, close in fact to the upper limit of the normal range. This patient experienced subacute hypoxaemia of a few month's duration. It is not known what the hypoxic ventilatory drive is in such cases, but it is unlikely to be significantly altered in such a brief period. There is no obvious explanation for the numerous enkephalin-containing cells in this case based upon a purely chemosensory rôle for the peptides.

The remaining four cases were examples of chronic carotid glomitis in which there were focal infiltrates of lymphocytes, and in case 9 plasma cells, into the stroma, especially around glomic blood vessels and nerves. The counts of enkephalin-containing cells in these cases were at the lower end of the normal range (Table 8.4). However, this condition is found only in elderly patients in whom there is atrophy and fibrosis of the glomic tissue. Low counts of labelled cells may be simply a reflection of the small numbers of chief cells per unit area available for counting. It is probable that this disease does not influence the enkephalins contained within the chief cells.

In conclusion, there is no unequivocal difference in the number of enkephalin-containing chief cells between normal and abnormal carotid bodies, merely a suggestion that they might alter in some cases. It is difficult to make comparisons when there is such a broad range of counts amongst normal organs which will mask any subtle changes in the population of cells. All that can be said at present is that in conditions such as chronic bronchitis and emphysema, where there are few dark cells, there are low counts of enkephalin-containing cells. Where there are many dark chief cells, as in the case of ventricular septal defect or in natives to the highlands of Ladakh (Khan et al. 1988), there are many cells positive for enkephalins. Enkephalins thus appear to act as markers for dark chief cells but whether they are involved in the chemosensory activity of the carotid body, or have some entirely different function, has yet to be ascertained.

References

Barnes PJ (1987) Regulatory peptides in the respiratory system. Experientia 43:832–839

Cuello AC, McQueen DS (1980) Substance P: a carotid body peptide. Neurosci Lett 17:215–219

Flenley D, Franklin D, Millar J (1970) The hypoxic drive to breathing in chronic bronchitis and emphysema. Clin Sci 38:503–518

Fried G, Meister B, Wikström M, Terenius L, Goldstein M (1989) Galanin-, neuropeptide Y- and enkephalin-like immunoreactivities in catecholamine-storing paraganglia of the fetal guinea pig and newborn pig. Cell Tissue Res 255:495–504

Gallagher PJ, Paxinos G, White SW (1985) The role of substance P in arterial chemoreflex control of ventilation. J Autonom Nerv Syst 12:195–210

Hansen JT, Brokaw J, Christie D, Karasek M (1982) Localization of enkephalin-like immunoreactivity in the cat carotid and aortic body chemoreceptors. Anat Rec 203:405–410

Hanson G, Jones L, Fidone S (1986) Physiological chemoreceptor stimulation decreases enkephalin and substance P in the carotid body. Peptides 7:767–769

Heath D, Quinzanini M, Rodella A, Albertini A, Ferrari R, Harris P (1988) Immunoreactivity to various peptides in the human carotid body. Res Commun Chem Pathol Pharmacol 62:289–293

Jago R, Heath D, Smith P (1982) Structure of the glomic arteries. J Pathol 138:205–218

Kameda Y (1989) Distribution of CGRP-, somatostatin-, galanin-, VIP-, and substance P-immunoreactive nerve fibres in the chicken carotid body. Cell Tissue Res 257:623–629

Khan Q, Heath D, Smith P, Norboo T (1988) The histology of the carotid bodies in highlanders from Ladakh. Int J Biometeorol 32:254–259

Khan Q, Smith P, Heath D (1990) The distribution of enkephalins in human carotid bodies showing cellular proliferation and chronic glomitis. Arch Pathol Lab Med 114:1232–1235

Kobayashi S, Uchida T, Ohashi T et al. (1983) Immunocytochemical demonstration of the co-storage of noradrenaline with met-enkephalin-arg[6]-phe[7] and met-enkephalin-arg[6]-gly[7]-leu[8] in the carotid body chief cells of the dog. Arch Histol Jpn 46:713–722

Kondo H, Iwanaga T, Nakajima T (1982) Immunocytochemical study on the localization of neuron-specific enolase and S-100 protein in the carotid body of rats. Cell Tissue Res 227:291–295

Kulik TJ, Johnson DE, Elde P, Locke JE (1983) Pulmonary vascular effects of bombesin and gastrin-releasing peptide in conscious newborn lambs. J Appl Physiol 55:1093–1097

Kummer W, Addicks K, Henkel H, Heym C (1985) Cholecystokinin-like immunoreactivity in cat extra-adrenal paraganglia. Neurosci Lett 55:207–210

Lundberg JM, Hökfelt T, Fahrenkrug J, Nilsson G, Terenius L (1979) Peptides in the cat carotid body (glomus caroticum): VIP-, enkephalin-, and substance P-like immunoreactivity. Acta Physiol Scand 107:279–281

McQueen DS (1980) Effects of substance P on carotid chemoreceptor activity in the cat. J Physiol (Lond) 302:31–47

McQueen DS, Ribeiro JA (1980) Inhibitory actions of methionine-enkephalin and morphine on the cat carotid chemoreceptors. Br J Pharmacol 71:297–305

McQueen DS, Ribeiro JA (1981) Effects of β-endorphin, vasoactive intestinal polypeptide and cholecystokinin octapeptide on cat carotid chemoreceptor activity. Q J Exp Physiol 66:273–284

Monti-Bloch L, Eyzaguirre C (1985) Effects of methionine-enkephalin and substance P on the chemosensory discharge of the cat carotid body. Brain Res 338:297–307

Pearse AGE (1969) The cytochemistry and ultrastructure of polypeptide hormone-producing cells of the APUD series and the embryologic, physiologic and pathologic implications of the concept. J Histochem Cytochem 17:303–313

Pokorski M, Lahiri S (1981) Effects of naloxone on carotid body chemoreception and ventilation in the cat. J Appl Physiol 51:1533–1538

Polak JM, Bloom SR (1980) Peripheral localization of regulatory peptides as a clue to their function. J Histochem Cytochem 28:918–924

Prabhakar NR, Runold M, Yamamoto Y, Lagercrantz H, von Euler C (1984) Effect of substance P antagonist on the hypoxia-induced carotid chemoreceptor activity. Acta Physiol Scand 121:301–303

Prabhakar NR, Mitra J, Cherniack NS (1987) Role of substance P in hypercapnic excitation of carotid chemoreceptors. J Appl Physiol 63:2418–2425

Prabhakar NR, Landis SC, Kumar GK, Mullikin-Kilpatrick D, Cherniack NS, Leeman S (1989) Substance P and neurokinin A in the cat carotid body: localization, exogenous effects and changes in content in response to arterial pO_2. Brain Res 481:205–214

Scheibner T, Reid DJC, Sullivan CE (1988) Distribution of substance P-immunoreactive structures in the developing cat carotid body. Brain Res 453:72–78

Smith P, Gosney J, Heath D, Burnett H (1990) The occurrence and distribution of certain polypeptides within the human carotid body. Cell Tissue Res 261:565–571

Stahlman MT, Kasselberg AG, Orth DN, Gray ME (1985) Ontogeny of neuroendocrine cells in human fetal lung. II. An immunohistochemical study. Lab Invest 52:52–60

Van Noorden S, Varndell IA (1987) Regulatory peptide immunocytochemistry at light- and electron-microscopical levels. Experientia 43:724–734

Varndell IM, Tapia FJ, De Mey J, Rush RA, Bloom SR, Polak JM (1982) Electron immunocytochemical localization of enkephalin-like material in catecholamine-containing cells of the carotid body, the adrenal medulla, and in phaeochromocytomas of man and other mammals. J Histochem Cytochem 30:682–690

Wharton J, Polak JM, Pearse AGE et al. (1980) Enkephalin-, VIP- and substance P-like immunoreactivity in the carotid body. Nature 284:269–271

The carotid bodies enlarge in response to hypoxaemia. This may take the form of an acute, reversible enlargement due to vascular engorgement. On the other hand it may be due to a cellular response to subacute or chronic hypoxia.

Vascular Engorgement

It is not known what changes take place in the human carotid body on acute exposure to hypoxia but reasonable deductions can be made by analogy with what occurs in rats kept in a decompression chamber. Such experiments show that the carotid bodies enlarge rapidly on acute exposure to diminished barometric pressure and just as rapidly return to normal size once the hypoxic stimulus is withdrawn (Heath et al. 1973; Laidler and Kay 1975a,b). In one experiment we studied the tissue volume of the carotid body by an application of Simpson's rule to histological sections (see Chapter 2) in three groups of ten adult male Wistar albino rats (Heath et al. 1973). The first group was kept for 5 weeks in a hypobaric chamber exposed to a barometric pressure of 380 mmHg, equivalent to a simulated altitude of 5500 m above sea level. The second was exposed to the same barometric pressure for 5 weeks and then allowed to recover in room air for a further period of 5 weeks. The third group acted as the control and was kept at normal barometric pressure throughout. In the control animals the mean carotid body volume (expressed in units of 10^6 μm^3) was 13.45 (Fig. 9.1), but after exposure to hypobaric hypoxia for only 5 weeks this volume rose to 47.81 (Fig. 9.2). After the hypoxic stimulus was withdrawn the tissue volume fell after only 5 weeks to 19.82. Similarly, when Laidler and Kay (1975a) exposed rats to a barometric pressure of 460 mmHg for between 25 and 96 days, the mean value of the total volume of the combined left and right carotid bodies rose from 47.16 $\times$ 10^6 μm^3 to 187.39 $\times$ 10^6 μm^3 (i.e. double the values resulting from the experiments of Heath et al. (1973) quoted above). Blessing and Wolff (1973) found that, when they subjected rats to a simulated high altitude of 7500 m for 3 months, the carotid bodies increased in volume from 32.8 $\times$ 10^6 μm^3 to 194.5 $\times$ 10^6 μm^3.

The reason for the rapid enlargement of the carotid bodies and its equally quick reversibility to almost normal levels is that it is due almost entirely to vascular engorgement, the cellular elements of the carotid body playing little or no part in it. Laidler and Kay (1975a) agreed with this interpretation, but were uncertain as to the functional significance of the increased vascularity. They thought it might be nothing more than a nonspecific reaction designed to increase blood flow and thus oxygen transport to a hypoxic organ, with increased metabolic activity. The studies of Hellström and Pequignot (1985) suggest that the vascularity of the organ in acute hypoxia is regulated through adrenergic β-receptors. They found that

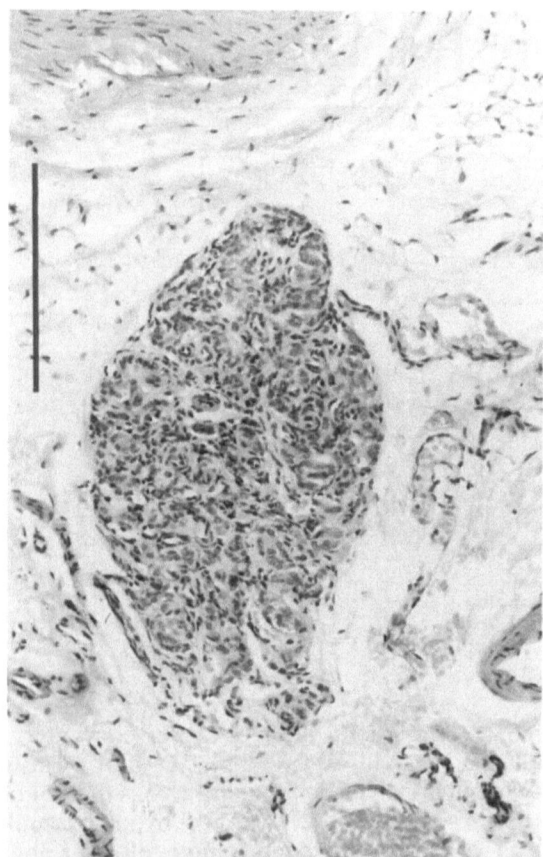

Fig. 9.1. Small, compact carotid body from a eupoxic Wistar albino rat at low altitude. (Haematoxylin–eosin (HE)) Scale line = 200 μm.

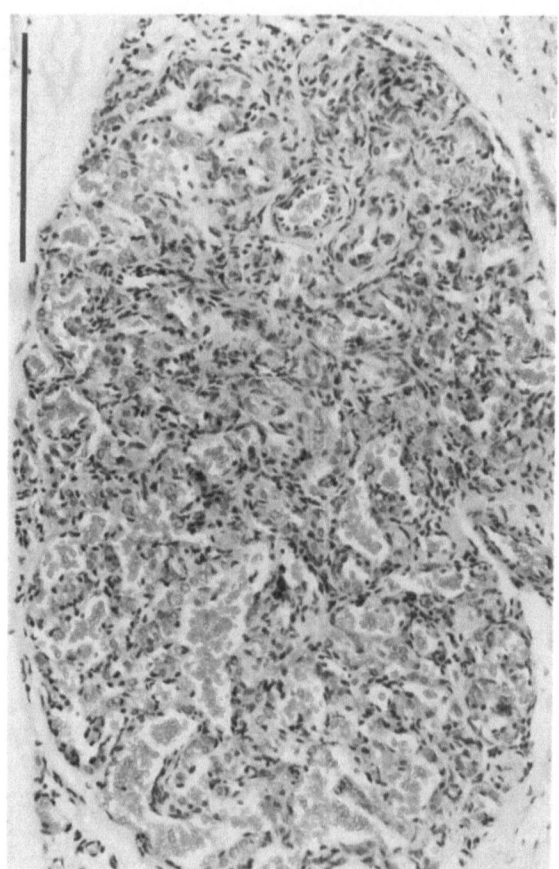

Fig. 9.2. Much enlarged carotid body from a Wistar albino rat subjected to a simulated altitude of 5500 m for 5 weeks. The hypobaric hypoxia has led to gross vascular engorgement. (HE) Scale line = 200 μm.

rats treated with propranolol adjusted more easily to hypoxia, and showed but slight enlargement of their carotid bodies, in which the vascular density remained normal. The location of adrenergic β-receptors in the carotid body is not known.

There seems to be no good reason why this rapid enlargement of the carotid body due to vascular engorgement should not take place during the periods of worsening of hypoxaemia that characterise the clinical course of chronic bronchitis and emphysema. Presumably recovery from bouts of superadded chest infections with associated alleviation from hypoxia will also bring about a rapid reversibility of vascular engorgement and a concomitant decrease in size of the carotid body.

Early Histological Changes in the Rat

Following the phase of vascular engorgement early structural changes can be detected in the cellular components of the carotid bodies. In humans these early organic changes will develop until they present characteristic histological appearances in patients with chronic obstructive lung disease and associated hypercarbia (see Chapter 10) and in native highlanders with associated hypocarbia (see Chapter 11). The microscopic features of these two groups of subjects are described in detail in Chapters 10 and 11 and we shall not repeat these accounts here. Here we shall be concerned with the initial structural responses of glomic tissue to hypoxaemia. To capture these very early cellular reactions we have to make recourse to the results of experimental work on laboratory animals. A meticulous study of the quantitative changes in the rat carotid body in chronic hypoxia with and without hypercarbia was made by Dhillon et al. (1984) and we may now consider their results, which are summarised in Table 9.1.

Table 9.1. Quantitative changes in the carotid body of the rat in chronic hypoxia with or without associated chronic hypercarbia (from Barer et al. (1986) based on studies of Dhillon et al. (1984))

Measurement for whole carotid body	Controls	Chronic hypoxia	Chronic hypoxia + hypercarbia
Number of chief cell nuclei $\times 10^3$	8.7±0.5	31.4±6.8	39.0 ±6
Total points counted on chief cell nuclei	454±70	1082±133	—
Total points counted on endothelial cell nuclei	122±27	524±38	—
Total points counted on connective tissue and sustentacular cell nuclei	4199±855	15 018±1710	—
Surface volume of blood vessels (mm^2)	2.57±0.38	26.22±9.61	18.96±2.1
Harmonic mean distance of glomic tissue from capillaries (10^3 mm)	1.13±0.08	0.64±0.12	0.48±0.03
Estimated number of dense-cored vesicles $\times 10^{10}$	0.09±0.1	0.28±0.08*	0.29±0.06*

Means ± s.e.m. All sections were significantly different from controls ($p<0.05$), except those marked with an asterisk.

The data in this table were obtained from control rats, from chronically hypoxic rats kept for 3 to 5 weeks in 10% oxygen in a normobaric chamber, and from chronically hypoxic and hyperbaric rats kept for 4 weeks in 10% (v/v) oxygen and 4% (v/v) carbon dioxide. Measurements were made on step sections throughout the carotid body by point counting with light microscopy or on electron micrographs (Dhillon et al. 1984). Compared with normal rats, chronically hypoxic rat carotid bodies contain more chief cell nuclei, more connective tissue and more sustentacular cell nuclei. There is a much greater blood vessel surface area, a shorter diffusion distance between capillaries and glomus cells (estimated as the harmonic mean distance), more endothelial cell nuclei and more dense-core vesicles. Barer et al. (1986) commented that the hyperplasia of chief cells, such that there is a two- or three-fold increase in the number of nuclei, might be held to be surprising because they are thought to arise from the neural crest, as we pointed out in Chapter 1. Nevertheless, unequivocal evidence was found for cell division in chronically hypoxic rats who were given vincristine to arrest mitosis in metaphase. The significance of this early hyperplasia of chief cells in response to hypoxaemia is intriguing and remains unexplained. The dopamine and noradrenaline contents of the carotid body are increased in chronic hypoxia, but in chronic hypercarbia only dopamine is raised (Barer et al. 1986; Table 9.2). The early increase in the number of sustentacular cells in the carotid bodies of rats is of interest in that these type II components are a prominent feature of the fully developed histological picture of carotid body hyperplasia.

Table 9.2. Dopamine and noradrenaline contents of the rat carotid body (pmol/carotid body±s.D.) in chronic hypoxia and hypercarbia (after Barer et al. (1986), based on data from Pallot and Al Neamy (1983) and Barer and Pallot (1984))

Condition	Dopamine content	Noradrenaline content
Normoxic controls	6.9±2.7*	8.3±2.7*
Chronic hypoxia (10% O_2 for 21 days)	85.8±31.8*	147±74.6*
Normoxic controls	4.8±2.1**	10.6±4.4***
Chronic hypercarbia (7% CO_2 for 21 days)	14.8±8.4**	14.6±6.3***

*$p<0.01$; **$p<0.02$; ***$p<0.05$.

Carotid Body Hyperplasia

The fully developed features of carotid body hyperplasia in man are described in the following two chapters. The condition is found in patients with chronic obstructive lung disease in which chronic normobaric hypoxia is associated with chronic hypercarbia (see Chapter 10). It occurs in patients with systemic hypertension in which there is no generalised hypoxaemia but in which the glomic tissues are rendered ischaemic by occlusive changes in the radicles of the glomic arteries (see Chapter 10). Finally it occurs in native highlanders in whom exposure to hypobaric hypoxia is associated with hypocarbia (see Chapter 11).

Dark Cell Prominence

Superimposed on the histological features of carotid body hyperplasia may be prominence of the dark variant of chief cells. This occurs diffusely throughout the glomic substance of native highlanders and we describe this in Chapter 11. It may also be seen as a secondary focal proliferation of dark cells, or their predecessors progenitor cells, in cases of chronic obstructive lung disease or systemic hypertension in which the typical features of carotid body hyperplasia are already established (Figs. 9.3, 9.4). The nodules may range from 40 to 400 μm in diameter and some of the cells in them may show pleomorphism. In two cases that we reported, the nodules were composed of dark cells in a woman of 80 years with systemic hypertension (Fig. 9.3) and of progenitor cells in a man of 72 years with panacinar emphysema complicated by chronic hypoxaemia (Fig. 9.4; Heath et al. 1984). In the case of systemic hypertension the dark cells accounted for 47% of

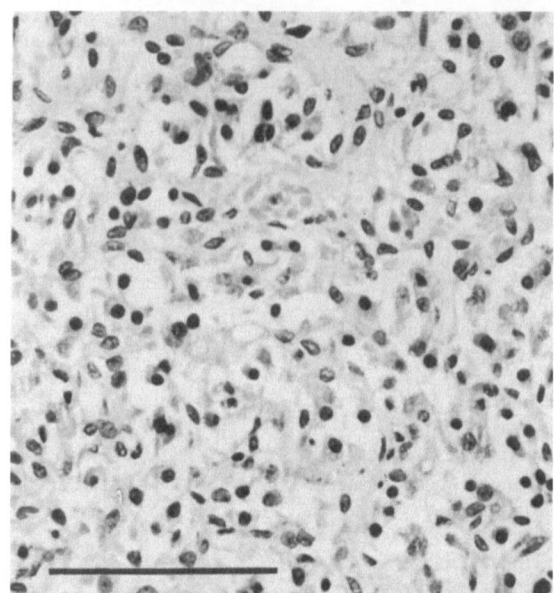

Fig. 9.4. Section of carotid body from a man of 72 years with hypoxic cor pulmonale due to panacinar emphysema. There is infiltration of the glomic tissue by progenitor cells. (HE) Scale line = 80 μm.

the cells in the areas of focal proliferation in the carotid body, while in the case of emphysema the dark cells accounted for 20% and the progenitor cells 42% in the localised nodules.

Prominence of dark cells may also arise de novo in the absence of established carotid body hyperplasia. Conditions in which focal dark cell proliferation has been reported are listed in Table 9.3. As an example we refer to the case of a woman of 62 years with a ventricular septal defect and recently reversed intracardiac shunt (Smith et al. 1986a). Her carotid bodies were exposed to hypoxaemia for only a few months. They were not enlarged but were abnormally cellular with a proliferation of

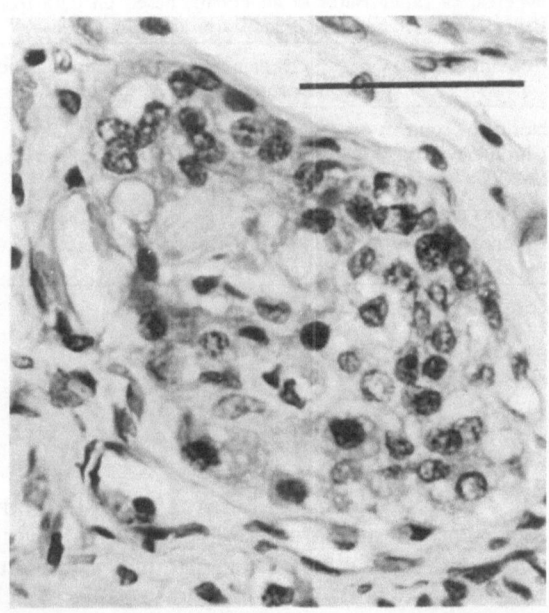

Fig. 9.3. Section of carotid body from a woman of 80 years with systemic hypertension. A nodule of dark cells is present. (HE) Scale line = 50 μm.

Table 9.3. Conditions in which focal dark- or progenitor-cell proliferation in the carotid body have been reported in association with hypoxia

Condition	Reference
Chronic obstructive lung disease and systemic hypertension	Heath et al. 1984
Ventricular septal defect with reversed intracardiac shunt	Smith et al. 1986a
Rabbits exposed to experimental normobaric hypoxia	Smith et al. 1986b
Hypoxic cor pulmonale	Heath et al. 1970

dark cells, many of which were abnormally large and showed ultrastructural features of metabolic activity (Fig. 9.5). A differential cell count of the glomic tissue revealed 21% of dark cells on the left and 14% on the right. Another reported case of cyanotic congenital heart disease in which the carotid bodies were examined was that of a girl of 8 years with an atrial septal defect (Lack 1978). In this instance the carotid bodies were heavier than normal, with enlargement of glomic lobules as in our case of ventricular septal defect. There was also hyperplasia of chief cells, with nuclei described as being hyperchromatic, which could be interpreted as representing dark cell hyperplasia.

The relation of the proliferation of dark and light variants of the chief cells to stimulation by hypoxia is illustrated by experimental work on rabbits (Smith et al. 1986b). Six Dutch rabbits were kept in a normobaric, hypoxic chamber to subject them to the same degree of hypoxia as experienced at Cerro de Pasco at high altitude (4330 m) in the Peruvian Andes. Three rabbits were killed after living in the chamber for 3 months, while the remaining three were subjected to hypoxia for 6 months. A differential cell count of glomic tissue on control rabbits showed 14% dark cells, but in the animals kept for 3 months in the hypoxic chamber the count rose to 31%. In these rabbits the lobular architecture was

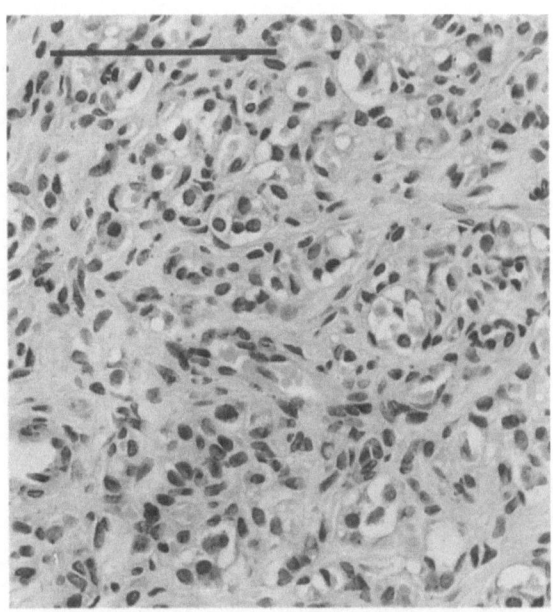

Fig. 9.6. Dutch rabbit subjected to hypoxia for 3 months. Much of the lobular architecture is obliterated by increased cellularity in which the dark variant of chief cells predominates. The nuclei are intensely haematoxyphilic. (HE) Scale line = 80 μm.

disturbed and almost obliterated by increased cellularity due to dark cells (Fig. 9.6). The characteristics of these cells were the oval nuclei with dense chromatin and sharply defined edges (Fig. 9.7). The cytoplasm showed intense vacuolation. In rabbits kept in the chamber for 6 months the dark cell count had fallen to 6%. In these animals the disruption of the glomic tissue remained but the nature of the increased cellularity had changed to consist of the light variant of chief cells (Fig. 9.8). At higher magnification these light cells showed the classic round outline with punctate heterochromatin (Fig. 9.9). These figures and appearances are consistent with the view that the initial cellular response of the carotid bodies to hypoxia in the rabbit is by dark cells which later change into light cells. In this respect native Peruvian rabbits showed a differential dark cell count of only 6% (Smith et al. 1986b). It seems likely that the same sequence of response takes place in the human carotid body. In assessing the significance of dark cell prominence in young animals and people it has to be kept in mind that this histological feature is characteristic of the young carotid body (see Chapter 4). The biochemical and physiological implications are that the cytoplasm of dark variants of chief cells is rich in leucine- and methionine-enkephalins, although the rôle of these peptides in chemoreception or an

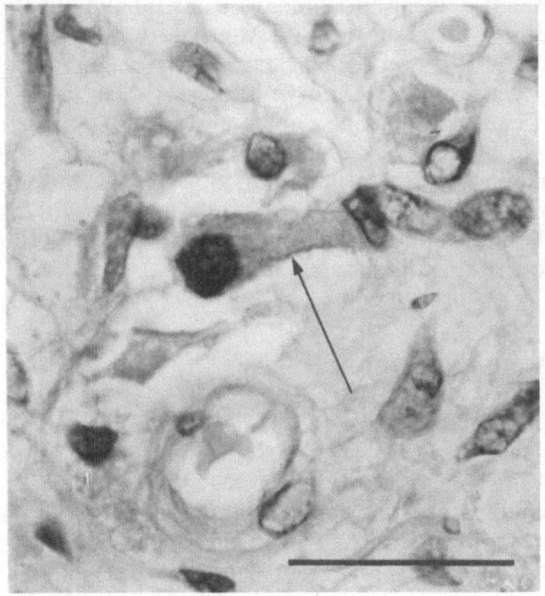

Fig. 9.5. Section of carotid body from a woman of 62 years with reversed intracardiac shunt through a ventricular septal defect causing several months of systemic hypoxaemia. The glomic tissue contains many dark cells with eccentric nuclei and long streamers of cytoplasm (*arrow*). (HE) Scale line = 20 μm.

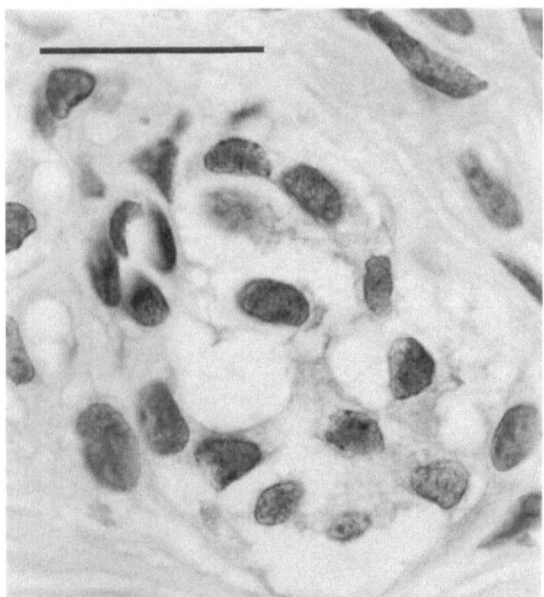

Fig. 9.7. Detail of Fig. 9.6 showing a cluster of dark cells in which the heterochromatin in the nuclei is densely packed. Their cytoplasm is intensely vacuolated. (HE) Scale line = 20 μm.

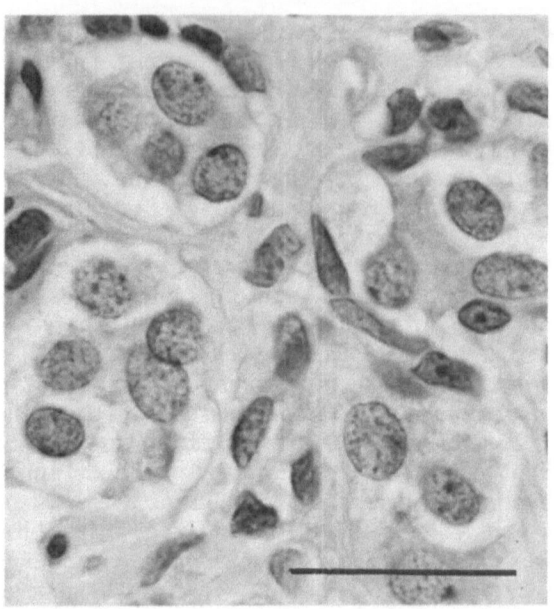

Fig. 9.9. Detail of Fig. 9.8 showing cytological features of the light cells. Their nuclei are large, round and pale with a finely stippled chromatin pattern and prominent nucleoli. Their cytoplasm is vacuolated. (HE) Scale line = 20 μm.

endocrine function is at present obscure (see Chapter 8). The appearance of cells packed with enkephalins is in itself difficult to interpret, for their very presence may represent, on the one hand, synthesis and secretion, or, on the other,

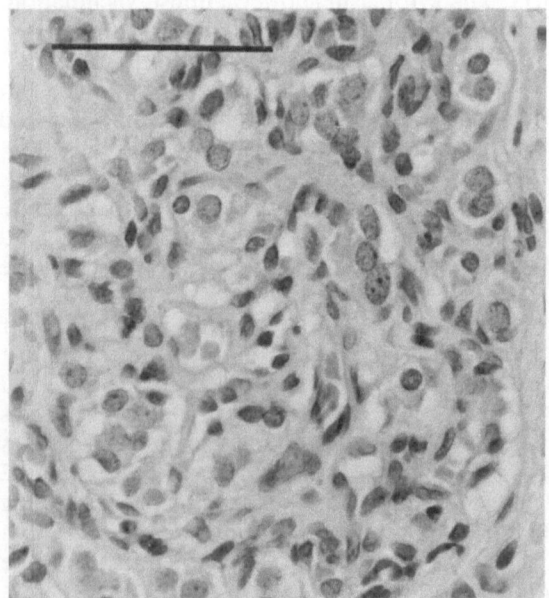

Fig. 9.8. Dutch rabbit subjected to hypoxia for 6 months. There is increased cellularity of the carotid body which consists in the main of the light variant of chief cells. (HE) Scale line = 50 μm.

increased storage. The enkephalins are members of the opiate peptides and as such might be thought to exert a depressant effect on chemoreceptor activity but this is by no means certain (see Chapter 8). Hence we do not know what effects a local enrichment of enkephalins in the glomic tissue have on the ventilatory response to hypoxia. It is not at all clear whether an enhanced or depressed ventilatory response is advantageous in chronic hypoxia, at high altitude for example (see Chapter 11). Finally, it has always to be kept in mind that the primary function of dark variants with their peptide-rich cytoplasm may not be chemoreception at all but another, as yet undiscovered, endocrine function.

Anaemic Hypoxaemia

Our studies suggest that the carotid bodies do not enlarge in response to anaemic hypoxaemia (Winson and Heath 1973). We obtained the combined carotid body weights at necropsy in a series of subjects who had been anaemic and related those weights to a representative haemoglobin level measured in the period before death. We studied 26 male and 9 female subjects who had been anaemic and 5 male and 10 female subjects who had not. Criteria for anaemia (de Gruchy 1972) were taken to be a haemoglobin level below

13.5 g/dl for men and less than 11.5 g/dl for women. A wide range of diagnoses accounted for the anaemia in both sexes. The carotid bodies were not enlarged in the subjects who had been anaemic, the mean carotid body weights in the anaemic men and women being, respectively, 21.8 mg and 13.3 mg, both figures being well within what we regard as a normal range (see Chapter 2). These findings do not confirm the experimental work of Tramezzani and his colleagues (1971), who reported that in cats made chronically anaemic by daily bleeding the carotid bodies enlarged. They also reported that in six cats removal of the carotid bodies led to a fall in the reticulocyte count and to anaemia. Subsequently, Lugliani et al. (1971) found no evidence in humans of a relation between the carotid bodies and the haemoglobin level, haematocrit reading, or reticulocyte count when they studied a series of patients undergoing bilateral resection of their carotid bodies as part of the treatment of chronic obstructive lung disease. They found there to be no significant change in haematocrit reading or reticulocyte count from preoperative values when studied for a period of up to three years postoperatively. Hence, the results of our study, and those of Lugliani and his colleagues, do not support the suggestion of Tramezzani et al. (1971) that the carotid bodies are concerned with the control of erythropoiesis. Chronic anaemia does not influence the size or histological appearance of carotid bodies either in normotensive Wistar rats or in spontaneously hypertensive rats of the Okamoto-Aoki strain (Habeck and Przybylski 1989). Functional studies have demonstrated that haemoglobin concentration has but a negligible effect on the activity of afferent carotid chemoreceptor fibres (Davies et al. 1981; Hatcher et al. 1978).

References

Barer GR, Pallot DJ (1984) Changes in catecholamine content of the rat carotid body in chronic hypoxia or hypercapnia and its modification by carotid body sympathectomy. J Physiol (Lond) 354:63P

Barer G, Wach R, Pallot D, Bee D (1986) Almitrine, hypoxia, systemic hypertension and the carotid body. In: Heath D (ed) Aspects of hypoxia. Liverpool University Press, Liverpool, pp 113–129

Blessing MH, Wolff H (1973) The carotid bodies at simulated high altitude. Pathol Microbiol 39:310–312

Davies RO, Nishino T, Lahiri S (1981) Sympathectomy does not alter the responses of carotid chemoreceptors to hypoxemia during carboxyhemoglobinemia or anemia. Neurosci Lett 21:159–163

de Gruchy GC (1972) Clinical haematology in medical practice. Blackwell, Oxford

Dhillon DP, Barer GR, Walsh M (1984) The enlarged carotid body of the chronically hypoxic and chronically hypoxic and hypercapnic rat: a morphometric analysis. Q J Exp Physiol 69:301–317

Habeck JO, Przybylski J (1989) Carotid and aortic bodies in chronically anemic normotensive and spontaneously hypertensive rats. J Auton Nerv Syst 28:219–226

Hatcher JD, Chiu LK, Jennings DB (1978) Anemia as a stimulus to aortic and carotid chemoreceptors in the cat. J Appl Physiol 44:692–702

Heath D, Edwards C, Harris P (1970) Post-mortem size and structure of the human carotid body. Thorax 25:129–140

Heath D, Edwards C, Winson M, Smith P (1973) Effects on the right ventricle, pulmonary vasculature, and carotid bodies of the rat on exposure to, and recovery from, simulated high altitude. Thorax 28:24–28

Heath D, Smith P, Jago R (1984) Dark cell proliferation in carotid body hyperplasia. J Pathol 142:39–49

Hellström S, Pequignot JM (1985) Beta-adrenergic blockade during long-term hypoxia. Effects on carotid body. A structural and biochemical study. In: 8th International Symposium on the peripheral arterial chemoreceptors. Oeiras, Portugal (abstract)

Lack EE (1978) Hyperplasia of vagal and carotid paraganglia in patients with chronic hypoxemia. Am J Pathol 91:497–516

Laidler P, Kay JM (1975a) A quantitative morphological study of the carotid bodies of rats living at a simulated altitude of 4300 metres. J Pathol 117:183–191

Laidler P, Kay JM (1975b) The effects of chronic hypoxia on the number and nuclear diameter of type I cells in the carotid bodies of rats. Am J Pathol 79:311–318

Lugliani R, Whipp BJ, Winter B, Tanaka KR, Wasserman K (1971) The role of the carotid body in erythropoiesis in man. N Engl J Med 285:1112–1114

Pallot DJ, Al Neamy KW (1983) The effects of hypoxia, hypercapnia and almitrine bismesylate on carotid body catecholamines. Eur J Respir Dis Suppl 126:203–207

Smith P, Hurst G, Heath D, Drewe R (1986a) The carotid bodies in a case of ventricular septal defect. Histopathology 10:831–840

Smith P, Heath D, Fitch R, Hurst G, Moore D, Weitzenblum E (1986b) Effects on the rabbit carotid body of stimulation by almitrine, natural high altitude, and experimental normobaric hypoxia. J Pathol 149:143–153

Tramezzani JH, Morita E, Chiocchio SR (1971) The carotid body as a neuroendocrine organ involved in the control of erythropoiesis. Proc Natl Acad Sci USA 68:52–55

Winson M, Heath D (1973) The carotid bodies in anemia. Arch Pathol Lab Med 96:58–60

10 Carotid Body Hyperplasia

Hyperplasia is the commonest form of histological abnormality which occurs in the carotid body, with the exception of chronic carotid glomitis (see Chapter 6). It is associated with a diversity of diseases, including pulmonary emphysema, bronchial asthma and systemic hypertension, but the underlying cause of hyperplasia in all of them is probably hypoxaemia. Hyperplasia may be suspected on naked-eye examination at necropsy but it has to be confirmed by microscopy. Unequivocal proof that glomic cells have proliferated is obtained by using morphometric techniques to determine the total number of cells present and comparing this figure with that from normal organs, as outlined in Chapter 3. This is a very time-consuming procedure suitable only for research projects on a limited number of cases. A less arduous and more rapid technique is a differential count of the various types of cell present, which may suggest rather than prove the existence of hyperplasia. Even a purely qualitative examination of the characteristic histological changes of carotid body hyperplasia may suggest the condition to the practised eye.

Macroscopic Appearances

The hyperplastic carotid body is enlarged and this increase in size can often be appreciated simply by examining it in situ on the carotid bifurcation. The enlarged carotid body in hypoxaemic cases of chronic bronchitis and emphysema is congested and mauve in colour, in contrast to the tan appearance of the normal organ. One has to be careful to distinguish normal anatomical variation from pathological enlargement for the normal carotid body may be double or bilobed (see Chapter 2). When a bilobed variant undergoes enlargement, it may be coarsely nodular, spuriously simulating nodular hyperplasia (Khan et al. 1988a). On the other hand, fine nodularity with a mulberry-like appearance is very suggestive of hyperplasia (Smith et al. 1982; Khan et al. 1988a).

The suspected increase in size is easily confirmed by weighing, but care must be taken to dissect away adherent flecks of adipose tissue first. In such a small organ incomplete dissection may inadvertently convert a normal carotid body into an apparently hyperplastic one on the basis of weight. The upper limit of normal combined carotid body weight may be regarded as 30 mg, a figure obtained from the upper 95% probability limit of weight from 57 cases (see Chapter 2). Limits calculated from 43 cases of carotid body hyperplasia (Smith et al. 1982) were found to overlap to a considerable degree with those from the controls (Table 10.1). Thus a combined weight below 30 mg does not necessarily indicate normality. In order to identify hyperplasia more precisely, histological assessment is necessary.

Histopathology

Hyperplasia of the carotid body can often be inferred from sections under the low power objective of the microscope because the glomic lobules are greatly enlarged. The normal appearance of several, small, discrete lobules separated by fibrous septa may be lost, since, as individual lobules enlarge, they fuse with their neighbours to

Table 10.1. A comparison of measurements of the carotid bodies from 57 normal subjects and 43 cases of sustentacular cell hyperplasia (after Smith et al. 1982)

Parameter	Group	Mean	Range of 95% probability	
			Lower	Upper
Combined weight (mg)	Normal	18	7	30
	Hyperplastic	32	3	61
Diameter of lobules (μm)	Normal	411	258	565
	Hyperplastic	586	297	896
Diameter of cell clusters (μm)	Normal	82	54	110
	Hyperplastic	82	56	109
Sustentacular cell count (%)	Normal	39	31	47
	Hyperplastic	55	43	67

form large lobules of irregular or elongated profile. The proportion of the carotid body occupied by fibrous tissue is correspondingly diminished. When these lobules are examined in more detail, the most prominent feature in the majority is a striking increase in the number of elongated cells (Fig. 10.1). These form thick rings, several cells in width, surrounding each core of chief cells (Figs. 10.1–10.3). The cores of chief cells are often reduced in size and appear to be compressed by the thick rings of long cells forming whorls around them (Figs. 10.3, 10.4) (Heath et al. 1982; Smith et al. 1982). Because the elongated cells are located at the periphery of cell clusters it has been assumed that they are sustentacular in type but the nuclei of many of them are plump and pale and more closely resemble those of Schwann cells (Fig. 10.5). Indeed, electron microscopy has confirmed

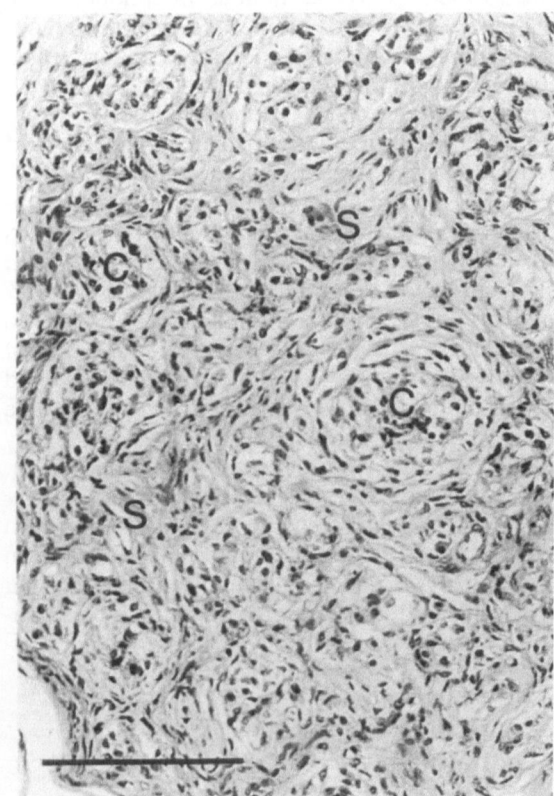

Fig. 10.1. Part of a glomic lobule from the enlarged carotid body of a 72-year-old man with centrilobular emphysema and severe hypoxaemia. The right ventricle weighed 198 g (normal <65 g) and the combined weight of the carotid bodies was 61 mg. The lobule is enlarged and shows thick whorls of sustentacular cells (*s*) surrounding the cores of chief cells (*c*). (Haematoxylin–eosin (HE)) Scale line = 100 μm.

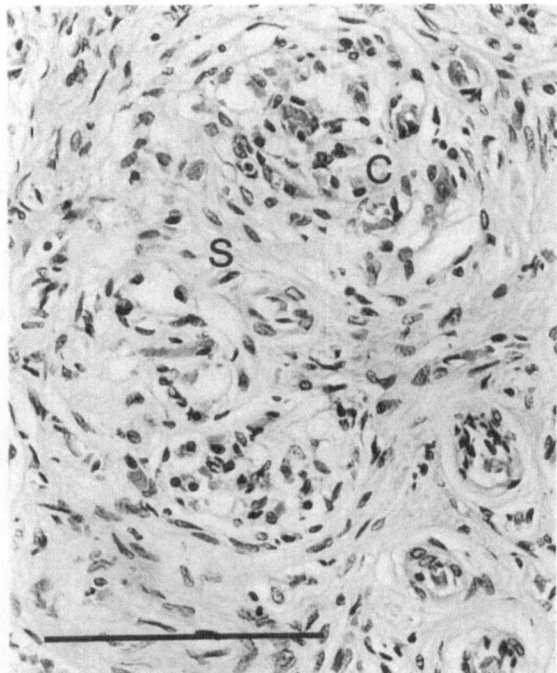

Fig. 10.2. Sustentacular cell hyperplasia in the carotid body of an 80-year-old man with systemic hypertension. The left ventricle and septum weighed 279 g (normal <190 g) and the combined carotid body weight was 50 mg. Numerous elongated sustentacular cells (*s*) form thick concentric rings around cores of chief cells (*c*). (HE) Scale line = 100 μm.

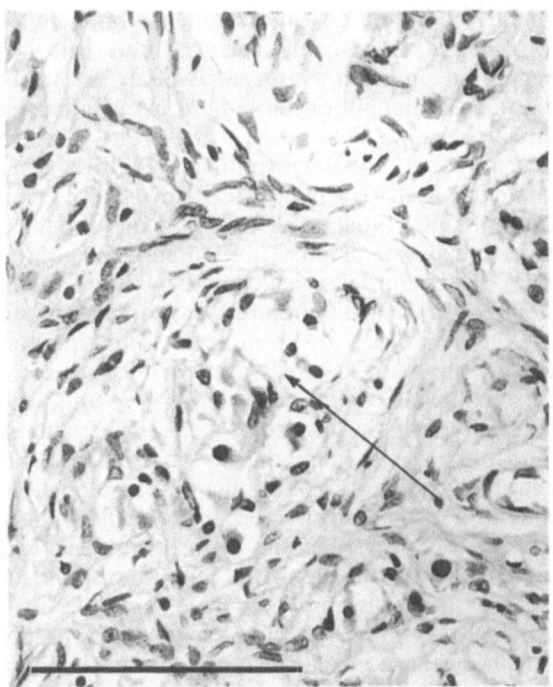

Fig. 10.3. Sustentacular cell hyperplasia in a woman of 80 years with systemic hypertension. A small, atrophic core of chief cells (*arrow*) is engulfed by a dense proliferation of elongated sustentacular cells. (HE) Scale line = 100 μm.

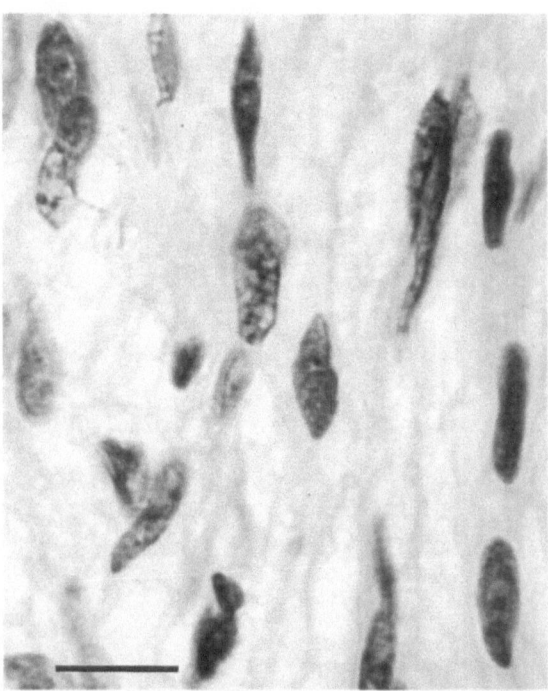

Fig. 10.5. Detail of elongated cells at the periphery of a cell cluster from a case of sustentacular cell hyperplasia secondary to pulmonary emphysema. The nuclei of the cells are variable in shape and staining intensity. Some are dark and narrow and may be sustentacular. Others are plump and stippled and may be Schwann cells. Their precise nature cannot be determined with certainty. (HE) Scale line = 10 μm.

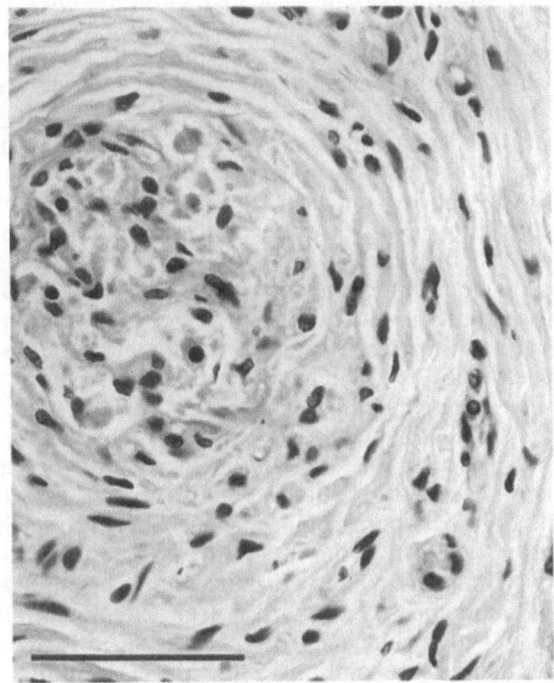

Fig. 10.4. Part of a cell cluster from a 60-year-old man with the Pickwickian syndrome. A central core of chief cells is surrounded by elongated cells which form concentric rings around it. (HE) Scale line = 50 μm.

that these peripheral elongated cells are closely associated with myelinated nerves. The more centrally situated ones closer to the cores ensheathe axons but also ramify amongst the chief cells and are thus sustentacular (Jago et al. 1984). Elongated cells comprise a mixture of both sustentacular and Schwann cells and by light microscopy it is impossible to distinguish between them with any certainty (Fig. 10.5). Both may merge imperceptibly with the much smaller proportion of fibroblasts at the periphery of the lobules. In other words the blanket term "sustentacular cell hyperplasia" is not wholly accurate but it has the advantage of conveying a visual impression of the histological appearance, provided its limitations are appreciated.

The proliferation of sustentacular cells is associated with an increase in density of nerve fibres which can be demonstrated using Bodian's silver protargol method (see Chapter 3). This stain reveals a few, thick, myelinated nerve fibres running together through the connective tissue between the lobules and in the surrounding stroma. In the interlobular septa the nerves

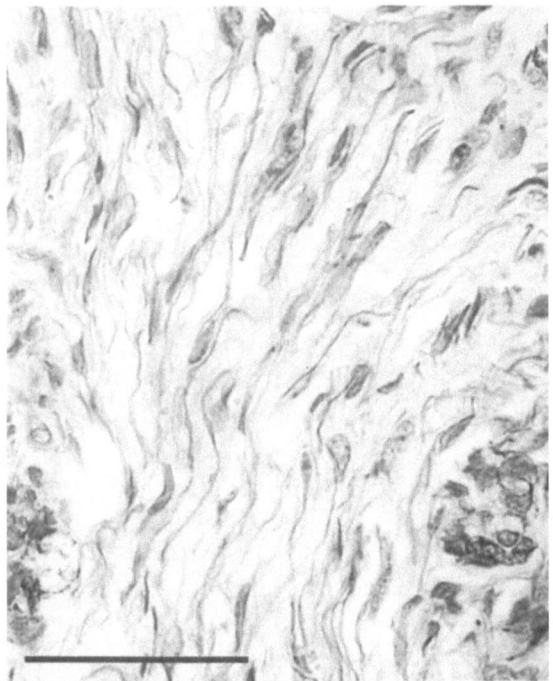

Fig. 10.6. Proliferation of nerve axons in an enlarged carotid body of a woman of 58 years with systemic hypertension. The axons are associated with hyperplasia of sustentacular cells surrounding cores of chief cells. (Bodian silver) Scale line = 50 μm.

quently it is not possible to define, and hence measure, a single axon. Instead, the quantity of nerve fibres present can be assessed by determining the total length of nerve fibre in each field of the microscope. This can be achieved by using an intercept method of morphometry with a Merz semicircle grid overlying the field (Aherne and Dunnill 1982; Fitch et al. 1985). Using this technique the density of nerve fibres, expressed as mm/mm^2 of section, was calculated by Fitch et al. (1985) to be 12.6 in five control carotid bodies and 44.3 in eight cases of sustentacular cell hyperplasia. In this study there was a strong statistical correlation between nerve density and the percentage of sustentacular cells. This almost four-fold increase in the density of axons makes it extremely unlikely that the numerous elongated cells in carotid body hyperplasia are fibroblasts and that the process is simply one of fibrosis. It is more likely that the elongated cells are sustentacular cells and Schwann cells, which proliferate to support the increased number of axons at the periphery of the cell clusters, a conclusion which is confirmed by electron microscopy (see Chapter 18).

While all these histological features of carotid body hyperplasia apply to middle-aged and elderly patients with chronic obstructive lung disease, they may appear in a slightly different form in young patients particularly. In this form hyperplasia involves the chief cells predominantly and thick whorls of sustentacular cells and Schwann cells are absent. The chief cells produce numerous clusters surrounded by thin rims of sustentacular cells as in the normal carotid body (Fig. 10.7). Sometimes the clusters are much smaller than usual, perhaps consisting of only three or four chief cells (Khan et al. 1988b). Dark variants of chief cells may be prominent but this is not invariable. This form of the condition may be termed "chief cell hyperplasia" to distinguish it from the commoner sustentacular cell type. Chief cell hyperplasia is typically seen in native highlanders (see Chapter 11) where the ambient hypoxia is associated with hypocarbia in contrast to the situation in chronic obstructive lung disease where the alveolar hypoxia is associated with hypercarbia. Chief cell hyperplasia is also found in children and young adults with hypoxaemia secondary to congenital heart disease or pulmonary cystic fibrosis (see Chapter 5). It may be that chief cell hyperplasia is an early stage of the disease and, with the passage of time, sustentacular cells with their axons continue to proliferate at the expense of chief cells to produce the histological picture of sustentacular cell hyperplasia. An alternative view (see below) is that the

become appreciably thinner and are frequently unmyelinated. Carotid body hyperplasia does not involve these large nerve fibres but rather the fine unmyelinated nerve fibres which tend to follow the same concentric course as the sustentacular cells around the cores of chief cells in the glomic clusters (Fig. 10.6; Fitch et al. 1985). These profuse rings of axons are closely associated with the sustentacular or Schwann cells with which they form a simple mesaxonal relationship (Jago et al. 1984). Fine axons ramify amongst the chief cells, although there are fewer of them than might be anticipated from their high density at the periphery of the clusters. As in the normal carotid body a variety of nerve endings can be seen on chief cells. In the main these are boutons, appearing as swellings at the terminations of axons, or calyces seen as a terminal U-shaped form. Sometimes the ends of axons seem to curve around the chief cell (Fitch et al. 1985). There is no apparent alteration in the proportions of the different types of nerve endings between normal and hyperplastic carotid bodies. The increase in nerve density in sustentacular cell hyperplasia has been quantified. Since nerve fibres are in the plane of section for only a small part of their total length and branch fre-

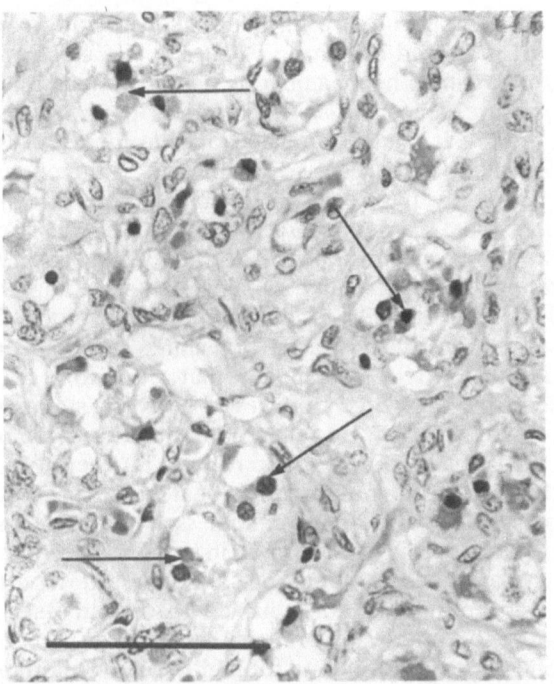

Fig. 10.7. Part of a glomic lobule from a man of 35 years who was domiciled at an altitude of 3500 m in Ladakh and killed in a traffic accident. There are numerous mini-clusters consisting of small groups of chief cells (*arrows*) with scanty sustentacular cells between them, typical of chief cell hyperplasia. (HE) Scale line = 50 μm.

two forms of hyperplasia are distinct entities and that their initiation depends at least upon the age of the individual at the time of commencement of the hypoxaemic stimulus.

Quantitative Histology

Sustentacular Cell Hyperplasia

The diameters of glomic lobules and cell clusters can be measured rapidly (see Chapter 3). The mean diameter of glomic lobules was measured in 43 pairs of carotid bodies with sustentacular cell hyperplasia and found to be considerably greater than in 57 controls (Table 10.1; Smith et al. 1982). There is, however, a large overlap of the 95% probability limits between the two groups (Table 10.1) so that this measurement alone will not always identify hyperplasia. The mean diameter of cell clusters is identical in the two groups with similar 95% limits (Table 10.1; Smith et al. 1982). Clusters comprise cores of chief cells plus rims of sustentacular cells which are much thicker in the

usual form of hyperplasia. Hence the cores must be smaller if the diameters of clusters are unchanged.

Differential cell counts reveal predictably that the percentage of sustentacular cells is greatly increased (Table 10.1). Furthermore, there is very little overlap of the 95% probability limits so that, in the 43 cases from which these data were calculated, only four were below the upper limit of 47% for normals (Smith et al. 1982). The sustentacular cell count must inevitably include all elongated cells in the lobule except those which can be easily distinguished, such as vascular smooth muscle, pericytes, endothelium or fibrocytes. A numerical definition of sustentacular cell hyperplasia may be expressed as a combined carotid body weight exceeding 30 mg, a lobular diameter exceeding 565 μm and a differential count of sustentacular cells greater than 47% (Smith et al. 1982). None of these individual components alone will identify hyperplasia in every case but when all three are present together its presence can be regarded as unequivocal.

The differential counts of chief cells are not included in the numerical definition of sustentacular cell hyperplasia because they are little different from those of controls. Thus, the proportions of light, dark and progenitor variants are, respectively, 39%, 4% and 2% in hyperplastic carotid bodies and 54%, 5% and 2% in controls (Smith et al. 1982). These differential cell counts are numerically related such that if one increases the others must decrease. However, even if the counts are calculated to include only chief cells, the values are similar, with proportions in normal and hyperplastic carotid bodies respectively of 89% and 87% for light cells, 8% and 9% for dark cells and 3% and 4% for progenitor cells. There is, therefore, no increase in the percentage of dark cells in sustentacular cell hyperplasia.

Chief Cell Hyperplasia

Comparisons of quantitative data between the two forms of carotid body hyperplasia are difficult because different morphometric methods have been employed. Lack (1977) studied necropsy specimens of carotid bodies which showed the histological picture of chief cell hyperplasia from 12 subjects who had been hypoxaemic in life. The mean combined weight of the carotid bodies was 48 mg, well above our criteria for the upper limit of normal. The sizes of lobules were expressed as maximum and minimum dimensions, the mean values being, respectively, 719 μm and 421 μm. These dimensions are equivalent to a mean lobule diameter of 570 μm, which is similar to that obtained in the 43 cases of sustentacular cell hyperplasia (Table 10.1). The number of lobules present on sections taken through the greatest diameter of the carotid bodies was also increased from 11–14 in controls to 16–20 in cases of cystic fibrosis or to 18–25 in congenital cardiac septal defects (Lack et al. 1985).

Although many of the cell clusters appear subjectively to be small, their diameters have been measured only by Khan et al. (1988b), who found that in two out of four native highlanders of Ladakh the mean diameters of cell clusters were 44 and 45 μm, approximately half the normal value. The sustentacular cell count was normal in three out of four, the exception being a man of 52 years who had a count of 58% in keeping with his age (see Chapter 4). Lack et al. (1985) stated that in their cases of cystic fibrosis and congenital heart disease the number of sustentacular cells was similar in 128 controls and 47 hypoxaemic patients under the age of 30 years.

Nature of Sustentacular Cell Hyperplasia

A proliferation of sustentacular cells in the carotid body is not unique to hyperplasia but occurs also as part of the ageing process (Hurst et al. 1985; see Chapter 4). It is important to distinguish between the two and this is usually possible because the lobules are commonly small and atrophic in the ageing carotid body but enlarged in hyperplasia. With increasing age, cell clusters tend to be few and surrounded by sustentacular cells many of which are not oriented circumferentially to the cores of chief cells and which show a patchy distribution. There is commonly an ingrowth of fibrous tissue from the stroma into the lobules. In hyperplasia the clusters are numerous and each is surrounded by predominantly circumferentially oriented sustentacular cells.

Sustentacular cell hyperplasia appears to be unique to humans. The carotid bodies in animal species which have to acclimatise to the hypobaric hypoxia of high altitude respond to hypoxaemia by hyperplasia of all cellular elements in the glomus (see Chapter 21). In species genetically adapted to high altitude none of the components responds to hypoxaemia (see Chapter 21). This human response by sustentacular cells is the same irrespective of whether it occurs in association with pulmonary emphysema or systemic hypertension. It is probable that sustentacular and Schwann cells proliferate in response to an increased need for

axonal support following an initial proliferation of nerves around cell clusters. The precise mechanism by which hypoxaemia or raised intravascular pressure leads to a proliferation of axons is obscure. It could be that axonal proliferation is provoked by a sustained stimulation of the nerve endings by hypoxaemia. Certainly there is evidence that the nerve endings on the chief cells are the primary transducers of the chemosensory reflex (Eyzaguirre and Zapata 1984; see Chapter 7). If this is so, nerves could respond to prolonged stimulation by increased branching and growth but this would presuppose that the axons are sensory afferents. However, silver staining provides no clue as to their nature and they could equally well be autonomic nerves. Furthermore, this theory does not explain axonal growth in cases of systemic hypertension, unless hyperplasia in these cases also has its origins in ischaemia secondary to intimal proliferation in glomic arteries. It has been suggested that the activity of sustentacular cells is not to support proliferating axons but to phagocytose cell debris formed during Wallerian degeneration of them. Within a few hours of axonal injury nerve endings retract from chief cells and become surrounded by processes of sustentacular cells (McDonald and Mitchell 1975). This explanation for sustentacular cell hyperplasia seems unlikely to us, for our ultrastructural studies of sustentacular cells in carotid body hyperplasia have not revealed residual bodies within their cytoplasm.

The axonal proliferation is curiously disorganised in that it forms dense, concentric rings around the cell clusters, from which only a few branches pass to contact the chief cells. This neural proliferation can be interpreted on the basis that the axons have lost the trophic influence of their target organ, the cores of chief cells, which are often much reduced in size and atrophic. If this loss of chief cells is an early event, the axons which normally supply them might be expected to proliferate in a vain attempt to find their target organ. The situation would be analogous to an amputation neuroma in which there is a profuse, uncoordinated proliferation of nerves in the region of the stump of an amputated limb. Indeed the sustentacular cell hyperplasia in emphysema has been described as showing "a neurinoma-like picture" (Habeck 1986). This hypothesis has the advantage of explaining the blunted hypoxic ventilatory drive which occurs in patients with chronic bronchitis and emphysema who have been hypoxaemic for many years (Flenley et al. 1970; Bradley et al. 1979; Flenley 1986).

At first sight this hypothesis fails to explain the

difference in structure of the carotid body between native highlanders and patients at sea level with chronic obstructive lung disease. Although both are hypoxic, the former show proliferation of chief cells whereas the latter are characterised by sustentacular cell hyperplasia. There are, however, some important differences between the two. Natives to high altitude are born into a hypoxic environment and their carotid bodies are hypoxaemic whilst they are growing. It is likely that these developing carotid bodies can respond to sustained stimulation by proliferation of chief cells, axons and sustentacular cells in an orderly, organised fashion. The same argument applies to the children and young adults, studied by Lack et al. (1985), in whom there was enlargement of carotid bodies in the absence of sustentacular cell hyperplasia. On the other hand, patients with chronic bronchitis and emphysema are usually middle-aged or elderly and hypoxaemia and hypercarbia develop after the age-changes of atrophy of chief cells and hyperplasia of sustentacular cells (see Chapter 4). It is possible that under chronic stimulation by hypoxia these atrophic chief cells are unable to respond by hyperplasia but the sustentacular cells, and presumably their axons, are already active and are able to respond by proliferation. However, because the chief cells are atrophic the axons are unable to establish adequate contact with their target organ and so they, and the sustentacular cells, continue to proliferate in an excessive and uncoordinated fashion to form a kind of neuroma. If this hypothesis is true, one would expect to find sustentacular cell hyperplasia when an ageing carotid body is chronically stimulated. Mini-clusters with only thin rims of sustentacular cells should be associated with hyperplasia of the young carotid body. As discussed below and summarised in Table 10.2 there is evidence that this is indeed the case.

Focal proliferation of dark or progenitor cells

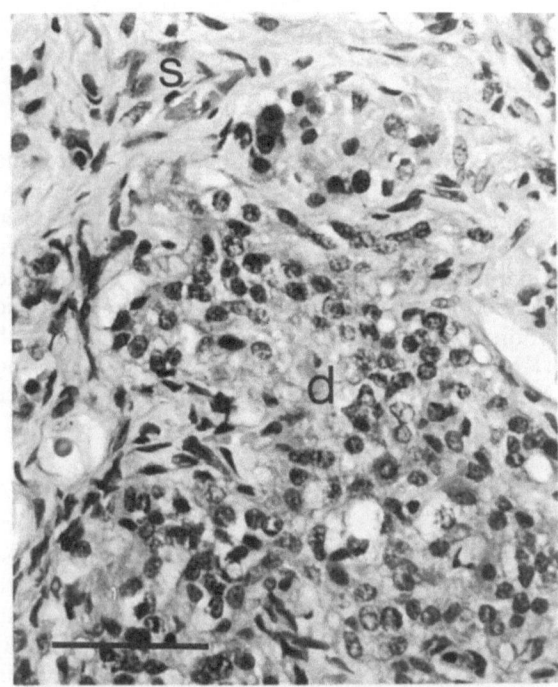

Fig. 10.8. Edge of a large focus of dark cell hyperplasia in a woman of 80 years with systemic hypertension. The proliferation of dark cells (*d*) is superimposed upon a background of sustentacular cell (*s*) hyperplasia. (HE) Scale line = 50 μm.

(Fig. 10.8) may occur on a background of generalised hyperplasia of sustentacular cells irrespective of whether this is due to chronic hypoxaemia or systemic hypertension (Heath et al. 1984; see Chapter 9). It may be caused by a sudden exacerbation of the long-standing stimulus for hyperplasia such as occurs when a patient with chronic obstructive lung disease develops a respiratory infection. The dark cells giving rise to these focal proliferations may be pleomorphic and dark cells have been reported in chemodectomas occurring at high altitude (see Chapter 11).

Table 10.2. Comparison of the histological type of carotid body hyperplasia with age and diagnosis

No. of cases	Age (yrs) Range	Age (yrs) Mean	Diagnosis	Form of hyperplasia	Reference
5	62–72	68	Emphysema	Sustentacular cell	Heath et al. 1982
1		60	Pickwickian syndrome	Sustentacular cell	Heath et al. 1982
5	58–80	75	Systemic hypertension	Sustentacular cell	Heath et al. 1982
30	1–30	22	Cystic fibrosis	Chief cell	Lack et al. 1985
17	<1–30	9	Congenital heart disease	Chief cell	Lack et al. 1985
4	4–52	30	Residence at high altitude	Chief cell; prominent dark cells[a]	Khan et al. 1988b
11	Not known	37	Bronchial asthma	Sustentacular cell; prominent dark cells	Bencini and Pulera 1991

[a] The oldest subject also showed moderate sustentacular cell hyperplasia.

Causes of Hyperplasia

Residence at High Altitude

Enlargement of the carotid bodies in Quechua Indians of the High Andes of Peru was the first example of carotid body hyperplasia to be described (Arias-Stella and Valcarcel 1973; see Chapter 11). Native highlanders are exposed to hypobaric hypoxia from birth and their carotid bodies typically show chief cell hyperplasia, with (Khan et al. 1988b) or without (Arias-Stella and Valcarcel 1976) a prominence of dark cells (see Frontispiece).

Chronic Bronchitis and Emphysema

These diseases together constitute the commonest cause of chronic obstructive lung disease and are frequently associated with carotid body hyperplasia. Hyperplasia is often found in those subjects with alveolar hypoxia, severe arterial oxygen desaturation, hypercarbia and systemic oedema, the so-called "blue-and-bloated" patient. The patients with chronic obstructive lung disease who develop enlargement of the carotid bodies are not those with any particular form of pulmonary emphysema nor those with the greatest loss of lung parenchyma, but patients with right ventricular hypertrophy associated with alveolar hypoxia and hypercarbia and muscularisation of pulmonary arterioles (Edwards et al. 1971). Combined carotid body weights as high as 84.8 mg (Heath et al. 1970) and 89.3 mg (Edwards et al. 1971) were found in cases of centrilobular emphysema with right ventricular hypertrophy. In cases of pulmonary emphysema in which the right ventricle was not hypertrophied, the combined carotid body weight ranged from 27.2 to 37.6 mg with a mean of 32.4 mg, whereas in cases with right ventricular hypertrophy the range was 32.2–89.3 mg with a mean of 56.2 mg (Edwards et al. 1971). Enlargement of lobules may also be pronounced in patients with chronic bronchitis and emphysema with individual lobules attaining diameters as high as 1500 μm and mean diameters of up to 1300 μm (Heath et al. 1982).

The histological picture of the lobules is typically one of sustentacular cell hyperplasia with reported differential cell counts of 55% of sustentacular cells in five cases of panacinar emphysema (Heath et al. 1982) and 62% in five cases of emphysema with right ventricular hypertrophy (Smith et al. 1982). As suggested above, sustentacular cell hyperplasia may be anticipated in patients whose carotid bodies have undergone age-changes prior to the onset of hypoxaemia. Such conditions are met in emphysema and in all of the cases cited above the patients were middle-aged or elderly with ages ranging from 54 to 87 years with a mean of 72 years. Habeck (1986) described extensive sustentacular cell hyperplasia in 13 patients over the age of 55 years with emphysema or pulmonary fibrosis but not in a boy of 15 with cystic fibrosis. An increase in weight of the carotid bodies, gross enlargement of lobules and sustentacular cell hyperplasia has also been described in a case of the Pickwickian syndrome (Heath et al. 1982). In such patients hypoxaemia is caused by alveolar hypoventilation brought about by gross obesity.

Cystic Fibrosis

This disease is characterised by abnormally viscous mucous plugs which obstruct bronchi commonly leading to chronic infection, pulmonary fibrosis, bronchiectasis and emphysema. In this condition the carotid bodies are abnormally heavy (Lack 1977; Lack et al. 1985). The dimensions and number of lobules are greater so that there is an increase in the total surface area of carotid bodies and in the ratio of the areas of glomic to stromal tissue. The lobules show the same proportions of chief cells and sustentacular cells as in normal carotid bodies indicating that the process of hyperplasia involves all cellular elements equally (Lack et al. 1985). The absence of sustentacular cell hyperplasia may be explained by the youth of the patients.

Cyanotic Congenital Heart Disease

Patients with congenital cardiac defects are commonly cyanotic and are hypoxaemic for the whole of their lives. This occurs in most cases of Fallot's tetralogy, or after the development of a reversed intracardiac shunt such as a large ventricular septal defect or patent ductus arteriosus (Harris and Heath 1986). Patients tend to be young and show chief cell hyperplasia (Lack et al. 1985).

Bronchial Asthma

Bronchospasm and the excessive production of viscid mucus in asthmatic patients bring about alveolar hypoxia and hypoxaemia, and hence hyperplasia of the carotid bodies might be expected in them. Glomic tissue for confirming this postulated histological change is now available only from centres where unilateral therapeutic glomectomy is still practised. This procedure in the treatment of bronchial asthma is said by its advocates (Bencini 1970) to reduce the dyspnoea and bronchospasm. It has been suggested that the carotid body reflexly influences the tone of smooth muscle in the bronchi and that glomectomy lowers this making the smooth muscle less responsive to bronchospastic stimuli (Bencini 1970). Carotid bodies thus resected have been studied by histology and morphometry (Bencini and Pulera 1991). Their 50 patients all suffered from episodes of dyspnoea but were submitted to operation only if the average daytime arterial oxygen tension was greater than 65 mmHg. Eleven of them had a history of bronchial asthma for up to five years and their carotid bodies were compared with controls obtained from ten subjects without cardiopulmonary disease coming to necropsy. The carotid bodies from the asthmatic subjects were not enlarged. The lobules were not increased in size and were irregular in shape, elongated and crowded closely together (Bencini and Pulera 1991). Large numbers of sustentacular cells were arranged as dense sheets at the centre of the lobules or in concentric rings around the clusters of chief cells. A differential cell count revealed that the proportion of sustentacular cells had increased from 33% in the controls to 68% in the asthmatics. Dark cells increased from 28% of all chief cells in the controls to 43% in the cases of asthma. Thus the histological picture is analogous but not identical with the sustentacular cell hyperplasia typical of pulmonary emphysema and appears to be produced by a prolonged stimulus of mild hypoxaemia.

Systemic Hypertension

The histology of the carotid body in some cases of systemic hypertension is indistinguishable both macroscopically and histologically from that in chronic bronchitis and emphysema (see Chapter 14). Hypertensive subjects coming to necropsy are likely to be midde-aged or elderly so that sustentacular cell hyperplasia develops in carotid bodies already showing age-changes and may be due to prolonged ischaemia secondary to intimal fibrosis of glomic arteries (Habeck et al. 1983).

The different histological types of carotid body hyperplasia are summarised in Table 10.2.

References

Aherne WA, Dunnill MS (1982) Morphometry. Edward Arnold, London, p 163

Arias-Stella J, Valcarcel J (1973) The human carotid body at high altitudes. Pathol Microbiol 39:292–297

Arias-Stella J, Valcarcel J (1976) Chief cell hyperplasia in the human carotid body at high altitudes. Physiologic and pathologic significance. Hum Pathol 7:361–376

Bencini A (1970) Reduction of reflex bronchotropic impulses as a result of carotid body surgery. Int Surg 54:415–423

Bencini C, Pulera N (1991) The carotid bodies in bronchial asthma. Histopathology 18:195–200

Bradley C, Fleetham J, Anthonisen N (1979) Ventilatory control in patients with hypoxaemia due to obstructive lung disease. Am Rev Respir Dis 120:21–30

Edwards C, Heath D, Harris P (1971) The carotid body in emphysema and left ventricular hypertrophy. J Pathol 104:1–13

Eyzaguirre C, Zapata P (1984) Perspectives in carotid body research. J Appl Physiol 57:931–957

Fitch R, Smith P, Heath D (1985) A quantitative study of nerve axons in carotid body hyperplasia. Arch Pathol Lab Med 109:234–237

Flenley D (1986) Long term oxygen therapy and the pulmonary circulation. In: Heath D (ed) Aspects of hypoxia. Liverpool University Press, Liverpool, pp 45–59

Flenley D, Franklin D, Millar J (1970) The hypoxic drive to breathing in chronic bronchitis and emphysema. Clin Sci 38:503–518

Habeck J-O (1986) Morphological findings at the carotid bodies of humans suffering from different types of systemic hypertension or severe lung diseases. Anat Anz Jena 162:17–27

Habeck J-O, Waller H, Protze J (1983) Pathological alterations of the arterial vessels of the carotid bodies in hypertensive humans. Dt Gesundh-Wesen 38:1970–1972

Harris P, Heath D (1986) The influence of hypertensive pulmonary vascular disease on the natural history of patients with congenital cardiac shunts. In: The human pulmonary circulation, 3rd edit. Churchill Livingstone, Edinburgh, pp 308–328

Heath D, Edwards C, Harris P (1970) Post-mortem size and structure of the human carotid body. Its relation to pulmonary disease and cardiac hypertrophy. Thorax 25:129–140

Heath D, Smith P, Jago R (1982) Hyperplasia of the carotid body. J Pathol 138:115–127

Heath D, Smith P, Jago R (1984) Dark cell proliferation in carotid body hyperplasia. J Pathol 142:39–49

Hurst G, Heath D, Smith P (1985) Histological changes associated with ageing of the human carotid body. J Pathol 147:181–187

Jago R, Smith P, Heath D (1984) Electron microscopy of carotid body hyperplasia. Arch Pathol Lab Med 108:717–722

Khan Q, Heath D, Smith P (1988a) Anatomical variations in human carotid bodies. J Clin Pathol 41:1196–1199

Khan Q, Heath D, Smith P, Norboo T (1988b) The histology of the carotid bodies in highlanders from Ladakh. Int J Biometeorol 32:254–259

Lack EE (1977) Carotid body hypertrophy in patients with cystic fibrosis and cyanotic congenital heart disease. Hum Pathol 8:39–51

Lack EE, Perez-Atayde AR, Young JB (1985) Carotid body hyperplasia in cystic fibrosis and cyanotic heart disease. A combined morphometric, ultrastructural, and biochemical study. Am J Pathol 119:301–314

McDonald DM, Mitchell RA (1975) The innervation of glomus cells, ganglion cells and blood vessels in the rat carotid body: a quantitative ultrastructural analysis. J Neurocytol 4:177–230

Smith P, Jago R, Heath D (1982) Anatomical variation and quantitative histology of the normal and enlarged carotid body. J Pathol 137:287–304

11   The Carotid Bodies at High Altitude

The initial hyperventilation on first ascent to high altitude is due to stimulation of the carotid bodies. Their contribution to respiration in normoxic man has been estimated as 15% (Barer et al. 1986) but this is much increased in response to hypobaric hypoxia. Not only is arterial chemoreceptor activity increased in lowlanders in response to the hypobaric hypoxia of high altitude but it is sustained. Under such circumstances it might be anticipated that the carotid bodies would undergo permanent enlargement in native highlanders. This was confirmed by Arias-Stella at a meeting of the American Association of Pathologists and Bacteriologists in San Francisco in 1969. He reported that the carotid bodies of Quechua Indians born and living in the Peruvian Andes are larger than those of mestizos living on the coast. This observation was so important as to form the basis for the study of the pathology of the carotid body beyond chemodectomas.

Carotid Bodies of Native Highlanders

This initial report was subsequently extended and published by Arias-Stella and Valcarcel (1973). They described the carotid bodies in two series of necropsies, one from sea level on mestizos from Lima and one from the Peruvian mining town of Cerro de Pasco at an altitude of 4330 m on native Quechua highlanders. The cases were matched for age and sex and mostly comprised accidental deaths with no significant cardiovascular or pulmonary pathology. The carotid bodies of the

highlanders were heavier (Fig. 11.1) and larger (Fig. 11.2) in each group and the differences became greater with increasing age. Subsequently Khan et al. (1988) were able to confirm this

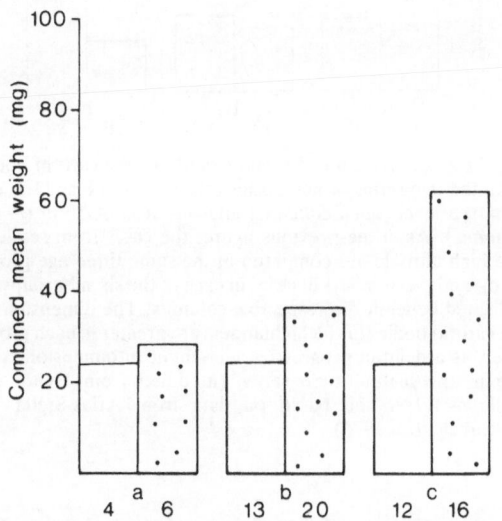

Fig. 11.1. Combined weights of the carotid bodies from two series of necropsies, one from mestizos at sea level (*open columns*) and one from native highlanders at 4330 m (*dotted columns*). The cases from sea level and high altitude are compared in the age groups: 10–20 years (a), 21–40 years (b) and 41–70 years (c). The numbers of cases in each of the six subgroups are indicated beneath the respective columns. The carotid bodies of the highlander are heavier in each group. In addition, there is a definitive progressive increment in weight with age in the high-altitude series (modified from Heath and Williams 1989, and based on data from Arias-Stella and Valcarcel 1973, 1976).

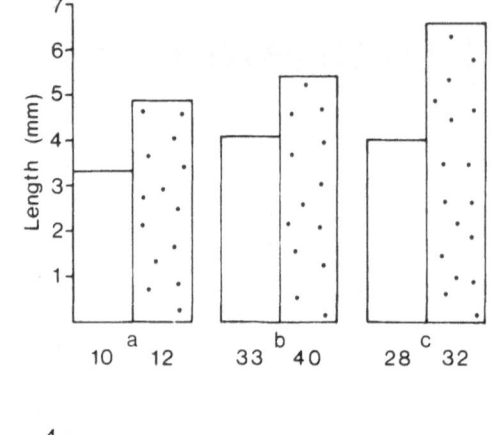

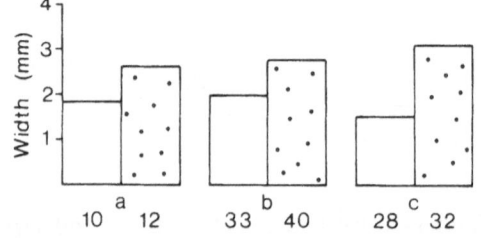

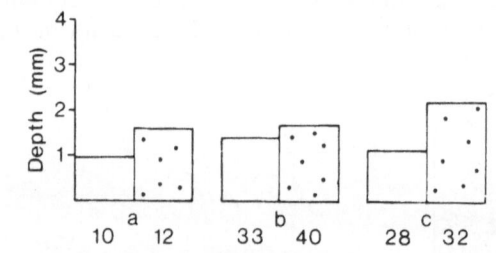

Fig. 11.2. a length, **b** width and **c** depth of the carotid bodies from the two series of necropsies referred to in Fig. 11.1, one from sea level (*open columns*) and one from 4330 m (*dotted columns*). As in the previous figure, the cases from sea level and high altitude are compared in the same three age groups (a–c), and the numbers of cases in each of the six subgroups are indicated beneath the respective columns. The dimensions of the carotid bodies of the highlanders are greater in each group. There is a definite progressive increment in dimensions with age in the high-altitude series (modified from Heath and Williams 1989, and based on data from Arias-Stella and Valcarcel 1973, 1976).

increase in carotid body weight in three adult highlanders domiciled at altitudes between 3300 and 4200 m in the vicinity of Leh in Ladakh. Mean normal combined carotid body weights from an unselected group of 42 domiciled residents at sea level had previously been shown to be 24.2 mg (Heath et al. 1970). In one young Ladakhi male of 29 years from Chusul (4200 m) the carotid bodies were of normal size and weight, the combined weight being 18 mg. In a man aged 35 years

from Choglamsar (3500 m) this weight had increased to 50 mg and in a third male aged 52 years from Stok (3300 m) it was 78 mg (Khan et al. 1988). Clearly in native highlanders from both the Andes and the Karakorams there is a progressive increase in the weight of the carotid bodies with age. This enlargement has an organic basis of proliferation of glomic cells unlike the readily reversible increases in size of these organs due to vascular engorgement, as seen in rats on acute exposure to hypobaric hypoxia (see Chapter 9). As such it seems likely that the chronic enlargement in the highlander is much more slowly reversible, if at all. The nature of the histological changes in the structure of the carotid bodies of highlanders remains surprisingly controversial.

Arias-Stella and Valcarcel (1976) found that in the enlarged carotid bodies of Quechua highlanders from the Andes the lobules were larger and more numerous and this was due to a proliferation of chief cells, type unspecified. The hyperplasia was said to be so dense and diffuse as to give the section a homogeneous appearance and the intervening bands of fibrous stroma were thinner. These authors attempted to measure what they termed "the functional area" of glomic tissue. They measured the area occupied by glomic tissue in sections through the centre of the carotid bodies from 51 lowlanders and 56 highlanders. In the lowlanders the mean functional area ranged from 857 to 1171 μm^2 but in the highlanders it was significantly greater, at a range of 600 to 1800 μm^2. The hyperplastic chief cells showed intense vacuolation. After fixation in formaldehyde vapour, the dense-core vesicles normally show a green-yellow natural fluorescence due to the biogenic amines. In the enlarged carotid bodies of Quechua highlanders such fluorescent granules were scarce or absent, strongly suggesting a discharge of biogenic amines from the chief cells under the influence of hypoxaemia.

The histological findings in the carotid bodies of four Ladakhi highlanders were rather different. Khan et al. (1988) were able to confirm that the lobules were enlarged compared to those of sea-level subjects but this was due to a proliferation of distinct clusters which were smaller than those found in lowlanders (Table 11.1). Another prominent feature was an increase in the differential count of the dark variant of chief cells (Heath et al. 1984; Table 11.1). Dark cells were very prominent in a boy of 4 years but even in the two older highlanders aged 35 and 52 years the incidence was 14%, about three times higher than the values anticipated for adult lowlanders. The dark cells showed a striking cytological appearance. There

Table 11.1. Diameters of lobules and clusters and differential cell counts in 4 Ladakhi highlanders (after Khan et al. 1988)

Age	Sex	Lobule diameter (μm)			Cluster diameter (μm)			Differential cell count (%)			
		n	Mean	CL (±)	n	Mean	CL	L	D	P	S
4	M	22	314	40.4	50	76	4.6	48	23	1	28
29	M	21	355	46.8	40	75	3.6	54	17	2	27
35	M	19	495	56.3	40	44	2.6	47	14	2	37
52	M	22	518	81.4	40	45	2.6	26	14	2	58
Sea level (Smith et al. 1982)		57	411	154	57	82	28	54	5	2	39

CL, 95% confidence limits; n, number of samples; L, light cells; D, dark cells; P, progenitor cells; S, sustentacular cells.

was a pronounced increase in the size of the nuclei, which contained much more heterochromatin arranged in a denser, coarser pattern. These dark cells were much larger than those seen in the carotid bodies of normal lowlanders and the copious basophilic cytoplasm formed long streamers (Fig. 11.3). The dense cytoplasm contrasted srikingly with that of the faintly staining, eosinophilic cytoplasm of the surrounding light chief cells. Its margins were sharply defined in contrast to the syncytium-like appearance of the adjacent light chief cells. The cytoplasm of the dark cells contained clear vesicles. In many cells these vesicles were small and of a bubbly appearance but in some they appeared to have fused to form large vesicles that had been discharged from the cell surface (Fig. 11.4). Progenitor cells were present but not prominent and the differential count revealed no increased incidence of this type of cell (Table 11.1). In the middle-aged highlander considerable numbers of sustentacular cells and focal collections of lymphocytes were present.

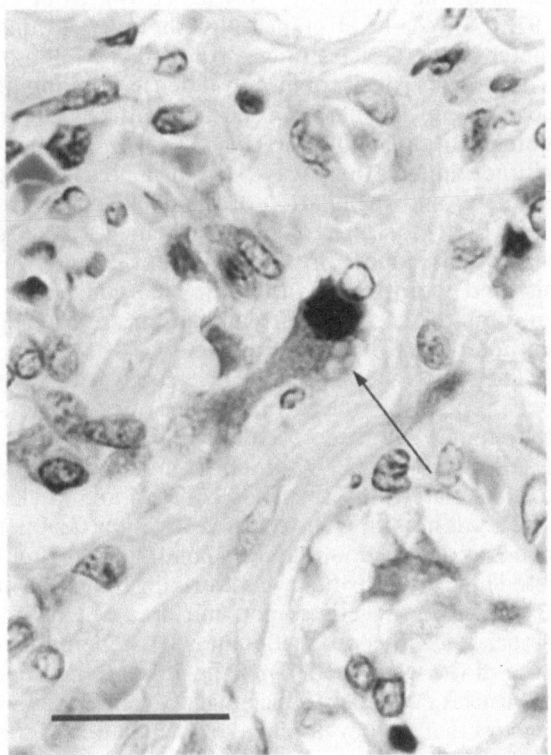

Fig. 11.3. Dark variant of chief cell of carotid body from a male Ladakhi highlander of 35 years with a large, dense nucleus and streamers of cytoplasm containing many small vesicles (*arrow*). (Haematoxylin–eosin (HE)) Scale line = 20 μm.

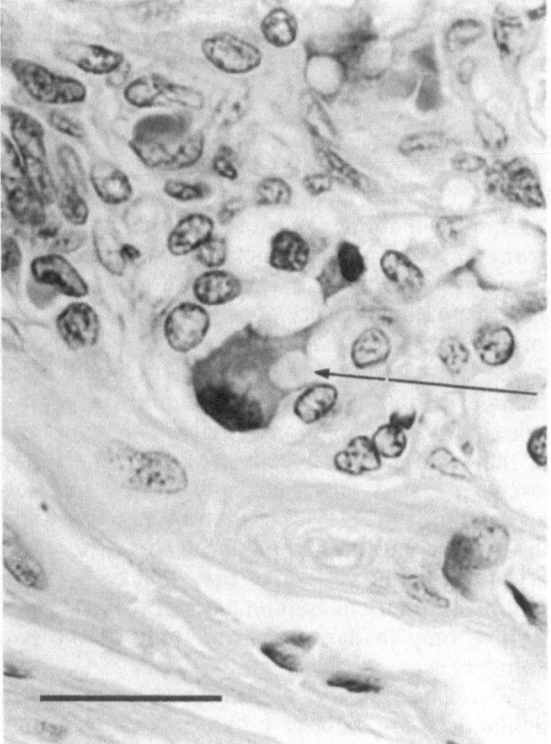

Fig. 11.4. Dark variant of chief cell of carotid body from the same case. The cell has voluminous cytoplasm containing vesicles some of which appear to have fused to form large vesicles that have discharged from the cell surface (*arrow*). (HE) Scale line = 20 μm.

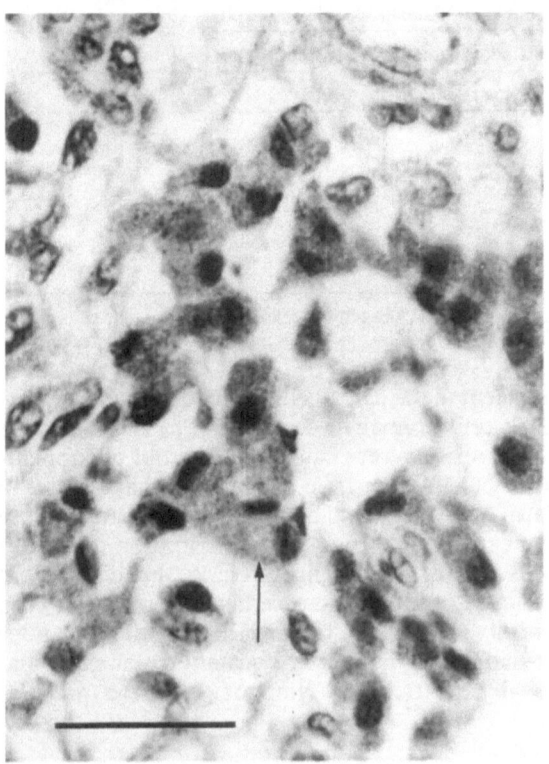

Fig. 11.5. A group of dark cells from the same case containing Met-enkephalin (*arrow*). Scale line = 20 μm.

Both of these histological features are consistent with age-changes in the carotid body (see Chapter 4).

After incubation with Met-enkephalin antisera, strong positive immunoreactivity was observed in the dark and progenitor variants of the chief cells (Fig. 11.5). The density of immunoreactivity was variable, from cells having intense deposits of gold in the cytoplasm to others with a much paler reactive cytoplasm. A few dark and progenitor cells showed no immunoreactivity. The cytoplasm of the enlarged dark cells with streamers and vesicles was particularly strongly positive for Met-enkephalin (Fig. 11.5). No immunoreactivity was observed with Met-enkephalin in the sustentacular and the light variant of the chief cells. Incubation with Leu-enkephalin antiserum was faintly positive only in dark cells and only in the two younger cases.

In conclusion, the carotid bodies of native highlanders are enlarged due to the sustained stimulation of the chemoreceptor tissue by hypobaric hypoxia. This results in the development of numerous mini-clusters of chief cells which group together and lead to an increase in size of lobules. There is a loss of formaldehyde-induced fluores-

cence, which suggests discharge of biogenic amines from the chief cells. There is a prominence of the dark variant of chief cells, which contain much Met-enkephalin in vesicles that accumulate to form larger intracytoplasmic cysts which then seem to be discharged from the cell. It is likely that the discharge of both the biogenic amines and peptides are associated functionally in some way with the chemoreceptor response to sustained hypobaric hypoxia.

Ultrastructure

So far as we are aware there are no electron microscopic studies of the carotid body of the native highlander and consequently we have to resort to investigations on animal species at high altitude on this matter. We found that in the carotid bodies of guinea-pigs born and bred in Cerro de Pasco (4330 m) in the Peruvian Andes there were striking and highly characteristic changes in the dense-core vesicles. The chief cells showed a pronounced increase in size and vacuolation of the dense-core vesicles (Edwards et al. 1972). In sea-level guinea-pigs the average diameter of the vesicles ranges from 100 to 150 nm. The vesicles consist of a central dense osmiophilic core with a very narrow clear halo adjacent to an outer limiting membrane (Fig. 11.6). In the high-altitude animals there was a striking increase in the diameter and vacuolation of the dense-core vesicles (Fig. 11.7). The central proteinaceous core was often reduced in size and density and was situated eccentrically within the enlarged vesicle. In some the core was virtually absent, producing vacuoles up to 350 nm in diameter.

The same vacuolation of dense-core vesicles has been reported in the carotid bodies of experimental animals subjected to simulated high altitude. Laidler and Kay (1978a) found it in rats subjected to a simulated altitude of 4300 m for 4 to 5 weeks. These authors in a further investigation (Laidler and Kay 1978b) found that in conditions of simulated high altitude there was a striking reduction in the number of vesicles per unit area but it was associated with a proportionately large increase in area of the whole cell. Thus the absolute number of vesicles per cell was unchanged. Animal studies suggest that the development of vacuolation of the dense-core vesicles is related to the duration of the exposure to hypoxia. Smith et al. (1986) therefore examined the carotid bodies of rabbits which were born and lived at Cerro de Pasco (4330 m) in the Andes, and rabbits which were subjected to a

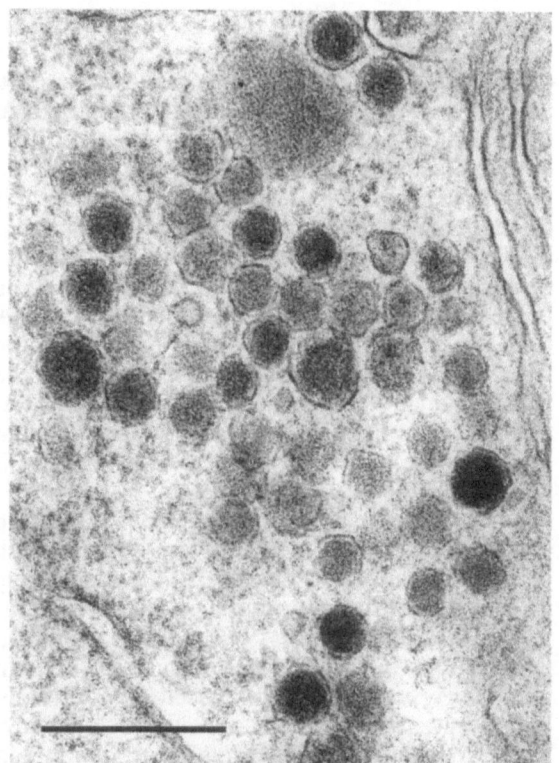

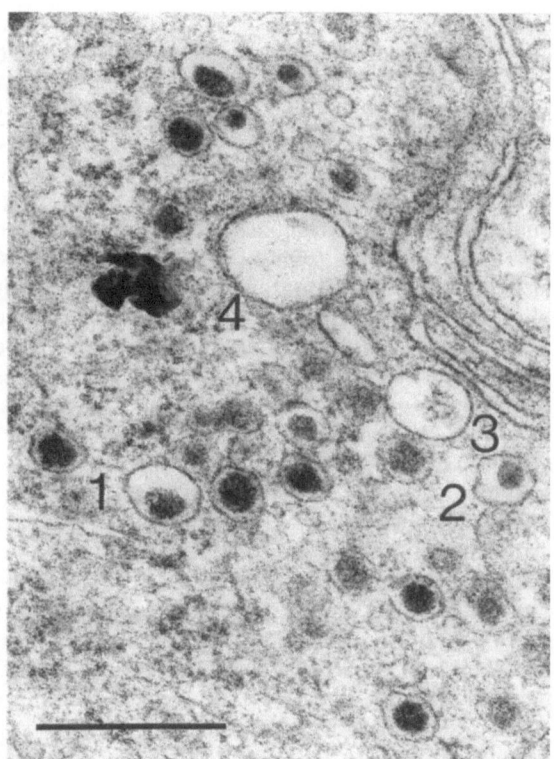

Fig. 11.6. Electron micrograph of carotid body from a low-altitude guinea-pig. It shows membrane-bound granules in the cytoplasm of a chief cell. In many granules there is a central dense osmiophilic core with a very narrow clear halo subjacent to an outer limiting membrane. In others there is no limiting membrane and the contents appear to merge into the surrounding cytoplasm. Scale line = 0.5 μm.

Fig. 11.7. Electron micrograph of carotid body from a high-altitude guinea-pig. It shows that some of the dense-core vesicles are distended by widening of the clear halo, which leaves the hitherto central osmiophilic core in an eccentric position (1). In others the core is much less dense than normal (2) or is represented by only faint remains (3). In some granules the core has been lost and the appearances are those of a microvacuole (4). Scale line = 0.5 μm.

comparable simulated high altitude for periods of either 3 or 6 months. Vacuolation was seen in all these groups of rabbits but was most pronounced in the native Peruvian animals, where it involved about half of the vesicles. Fewer vacuoles were found in the rabbits exposed to simulated high altitude for 6 months and were rare in the animals exposed to the same degree of hypobaric hypoxia for only 3 months.

The functional significance of vacuolation of the dense-core vesicles is unknown but certain observations and speculations can be made (see p. 81 in Smith 1986). It is unlikely to represent discharge of catecholamines in response to hypoxia, since in the rabbit at least the change takes some 6 months to develop and is not seen on acute exposure to hypoxia. Conceivably the change may represent storage of catecholamines consistent with the increased levels of dopamine which have been demonstrated in the chief cells of hypoxic animals. The process could represent secretion of a soluble peptide, the diminution in size of the core indicat-

ing loss of proteinaceous substance, and the distension of the surrounding "halo" may be due to imbibition of fluid following an increase in osmotic pressure of the granule contents. Other workers have been unable to confirm the existence of this vacuolation in the dense-core vesicles in rabbits living on the Bolivian altiplano at a comparable altitude (Møller et al. 1974) and in rats exposed to the prolonged hypoxia of a simulated altitude of 7000 m (Blessing and Kaldeweide 1975).

The Blunted Ventilatory Response to Hypoxia

The enlarged carotid body with its pathological changes seen at levels of both light and electron microscopy is associated with a blunted ventilatory response in permanent residents at high altitude. When a human from sea level ascends into mountains he or she hyperventilates. The native highlander shows pronounced hypoventilation compared to such newcomers, although there is still hyperventilation compared to lowlanders living at sea level. The blunted ventilatory response to hypoxia has been reported in native highlanders of both the Andes and of Himalaya (Chiodi 1957; Lahiri and Milledge 1965; Severinghaus et al. 1966; Lefrançois et al. 1968; Sørensen and Severinghaus 1968a). Genetic factors are unlikely to be a common denominator in this blunted response. Both the Quechua people of the Andes and the Sherpas of the Himalaya are mongoloid but genetically unrelated Caucasians living permanently in Kashmir at 3500 m (Ramaswamy 1962) and in Colorado at 3100 m (Forster et al. 1969) also show a blunted ventilatory response to hypoxia.

It seems more likely that the blunted response is acquired and this takes a long time to occur. Thus it develops after about 12 years' residence at high altitude, manifesting itself in teenage highlanders and becoming universal in adults. In some highland children, however, a diminution of ventilatory response to hypoxia appears only two or three years after birth. Once these highlanders have lost their sensitivity to hypoxia they fail to regain it even after prolonged subsequent residence at sea level (Sørensen and Severinghaus 1968b; Lahiri et al. 1969). In the same way the offspring of native highlanders born at sea level show the same high ventilatory response to low arterial oxygen tension as that of natural lowlanders. Clearly the carotid bodies are very susceptible to the level of arterial oxygen tension during the first few years of life and in some way this diminution imprints itself on the future functioning of the chemoreceptor. There is some evidence for the fact that after prolonged residence at high altitude for years even adults may exhibit an attenuation of the ventilatory response to hypoxia. Some authorities such as Chiodi (1957) believe that the loss of sensitivity to hypoxaemia is often matched by one for carbon dioxide tension. Probably the most characteristic feature of this blunted ventilatory response to the hypoxia of high altitude is its irreversibility. It is not restored even after prolonged residence at sea level. The same loss of sensitivity to arterial oxygen tension is found in many patients with chronic obstructive lung disease but an even greater number seem to be normal in this respect (Flenley and Millar 1967).

The sharing of this insensitivity of chemoreceptors in native highlanders and some "blue bloaters" leads one to seek out shared structural features with which the functional changes may be related. In both there is a hyperplasia of chief cells surrounded by sustentacular cells. One could postulate that the hyperplastic chief cells may be defective per se in their physical response to lowered arterial oxygen tension compared to normal chief cells. Alternatively the proliferation of sustentacular cells might be considered to compress the cores of chief cells, compromising their chemoreceptor function. We return to this point below, where we consider effects of the sustentacular cell proliferation of increasing age and falling ventilatory response to hypobaric hypoxia. A third factor that has to kept to the forefront is the prominence of the dark variant of chief cells of the native highlander with its cytoplasm rich in Met- and Leu-enkephalins. The basic difficulty is that one does not know the rôle of the enkephalins, if any, in the process of chemoreception.

Arias-Stella and Valcarcel (1976) related the augmented carotid body size and weight with increasing age of highlanders to progressive insensitivity of those chemoreceptors. They refer to the studies of Sime (1973), who was able to demonstrate that there is a progressive fall in pulmonary ventilation with age at high altitude. He submitted highlanders at 4330 m to a steady-state exposure to various levels of arterial oxygen tension ranging from hyperoxia to hypoxia, and in isocapnic as well as in hypercapnic conditions. Subjects sleeping naturally were also studied. His studies showed that highlanders progress irreversibly with age from hyperventilation to hypoventilation, irrespective of whether they are tested under basal, or acute hypoxic or hyperoxic conditions. The progressive hypoventilation with age in native highlanders may be related to the progressive proliferation of sustentacular cells around cores of chief cells as an ageing phenomenon (Hurst et al. 1985; see Chapter 4). Earlier in this chapter we referred to this process occurring in a middle-aged highlander from Ladakh (Fig. 11.8).

No state comparable to a "blunted" ventilatory response to hypoxia has been found in animals living at high altitude. Llamas, goats and sheep have normal hypoxic responses, although cattle may show some depression of ventilation with time during hypoxic exposure (Severinghaus 1972). Attempts to produce hypoxic blunting in the laboratory have had limited success (Barer et al. 1986, p. 121). Temporary blunting which persists for a few days only on return to normoxia has been produced in rats (Barer et al. 1976) and cats (Tenney and Ou 1977; Lahiri et al. 1983). In contrast, highlanders who return to sea level retain their poor response to hypoxia and it is disputed as to whether there is a changed response in patients whose congenital cardiac defects have

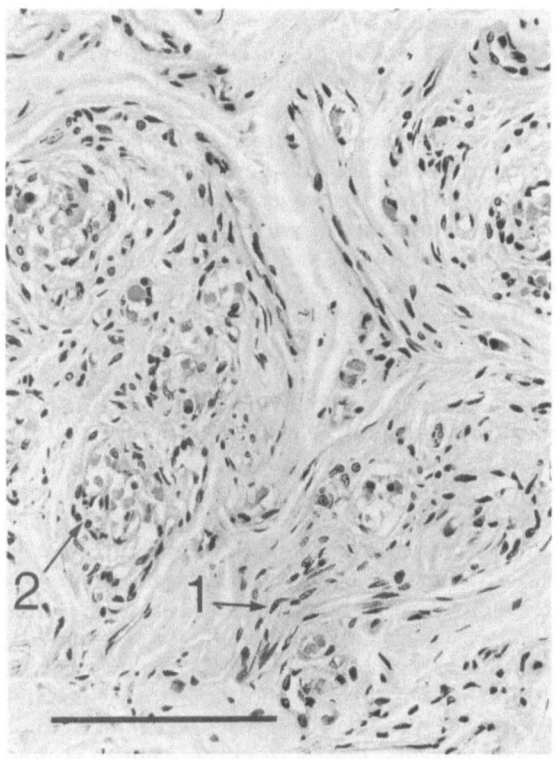

Fig. 11.8. Section of carotid body from a Ladakhi highlander aged 52 years showing proliferation of elongated sustentacular cells (*arrow 1*) that have encroached on the central cores of chief cells (*arrow 2*). (HE) Scale line = 100 μm.

been corrected by surgery. Some athletes are also said to have a blunted ventilatory response to hypoxia (Barer et al. 1986, p. 121).

The chronically hypoxic rats of Tenney and Ou (1977) were found to have a depressed ventilatory response to hypoxia which was restored to normal after section of the brain at the mid-collicular level and became enhanced after decortication. Thus the peripheral chemoreceptor information is subject to inhibition or accentuation within the central nervous system. Lahiri et al. (1983) also produced blunting in cats exposed to chronic hypoxia. They reported impulse traffic in the carotid nerve and found it increased after the nerve was sectioned, more so than in normal cats. It is known that the carotid nerve contains some efferent fibres which are believed to be inhibitory to the chemoreceptors. These workers suggest that centripetal impulses might have reduced carotid body activity in their cats. There is no certainty that these changes in chronically hypoxic rats and cats bear any relationship to those found in long-term human residents at high altitude (Barer et al. 1986, p. 123).

Loss of Ventilatory Response and Well-being at High Altitude

The rôle that the blunting of chemoreceptor response to hypoxia in native highlanders might play in adjustment to life in the mountains is a matter of considerable controversy. There is no doubt that a brisk ventilatory response to hypoxia is one of the most important components of acclimatisation of the lowlander ascending into high altitude. This is the classic manifestation of the response of the carotid body of the Caucasian from sea level. Milledge (1987) argued that two criteria should be employed to assess the value of any change in the ventilatory response to hypoxia. The first is the ability to acclimatise rapidly, including freedom from acute mountain sickness. The second is the ability to perform well at extreme altitudes.

There are many reports to confirm that a brisk ventilatory response protects against acute mountain sickness (Hackett et al. 1982). Recent studies suggest that an intense ventilatory response to extreme high altitude is important in achieving a good performance in mountain climbing (Schoene et al. 1984). Indeed a high ventilatory response to hypoxia at sea level has been suggested to be an indicator for a climber's capability at high altitude. On the other hand lowlanders with a brisk ventilatory response are at a greater risk of developing periodic breathing at a lower altitude, with all the disadvantages and dangers of developing periods of severe hypoxaemia during the phases of apnoea.

Most of this would seem to indicate that carotid bodies initiating and maintaining a high ventilatory response to hypoxia at high altitude provide all the advantages. However, Milledge (1987) pointed out the difficulty of accepting that Sherpas and other highlanders are anything but well adjusted to their mountain environment. They can usually carry more and climb faster while at the same time ventilating less and being less dyspnoeic than lowlanders climbing with them. It also has to be kept in mind that the depressed carotid body of the Sherpa does not lend itself so readily to the onset of nocturnal periodic breathing at high altitude with its disadvantage of repeated episodes of very low levels of systemic arterial oxygen saturation. Before taking too strong a stand as to the advantage or otherwise of carotid bodies able to initiate or maintain a high ventilatory response to hypoxia at high altitude, it is advisable to reflect on the fact that Mount Everest was conquered for the first time by two climbers of very different physiological background. Sir Edmund Hillary was a Caucasian with low-altitude carotid bodies likely to sustain a high ventilatory response to hypoxia. On the other hand Tensing Sherpa was a native highlander with high-altitude carotid bodies manifesting a depressed ventilatory response to hypoxia.

Chemodectomas at High Altitude

The carotid body tumour is considered in detail in Chapter 19 but reference needs to be made to it here because of its reported increased incidence at high altitude. Morfit et al. (1953) reported a series of 12 patients with carotid body tumours seen over a period of 14 years from a University of Colorado tumour clinic, but the authors did not associate this high prevalence with the high altitude of the city (1610 m). However, 20 years later Saldaña et al. (1973) reported a series of cases of chemodec-toma in 25 Peruvian adults all but two of whom had been born, and lived, at altitudes between 2105 and 4350 m above sea level, whereas only a third of the Peruvian population live at or above an altitude of 3000 m. These authors believe that carotid body tumours are ten times more frequent at high altitude than at sea level. In their series they found that females predominated over males by about six to one and left-sided tumours were three times more prevalent. Histologically the tumour cells were arranged in classic "Zellballen" of chief cells but it is of special interest that Saldaña et al. (1973) reported a predominance of dark cells. The dark variant of the chief cell is known to become prominent in diseases and environmental conditions predisposing to chronic hypoxaemia (see Chapter 9). When the effects of hypobaric hypoxia inherent in life at high altitude are buttressed by alveolar hypoxia brought about by lung disease, the hypoxic stimulus thought to be responsible for carotid body hyperplasia and perhaps subsequent development of chemodec-toma is likely to be enhanced. In this respect Saldaña et al. (1973) reported the occurrence of bilateral carotid body tumours in a 29-year-old Quechua Indian who had worked at an altitude of 3870 m as a miner in the Andes, where he developed silicosis. Perhaps stimulated by the proposed association of increased numbers of carotid body tumours at high altitude, Gaylis and Mieny (1977) subsequently reported that no fewer than 27 chemodectomas were encountered in 25 patients seen in a surgical unit in Johannesburg Hospital at an altitude of 1800 m. This is a moderate degree of altitude which is lower than

that considered by Saldaña and his colleagues to be of significance. Rodriguez-Cuevas et al. (1986) reported that over a period of 21 years at an oncology head and neck clinic in Mexico City (2380 m) 40 patients were diagnosed as having paragangliomas, 87% of them being carotid body tumours. All of them were residents of Mexico City or neighbouring communities on a plateau with a mean altitude of 2000 to 2500 m above sea level. They noted that no fewer than 38 (95%) of their patients were women who accounted for only 52% of the patients attending the clinic. This propensity of the tumour for women is reminiscent of similar findings by Saldaña et al. (1973). Rodri-guez-Cuevas et al. (1986) noted that, in the reported series of patients at high altitude with paragangliomas, no malignant tumours had been reported. They were of the opinion that at high altitude carotid body tumours appeared later in life and they thought it likely that they might represent carotid body hyperplasia rather than neoplasia. Pacheco-Ojeda et al. (1988) reported a high prevalence of carotid body tumours at high altitude in Ecuador. The hospital from which this report came provided care to workers and to some Indian peasants, mostly from the Andean region, living at altitudes of between 2000 and 4000 m. Twenty carotid body tumours were operated on in 19 patients over a period of five years. The series comprised 5 men and 14 women. Nine were from Quito (2800 m) and 10 were from other Andean provinces; all of the patients lived at high altitude. It has to be kept in mind that, while the increased prevalence of chemodectomas in the Andes may be a reflection of chronic stimulation by hypobaric hypoxia, an alternative explanation may be an inherent genetic susceptibility to the tumour by Quechua and Aymara Indians.

Albores-Saavedra and Durán (1968) reported the association of chemodectoma with thyroid carcinoma in two patients at high altitude in Mexico City (2380 m). In these cases the carcino-mas were of papillary and follicular type, in contradistinction to the medullary carcinoma of thyroid, which occurs in association with phaeo-chromocytoma. The carotid body and the adrenal medulla share a common origin from neural ecto-derm and a common function in secretion of catecholamines. Two of the patients from the Andes with chemodectomas, reported by Saldaña and Salem (1970) and Saldaña et al. (1973), died of metastasising thyroid carcinoma.

It has been claimed that cattle as well as man show the same tendency to develop chemodecto-mas at high altitude. Arias-Stella and Bustos (1976) stated that as many as 40% of cattle living at an elevation of 4330 m in the vicinity of Cerro de Pasco in the Andes of Central Peru develop histological changes in their carotid bodies suggestive of chemodectoma. Both man and cattle have to acclimatise to high altitude and during the process over a long period their carotid bodies enlarge as described earlier in this chapter. In contrast, indigenous mountain species such as llamas and alpacas which show genetic adaptation to life in the high mountains (Heath and Williams 1989, p 331) do not develop chemodectomas.

References

Albores-Saavedra J, Durán ME (1968) Association of thyroid carcinoma and chemodectoma. Am J Surg 116:887–890

Arias-Stella J (1969) Human carotid body at high altitudes. Item 150 in the sixty-ninth programme and abstracts of the American Association of Pathologists and Bacteriologists, San Francisco, California, USA

Arias-Stella J, Bustos F (1976) Chronic hypoxia and chemodectomas in bovines at high altitudes. Arch Pathol Lab Med 100:636–639

Arias-Stella J, Valcarcel J (1973) The human carotid body at high altitudes. Pathol Microbiol 39:292–297

Arias-Stella J, Valcarcel J (1976) Chief cell hyperplasia in the human carotid body at high altitudes. Physiologic and pathologic significance. Hum Pathol 7:361–373

Barer GR, Edwards CW, Jolly AI (1976) Changes in the carotid body and the ventilatory response to hypoxia in chronically hypoxic rats. Clin Sci 50:311–313

Barer G, Wach R, Pallot D, Bee D (1986) Almitrine, hypoxia, systemic hypertension and the carotid body. In: Heath D (ed) Aspects of hypoxia. Liverpool University Press, Liverpool, pp 113–129

Blessing MH, Kaldeweide J (1975) Light and electron microscopic observations on the carotid bodies of rats following adaptation to high altitude. Virchows Arch [B] 18:315–329

Chiodi H (1957) Respiratory adaptations to chronic high altitude hypoxia. J Appl Physiol 10:81–87

Edwards C, Heath D, Harris P (1972) Ultrastructure of the carotid body in high-altitude guinea-pigs. J Pathol 107:131–136

Flenley DC, Millar JS (1967) Ventilatory response to oxygen and carbon dioxide in chronic respiratory failure. Clin Sci 33:319–334

Forster HV, Dempsey JA, Birnbaum ML, Reddan WG, Thoden JS, Grover RF, Rankin J (1969) Comparison of ventilatory responses to hypoxic and hypercapnic stimuli in altitude-sojourning lowlanders, lowlanders residing at altitude and native altitude residents. Fedn Proc Fedn Am Soc Exp Biol 28:1274–1279

Gaylis H, Mieny CJ (1977) The incidence of malignancy in carotid body tumours. Br J Surg 64:885–889

Hackett PH, Rennie D, Hofmeister SE, Grover RF, Grover EB, Reeves JT (1982) Fluid retention and relative hypoventilation in acute mountain sickness. Respiration 43:321–329

Heath D, Williams DR (1989) Carotid bodies of native highlanders. In: High-altitude medicine and pathology. Butterworths, London, pp 74–87

Heath D, Edwards C, Harris P (1970) Post-mortem size and structure of the human carotid body. Its relation to pulmonary disease with cardiac hypertrophy. Thorax 25:129–140

Heath D, Smith P, Jago R (1984) Dark cell proliferation in carotid body hyperplasia. J Pathol 142:39–49

Hurst G, Heath D, Smith P (1985) Histological changes associated with ageing of the human carotid body. J Pathol 147:181–187

Khan Q, Heath D, Smith P, Norboo T (1988) The histology of the carotid bodies in highlanders from Ladakh. Int J Biometeorol 32:254–259

Lahiri S, Milledge JS (1965) Sherpa physiology. Nature 207:610–612

Lahiri S, Kao FF, Velasquez T, Martinez C, Pezzia W (1969) Irreversible blunted respiratory sensitivity to hypoxia in high altitude natives. Respir Physiol 6:360–374

Lahiri S, Smatresk NJ, Pokorski M, Bernard P, Mokashi A, McGregor A (1983) Altered structure and function of the carotid body in chronically hypoxic cat. In: Pallot DJ (ed) The peripheral arterial chemoreceptors. Croom Helm, London and Canberra/Oxford University Press, New York, pp 303–309

Laidler P, Kay JM (1978a) Ultrastructure of carotid body in rats living at a simulated altitude of 4300 metres. J Pathol 124:27–33

Laidler P, Kay JM (1978b) A quantitative study of some ultrastructural features of the type I cells in the carotid bodies of rats living at a simulated altitude of 4300 metres. J Neurocytol 7:183–192

Lefrançois R, Gautier H, Pasquis P (1968) Ventilatory oxygen drive in acute and chronic hypoxia. Respir Physiol 4:217–228

Milledge JS (1987) The ventilatory response to hypoxia: how much is good for a mountaineer? Postgrad Med J 63:169–172

Møller M, Møllgard K, Sørensen SC (1974) The ultrastructure of the carotid body in chronically hypoxic rabbits. J Physiol (Lond) 238:447–453

Morfit HH, Swan H, Taylor ER (1953) Carotid body tumors. Arch Surg 67:194–214

Pachecho-Ojeda L, Durango E, Rodriquez C, Vivar N (1988) Carotid body tumors at high altitudes: Quito, Ecuador, 1987. World J Surg 12:856–860

Ramaswamy SS (1962) In: Bhatia SP (ed) International symposium on problems of high altitude. Indian Armed Forces Medical Services, New Delhi, p 74

Rodriguez-Cuevas H, Lau I, Rodriguez HP (1986) High-altitude paragangliomas diagnostic and therapeutic considerations. Cancer 57:672–676

Saldaña MJ, Salem LE (1970) High altitude hypoxia and chemodectomas. Am J Pathol 59:91a–92a

Saldaña MJ, Salem E, Travezan R (1973) High altitude hypoxia and chemodectomas. Hum Pathol 4:251–263

Schoene RB, Lahiri S, Hackett PH, Peters RM Jr, Milledge JS, Pizzo CJ, Sarnquist FH, Boyer SJ, Graber DJ, Maret KH, West JB (1984) Relationship of hypoxic ventilatory response to exercise performance on Mount Everest. J Appl Physiol 56:1478–1483

Severinghaus JW (1972) Hypoxic respiratory drive and its loss during chronic hypoxia. Clin Physiol Tokyo 2:57–79

Severinghaus JW, Bainton CP, Carcelan A (1966) Respiratory insensitivity to hypoxia in chronically hypoxic man. Respir Physiol 1:308–334

Sime F (1973) Ventilacion humana en hipoxia cronica. Etiopatogenia de la Enfermedad de Monge a desadaptacion cronica a la altura. Doctoral thesis. Universidad Peruana Cayetano Heredia, Lima, Peru

Smith P (1986) Electron microscopy of the abnormal carotid body. In: Heath D (ed) Aspects of hypoxia. Liverpool University Press, Liverpool, pp 77–96

Smith, P, Jago R, Heath D (1982) Anatomical variation and quantitative histology of the normal and enlarged carotid body. J Pathol 137:287–304

Smith P, Heath D, Fitch R, Hurst G, Moore D, Weitzenblum E (1986) Effects on the rabbit carotid body of stimulation by almitrine, natural high altitude, and experimental normobaric hypoxia. J Pathol 149:143–153

Sørensen SC, Severinghaus JW (1968a) Irreversible respiratory insensitivity to acute hypoxia in man born at high altitude. J Appl Physiol 25:217–223

Sørensen SC, Severinghaus JW (1968b) Respiratory sensitivity to acute hypoxia in man born at sea level living at high altitude. J Appl Physiol 25:211–216

Tenney SM, Ou LC (1977) Hypoxic ventilatory response of cats at high altitude, an interpretation of blunting. Respir Physiol 30:185–199

12 Ventilation and the Carotid Bodies

Respiration is the process by which the tissues of the body are supplied with oxygen while carbon dioxide is removed from them. It entails four distinct stages of conduction from the atmosphere to the cells. (a) In ventilation, air flows through the trachea and bronchial tree to the alveolar spaces of the lung. (b) Pulmonary diffusion is the stage by which oxygen in the alveoli comes into contact with the alveolar-capillary walls and passes through them to reach the blood. (c) The gas is then carried by blood transport from the capillaries of the lung to those of the tissues. (d) Finally, in the fourth stage of tissue diffusion, oxygen passes from the systemic capillaries to the respiratory enzymes of the intracellular mitochondria, where it is utilised.

Since the main function of respiration is to provide the tissues with oxygen and to eliminate carbon dioxide from them, it is to be expected that the arterial tensions of these respiratory gases will exert a powerful influence on ventilation. So far as hypoxaemia is concerned and to some extent for hypercarbia this is achieved through the agency of the carotid bodies and to a far lesser extent that of the aortic bodies, for these peripheral arterial chemoreceptors sense the levels of oxygen and carbon dioxide in the blood and relay the information to the respiratory areas in the brain. Central chemoreceptors in the medulla are of importance in sensing the levels of carbon dioxide. The neurons concerned, which generate and maintain the central rhythm of ventilation, are situated in the medulla and pons and are cited by Bradley (1981) as comprising the nucleus ambiguus, and nucleus retroambigualis in the ventrolateral part of the medulla, and the nucleus of the tractus solitarius in the dorsal part of the medulla.

Homeostasis of arterial oxygen tension depends on the carotid bodies but the response to hypercarbia is more dependent on other receptors which are found in the brain stem. Just under the ventral surface of the medulla there are receptors to levels of carbon dioxide. These are the so-called L area near the 12th cranial nerve, the M (Mitchell) area near the 7th to 10th cranial nerves, and the intermediate S area (Bradley 1981). These areas respond rapidly to an increased hydrogen ion concentration in the cerebrospinal fluid. Respiration is stimulated and the hyperventilation reduces the partial pressure of carbon dioxide, thereby returning the pH of the fluid to normal. On initial exposure to high altitude there is stimulation of the carotid bodies which thereby reduces the partial pressure of carbon dioxide in the cerebrospinal fluid. This brings about a relative alkalinity of the fluid as part of the general respiratory alkalosis and tends to inhibit ventilation by its action on the medullary hydrogen ion receptors. Some years ago Severinghaus et al. (1963) and Severinghaus and Mitchell (1964) presented physiological evidence claiming that this cerebrospinal alkalosis does not occur, due to active transport of bicarbonate ions out of the fluid; this brings about a relative increase in its acidity which stimulates ventilation and thus aids acclimatisation. The concept of the rôle of the pH of the cerebrospinal fluid as a homeostatic mechanism was soon challenged and found to be false. Dempsey et al. (1974) and Forster et al. (1975) found that the pH of lumbar cerebrospinal fluid in men at high altitude was more alkaline than in those at sea level. It would appear that the strong hypoxic peripheral chemoreceptor drive is in fact partly offset by alkaline inhibition of the medullary chemoreceptors.

The Contribution of the Carotid Bodies to Ventilation in Healthy Humans at Sea Level

The peripheral arterial chemoreceptors contribute only about 15% to the level of ventilation of normoxic humans at sea level (Jeyaranjan et al. 1987) but more during exercise and hypoxia. Carotid body drive is of particular importance in the "behavioural" control of breathing during daily activity. This term covers ventilatory control when there is no homeostasis of arterial gas tensions during, for example, speaking, eating, using respiratory muscles and during postural control or weight-lifting (Barer et al. 1986). Bilateral carotid endarterectomy effectively destroys carotid chemoreceptor function and Lugliani et al. (1971) could find no significant difference in resting ventilation or blood gases between asthmatic patients with bilateral carotid body resections and asthmatic controls.

An assessment of the level of the contribution of the carotid bodies to ventilation is based on the degree of suppression of ventilatory drive by the inhalation of oxygen to produce temporary hyperoxia. In this "oxygen test" the transient fall in ventilation, breath-by-breath, produced by inhalation of one or two breaths of pure oxygen is measured (Dejours 1957; Dejours et al. 1958). The onset of oxygen-induced ventilatory depression occurs after a delay of some 8–10 seconds, this being consistent with the lung-to-carotid body circulation time. Dejours et al. (1958) found that this test indicated a contribution to ventilation of the peripheral chemoreceptors in healthy humans at sea level of about 10%.

Exercise

The contribution from the carotid bodies is much enhanced during exercise. The oxygen test reveals this greater contribution to ventilation following a shortened circulation time of 5 seconds. During steady-state moderate exercise below anaerobic threshold the peripheral chemoreceptors contribute at least 20% of the ventilation, and in heavy exercise above anaerobic threshold the contribution is of the order of 25% (Wasserman 1976; Wasserman et al. 1979). The drop in ventilation following oxygen administration is transient, returning to normoxic levels. This is thought to be due to offsetting of the suppressed peripheral chemoreceptor drive by acidic stimulation of central respiratory centres by the secondary increase in carbon dioxide tension subsequent to the fall in ventilation, amongst other mechanisms. The importance of this central control is manifested by subjects who have been subjected to therapeutic glomectomy, for they maintain a normal ventilatory response during steady-state mild to moderate exercise (Wasserman et al. 1979). In contrast, during heavy exercise with hyperventilation, an oxygen-induced drop in ventilation is persistent, indicating that the peripheral chemoreceptors are the virtually exclusive mediator of this hyperventilatory component of ventilation. The metabolic acidosis of heavy exercise due to lactic acid accumulation stimulates the peripheral chemoreceptors causing hyperventilation, which constrains the decrease in blood pH.

Children

The maximum percentage reduction in ventilation seen during a switch to hyperoxia is sometimes called the peripheral chemoreceptor tone. The coupling of this tone assessed during steady-state exercise and ventilatory response at the onset of exercise is different between children and adults. Ventilatory responses in the form of minute ventilation and carbon dioxide output are significantly faster in young children than in teenagers and adults (Springer et al. 1989). It is of interest to recall in this respect that dark cell variants of chief cells may account for 15% of the differential count of glomic cells at this age, roughly three times the number found in adult life (see Chapter 5). The responses become even faster in children under hypoxic conditions, as in the adult, but the magnitude of the increase is not disproportionate in the child. This is surprising because the contribution to ventilatory drive during hypoxia is much greater in children. In other words the high level of peripheral chemoreceptor tone is not paralleled by an increase of the same magnitude in the response of minute ventilation at the onset of exercise.

Ventilation at High Altitude

The contribution of the carotid bodies to ventilation is greatly increased by exposure to hypoxia, be this associated with hypocarbia as in ascent to high altitude or with hypercarbia as in chronic obstructive lung disease. In hypoxia induced acutely it is necessary to lower arterial oxygen tension to about 55 mmHg before increased ventilation occurs (Cumming and Semple 1980).

In the case of lowlanders ascending to high altitude, once acclimatisation has taken place hyperventilation will be provoked by an arterial oxygen tension as high as 90 mmHg. The hypoxic drive from the peripheral arterial chemoreceptors is maintained in the newly acclimatised for many weeks on exposure to high altitude (Michel and Milledge 1963; Tenney et al. 1964). Lowlanders who live for years at high altitude do not lose their hypoxic drive due to the sensitivity of their peripheral arterial chemoreception to a diminished arterial oxygen tension. In the newcomer to high altitude, hyperventilation occurs within a few hours of arrival and increases rapidly during the first week. It comes to exceed that of native highlanders by some 20% (Lenfant and Sullivan 1971). With increasing duration of residence at high altitude this difference between the sojourner and the native diminishes.

So far as native highlanders are concerned Hurtado (1964) found that the mean total ventilation for 103 persons at sea level was 7.77 l/min (s.d. 1.24), but for 80 highlanders at 4540 m was 9.49 l/min (s.d. 1.77). Expressed per square metre of body surface area the figures were, respectively, 4.56 and 6.19. In general, the native highlander ventilates some 25%–35% above the value for humans at sea level.

In sharp contrast to the sustained hypoxic drive of lowlanders ascending into high altitude, the progressive increase in carotid body size and weight in highlanders is associated in adult life with a progressive insensitivity of these chemoreceptors (see Chapter 11). Normal hypoxic ventilatory drive develops in the native highlander up to the age of 8 years before becoming substantially lost during adult life (Lahiri et al. 1976). The blunted ventilatory response is acquired rather than genetic in nature (Lahiri 1971), and occurs not only in the Quechuas of the Andes but also in the genetically unrelated Sherpas in Himalaya and Caucasians living permanently in Kashmir and Colorado. The progressive fall in ventilation continues throughout adult life at high altitude, at the same time as the carotid bodies are increasing in size and weight. Sime (1973) submitted high-

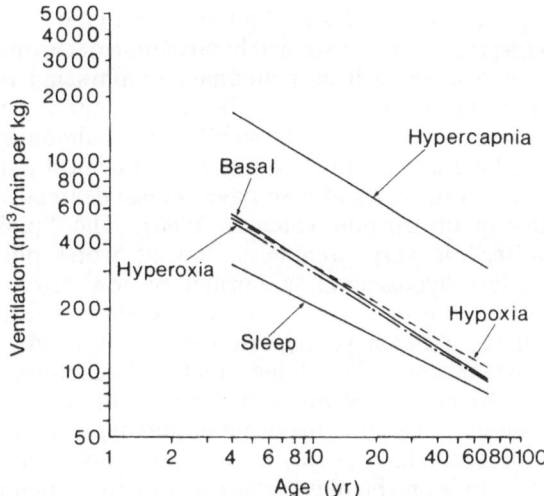

Fig. 12.1. The progression from hyperventilation to hypoventilation with increasing age in native highlanders. Basal and sleeping conditions are shown, together with the effects on the stimuli of hypercapnia, hypoxia and hyperoxia. The diagram is based on data of Sime (1973).

landers at 4330 m to steady-state exposure to various levels of arterial oxygen tension ranging from hyperoxia to hypoxia, in isocapnic as well as in hypercapnic conditions. Subjects sleeping naturally were also studied. Sime's (1973) studies showed that highlanders progress irreversibly with age from hyperventilation to hypoventilation. This applies equally whether the subjects are tested under basal, acute hypoxic or acute hyperoxic conditions (Fig. 12.1). In comparison to the levels obtained under isocapnia, hypercarbia led to hyperventilation and sleep to hypoventilation.

It is tempting to try to correlate this changing picture of the level of ventilation with the histological changes in the carotid body known to occur with increasing age. Prominence of the dark variants of chief cells with a cytoplasm rich in methionine- and leucine-enkephalins is found in the young both at sea level (Hurst et al. 1985) and at high altitude (Khan et al. 1988). One may speculate that in some way a profusion of dark cells is associated with a high level of ventilation. As we have seen in the preceding chapter, in one middle-aged highlander from Ladakh we found a proliferation of sustentacular cells (Khan et al. 1988) and this may be a histological association of progressive loss of ventilation.

Chronic Obstructive Lung Disease

The other situation apart from life at high altitude where the carotid bodies might be anticipated to make an increased contribution to ventilation is in a response to the sustained hypoxaemia of chronic lung disease such as pulmonary emphysema or cystic fibrosis. It has long been recognised that patients with chronic bronchitis and pulmonary emphysema may present one of two clinical patterns. Both types of case have similar degrees of airway obstruction (Flenley 1986). The "pink puffer" is very breathless, has at worst only modest hypoxaemia, a normal or low arterial carbon dioxide tension, without evidence of heart failure, pulmonary hypertension or secondary polycythaemia. The "blue bloater" is characterised by central cyanosis, oedema of the ankles, pronounced arterial hypoxaemia and hypercarbia, with secondary polycythaemia and pulmonary hypertension. Flenley (1986) was of the opinion that the pathological basis for these two different clinical patterns is far from clear. It has commonly been accepted that "pink puffers" suffer from pulmonary emphysema, commonly of the panacinar type, with little bronchitis, whereas "blue bloaters" have centrilobular emphysema (affecting respiratory bronchioles), with a pronounced bronchitic element. This view was supported by Thurlbeck (1976).

The two clinical patterns appear to be related to the different patterns of matching of ventilation and perfusion of the lung rather than depending on the underlying pathology per se. The determination of ventilation and perfusion by the partitioning of gases of different solubilities in patients with chronic obstructive lung disease (Wagner et al. 1977) commonly shows that "pink puffers" have increased ventilation of poorly perfused alveoli, whereas "blue bloaters" have increased perfusion of poorly ventilated spaces. However, as Flenley (1986) points out, in the original study of Wagner et al. (1977) whereas seven out of eight "pink puffers" conformed to this abnormality of respiration only four of 12 "blue bloaters" showed the anticipated abnormality of gas exchange.

Such indications that neither pathological lesions in the lungs nor physiological abnormalities in gas exchange within these diseased lungs appear to explain the clinical differences between "pink puffers" and "blue bloaters" suggest that other disorders of ventilatory control may be of prime importance. In this respect the hypoxic drive to breathing based on the carotid chemoreceptor has to be considered. Flenley and his colleagues (1970) found that the hypoxic drive to breathing was diminished in some "blue bloaters". Flenley (1986) reported that his group had devised a method of measuring hypoxic drive

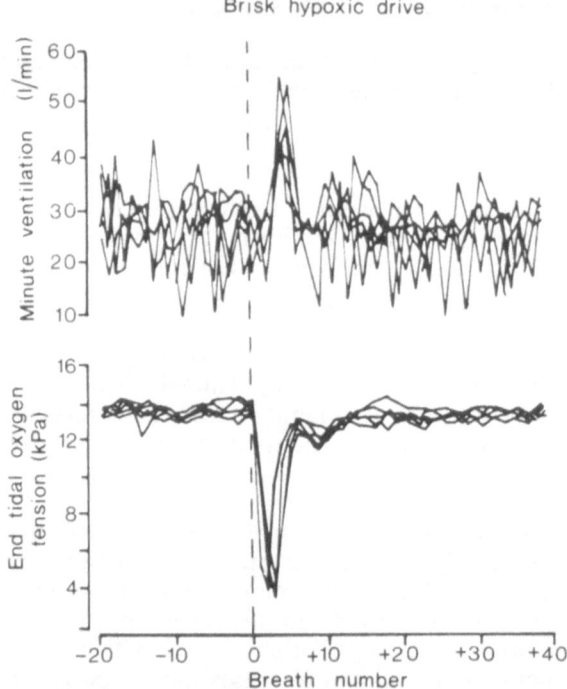

Fig. 12.2. Minute ventilation, as recorded for each breath expressed as litres per minute, during steady treadmill walking in a normal healthy subject, who had a brisk hypoxic drive to breathing, before, during and after sudden administration of three breaths of nitrogen. The end tidal oxygen tension, as determined by mass spectrometer, is shown below the ventilation trace. 1 kPa = 7.5 mmHg.

which related the minute ventilation calculated by computer for each breath, during steady-state treadmill exercise, to the simultaneously measured end tidal oxygen tension when the subject suddenly received three breaths of nitrogen, unknowingly during the steady-state exercise. They found that there was a seven-fold variability in the sensitivity of the hypoxic drive, which they believed to be mediated by the carotid chemoreceptors in apparently healthy human subjects (Figs. 12.2 and 12.3). Such observations led them to the conclusion that patients with the blue-and-bloated pattern of chronic bronchitis and emphysema may possibly have a pre-existing reduction in their hypoxic drive to breathing, whereas those with the pink-and-puffing pattern have a brisk hypoxic drive to breathing, this again possibly being present before they develop the disease (Flenley 1986). This hypothesis clearly has to be supported by long-term studies following patients in whom the hypoxic drive has been measured in health, and who subsequently develop chronic bronchitis and emphysema.

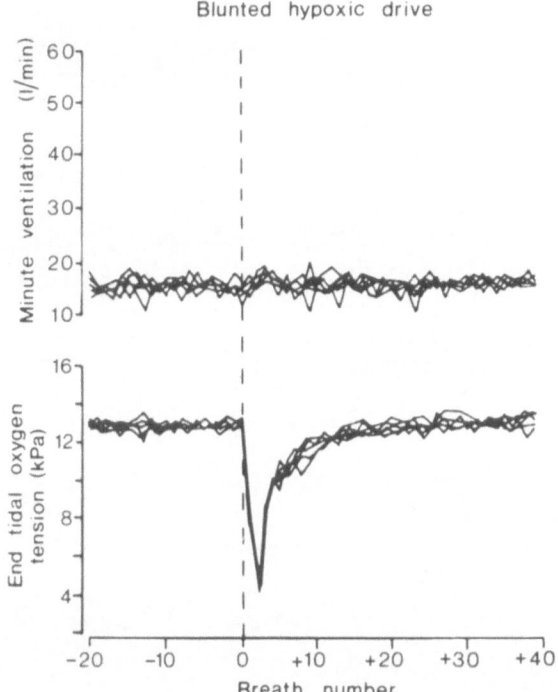

Fig. 12.3. The same traces in another healthy normal subject who had a blunted hypoxic drive to ventilation. 1 kPa = 7.5 mmHg.

References

Barer G, Wach R, Pallot D, Bee D (1986) Almitrine, hypoxia, systemic hypertension and the carotid body. In: Heath D (ed) Aspects of hypoxia. Liverpool University Press, Liverpool, pp 113–129

Bradley G (1981) Control of ventilation. In: Scadding JG, Cumming G (eds) Scientific foundations of respiratory medicine. Heinemann, London, pp 162–172

Cumming G, Semple SJ (1980) The effects of hypoxia and hypercapnia. In: Disorders of the respiratory system, 2nd edit. Blackwell Scientific, Oxford, pp 278–308

Dejours P (1957) Intérêt méthodologique de l'étude d'un organisme vivant à la phase initiale de rupture d'un equilibre physiologique. CR Acad Sci Paris 245:1946–1948

Dejours P, Labrousse Y, Raynaud J, Girard F, Teillac A (1958) Stimulus oxygène de la ventilation au repos et au cours de l'exercice musculaire à basse altitude (50 m) chez l'homme. Rev Fr Etud Clin Biol 3:105–123

Dempsey JA, Forster HV, Do Pico GA (1974) Ventilatory acclimatization to moderate hypoxaemia in man. The role of spinal fluid (H+). J Clin Invest 53:1091–1100

Flenley D (1986) Long-term oxygen therapy and the pulmonary circulation. In: Heath D (ed) Aspects of hypoxia. Liverpool University Press, Liverpool, pp 45–59

Flenley DC, Franklin DH, Millar JS (1970) The hypoxic drive to breathing in chronic bronchitis and emphysema. Clin Sci 38:503–518

Forster HV, Dempsey JA, Chosy LW (1975) Incomplete compensation of CSF (H+) in man during acclimatization to high altitude (4300 m). J Appl Physiol 38:1067–1072

Hurst G, Heath D, Smith P (1985) Histological changes associated with ageing of the human carotid body. J Pathol 147:181–187

Hurtado A (1964) Some physiologic and clinical aspects of life at high altitudes. In: Cander L, Moyer JH (eds) Aging of the lung. Grune and Stratton, New York, pp 257–278

Jeyaranjan R, Goode R, Beamish S, Duffin J (1987) The contribution of peripheral chemoreceptors to ventilation during heavy exercise. Resp Physiol 68:203–213

Khan Q, Heath D, Smith P, Norboo T (1988) The histology of the carotid bodies in highlanders from Ladakh. Int J Biometeorol 32:254–259

Lahiri S (1971) Genetic aspects of the blunted chemoreflex ventilatory response to hypoxia in high altitude adaptation. In: Porter R, Knight J (eds) High altitude physiology: cardiac and respiratory aspects, Ciba Foundation Symposium. Churchill Livingstone, Edinburgh, pp 103–112

Lahiri S, Delaney RG, Brody JS, Simpeer M, Velasquez T, Motoyama EK, Polgar C (1976) Relative role of environmental and genetic factors in respiratory adaptation to high altitude. Nature 261:133–135

Lenfant C, Sullivan K (1971) Adaptation to high altitude. N Engl J Med 284:1298–1309

Lugliani R, Whipp BJ, Seard C, Wasserman K (1971) Effects of bilateral carotid body resection on ventilatory control at rest and during exercise in man. N Engl J Med 285:1105–1111

Michel CC, Milledge JS (1963) Respiratory regulation in man during acclimatization to high altitude. J Physiol (Lond) 168:631–643

Severinghaus JW, Mitchell RA (1964) The role of cerebrospinal fluid in the respiratory acclimatization to high altitude in man. In: Weihe WH (ed) The physiological effects of high altitude. Pergamon Press, Oxford, pp 273–288

Severinghaus JW, Mitchell RA, Richardson BW, Singer MM (1963) Respiratory control at high altitude suggesting active transport regulation of cerebrospinal fluid pH. J Appl Physiol 18:1155–1166

Sime F (1973) Ventilacion humana en hipoxia cronica. Etiopatogenia de la Enfermedad de Monge a desadaptacion cronica a la altura. Doctoral thesis. Unversidad Peruana Cayetano Heredia, Lima, Peru

Springer C, Barstow TJ, Cooper DN (1989) Effect of hypoxia on ventilatory control during exercise in children and adults. Pediatr Res 25:285–290

Tenney SM, Remmers JE, Mithoefer JC (1964) Hypoxic-hypercapneic interaction at high altitude. In: Weihe WH (ed) The physiological effects of high altitude. Pergamon Press, Oxford, pp 263–272

Thurlbeck WM (1976). Pathophysiology: clinicopathologic correlations. In: Bennington JL (ed) Major problems in pathology, vol. V. W. B. Saunders, London, pp 340–344.

Wagner PD, Dantzker DR, Dueck R, Clausen JL, West JB (1977) Ventilation-perfusion inequality in chronic obstructive pulmonary disease. J Clin Invest 59:203–216

Wasserman K (1976) Testing regulation of ventilation with exercise. Chest 70:S173–S178

Wasserman K, Whipp BJ, Casaburi R, Golden M, Beaver WL (1979) Ventilatory control during exercise in man. Bull Eur Physiopathol Respir 15:27–51

13   Almitrine

Patients with chronic obstructive airways disease frequently exhibit alveolar hypoventilation with associated hypoxaemia and hypercarbia. They are commonly in a state of chronic respiratory failure and acute exacerbations brought about by respiratory infections may further impair their blood gas levels. Centrally acting respiratory stimulants such as doxapram have been used in an attempt to improve blood gases, but with limited clinical success and many side effects including stimulation of the central nervous system with ensuing convulsions, gastrointestinal upsets, systemic hypertension, bronchospasm and increased consumption of oxygen due to overactivity of skeletal muscle (Powles et al. 1983). Long-term domiciliary oxygen therapy is more successful but this is both expensive and onerous for the patients. A third theoretical possibility to improve oxygenation of the blood is to stimulate the carotid bodies with a drug which could directly improve ventilation without the risk of central nervous system disturbances. Such a drug is almitrine bismesylate.

Almitrine has the chemical name of 1-(4', 6'-diallylamino-2'-triazinyl)-4-(bis-4,4'-fluorobenzydryl)piperazine bismethane sulphonate (Laubie and Schmitt 1980) and its molecular structure is shown in Fig. 13.1. The bismethane sulphonate moiety is to increase its solubility. The drug is available in France under the trade name Vectarion® and is still undergoing clinical trials in several countries. All the studies on animals and man outlined below indicate that almitrine acts by stimulating the carotid and, to a lesser extent, the aortic bodies with a negligible effect on the central nervous system. It is rapidly absorbed when administered either intravenously or orally, reaching a maximum plasma concentration within 2–3 hours (Campbell et al. 1983). The subsequent half-life of elimination from the plasma varies from 4 to 50 hours. Very little of the drug is excreted by the kidneys, the majority being found in the faeces, so that the probable route of excretion is in the bile. Almitrine is extensively metabolised, with less than 20% being excreted in unchanged form. All the metabolites that have been tested have no effect upon ventilation and so it is assumed that

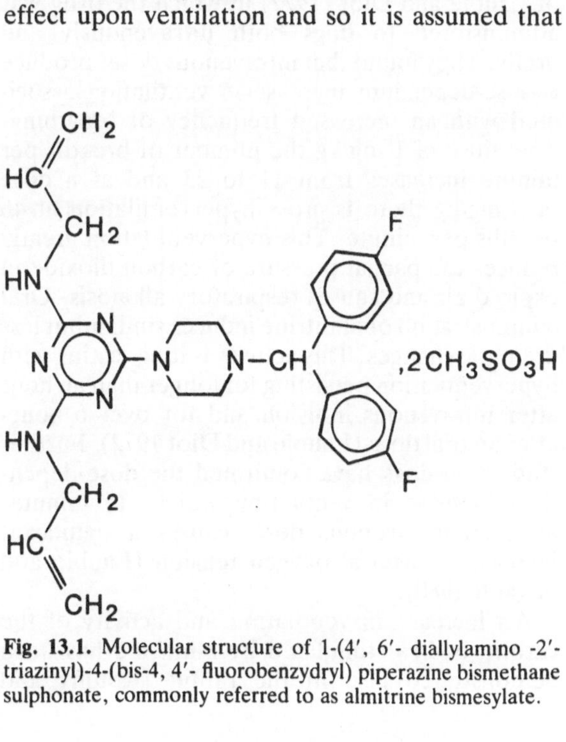

Fig. 13.1. Molecular structure of 1-(4', 6'- diallylamino -2'- triazinyl)-4-(bis-4, 4'- fluorobenzydryl) piperazine bismethane sulphonate, commonly referred to as almitrine bismesylate.

the pharmacological effects of almitrine are due to unmetabolised drug (Campbell et al. 1983).

The long half-life is explained by an avid binding to plasma proteins. In the tissues, concentrations higher than in the plasma may be attained and organs in which this occurs particularly include muscle, liver, lung, adrenal, kidney, carotid artery and carotid body (Gordon et al. 1987). Peak plasma levels after a single oral dose of 100 mg generally range from 150 to 200 ng/ml, but there is a wide intersubject variation which can be as much as ten-fold. With repeated oral doses the drug tends to accumulate in the body, with resulting high peak plasma levels ranging from 20 to 640 ng/ml after 3 months and from 83 to 904 ng/ml after a year (Campbell et al. 1983). In patients with chronic obstructive airways disease the plasma half-life of almitrine is similar to that in healthy individuals after a single dose. With long-term therapy, however, the plasma half-life is very much longer, up to 23.3 days (Evans et al. 1986).

Effects on Chemoreception in Experimental Animals

Early research into the effects of almitrine on experimental animals demonstrated its powerful stimulatory effect on the peripheral chemoreceptors. The first study to be published was that of Laubie and Diot (1972) in which the drug was administered to dogs both intravenously and orally. They found that intravenous doses produce a dose-dependent increase in ventilation associated with an increased frequency of breathing. At a dose of 1 mg/kg the number of breaths per minute increases from 11 to 23 and at a dose of 3 mg/kg there is gross hyperventilation at 46 breaths per minute. This hyperventilation greatly reduces the partial pressure of carbon dioxide in expired air and causes respiratory alkalosis. Oral administration of almitrine induces similar but less dramatic changes. This activity is long-lasting with hyperventilation persisting for longer than an hour after intravenous infusion and for over 6 hours after an oral dose (Laubie and Diot 1972). Further studies on dogs have confirmed the dose-dependent increase in ventilation which, 15 minutes after an intravenous dose, causes a significant increase in arterial oxygen tension (Laubie and Schmitt 1980).

An increase in ventilation and activity of the carotid bodies stimulated by almitrine has been demonstrated also in the rabbit (Roumy and Leitner 1981), the cat (Gautier and Bonora 1982) and the rat (Dhillon and Barer 1982). In rats subjected to chronic hypoxia in a chamber containing 10% oxygen for 4 weeks almitrine results in further increases in ventilation above those already present due to the hypoxic stimulation of their carotid bodies (Dhillon and Barer 1982). Rats have also been subjected to a hypoxic and hypercarbic atmosphere simulating the alveolar gases in patients with chronic bronchitis and emphysema. Almitrine significantly improves the arterial oxygen tension in these animals but has a minimal effect on arterial carbon dioxide tension despite increased ventilation, the latter effect being attributable to increased storage of carbon dioxide in the body fluids (Dhillon and Barer 1982).

Neurophysiological experiments have demonstrated that this stimulatory action of almitrine is due to a direct influence on the peripheral chemoreceptors. Sectioning the carotid sinus nerves greatly reduces the effect of almitrine on ventilation and cutting both sinus and vagus nerves abolishes it altogether (Laubie and Diot 1972; Laubie and Schmitt 1980). Thus, isolating both the carotid and aortic bodies from their afferent and sympathetic neural connections renders almitrine inactive, suggesting that its action is exclusively on these organs. Furthermore, injecting the drug into the vertebral artery induces no change in ventilation (Laubie and Diot 1972). Ablation of the nucleus tractus solitarius (see Chapter 12) abolishes all activity of almitrine. This centre in the brain-stem is the site of the first synapses of vagal efferent nerve fibres and also of baroreceptor and chemoreceptor afferent fibres from the carotid sinus nerve (Laubie and Schmitt 1980). Recording the electrical impulses along the carotid sinus nerve demonstrates that almitrine causes large and sustained increases in chemoreceptor discharges (Laubie and Schmitt 1980; Bisgard 1981; O'Regan et al. 1987). Reducing the concentration of oxygen in inspired air causes a further increase in the rate of discharge which is greater than the additive effects of hypoxia and almitrine alone (Bisgard 1981). Thus almitrine appears to act by sensitising the carotid bodies to stimulation by hypoxaemia. This effect is independent of the depressant action of dopamine, since its application merely depresses the baseline level of firing and not the magnitude of the increase induced by almitrine. Similarly, the dopamine antagonist domperidone simply raises the baseline level of discharge. It can be concluded that the action of almitrine on the carotid body does not involve dopamine receptors (Bisgard 1981).

Effects in Healthy Man

The contribution of the carotid bodies to resting ventilatory drive in man is only 10%–15% although this value rises considerably during exercise or hypoxaemia (see Chapter 12). It is not surprising, therefore, that almitrine induces virtually no effect upon minute ventilation or blood gases in resting, healthy subjects breathing air (Guillerm and Radziszewski 1974; Stradling et al. 1982; Stanley et al. 1983a,b). In one study, almitrine at a dose of 5 mg/kg induced a significant increase in ventilation, but this dose is higher than that recommended for therapeutic use (Guillerm and Radziszewski 1974).

When healthy volunteers breathe hypoxic gas mixtures, almitrine stimulates a large increase in ventilation in addition to that elicited by hypoxia alone (Stradling et al. 1982; Stanley et al. 1983a,b). With intravenous infusions of almitrine the increase in minute ventilation is dependent upon the dose administered (Stanley et al. 1983a). A dose-dependent effect is also evident after oral dosing such that a dose of 50 mg causes a 78% elevation and this increases to 120% when the dose is doubled (Stanley 1981). Almitrine has no effect on metabolic rate, heart rate or systemic blood pressure (Stanley et al. 1983b). Hypercarbia has a negligible effect upon the ventilatory response to almitrine (Stradling et al. 1982; Stanley et al. 1983a) although high oral doses may induce a small but significant increase in ventilation as the partial pressure of carbon dioxide in the alveolar spaces rises (Stanley et al. 1983b). This is further evidence that almitrine acts on the peripheral chemoreceptors, since hypercarbia is mainly a central nervous stimulant of ventilation. The small stimulatory effect of almitrine in hypercarbia may also be mediated by the peripheral chemoreceptors, in which a slight residual sensitivity to carbon dioxide persists even in normoxia (Stanley et al. 1983b).

The most cogent evidence to support the argument that almitrine stimulates the carotid bodies comes from studies on patients with chronic bronchitis and emphysema who have had both carotid bodies removed to alleviate symptoms of dyspnoea (de Backer et al. 1985). In eight of such patients almitrine failed to induce any improvement in ventilation or oxygenation of the blood. Two patients were studied before and after the operation. With their carotid bodies intact, almitrine significantly improved gas exchange but after bilateral glomectomy it was without any effect, indicating that almitrine acts entirely on the carotid bodies and not the aortic bodies or central nervous system in humans (de Backer et al. 1985).

Short-Term Studies on Hypoxaemic Patients

The many clinicians who have studied the acute effects of almitrine on patients with chronic obstructive airways disease are unanimous in their opinion that the drug improves oxygenation of the blood. The results of some of these studies are summarised in Table 13.1. In all but three of these the mean arterial oxygen tension in each patient before almitrine administration was 60 mmHg or less, indicating a state of chronic respiratory failure. Following either intravenous infusion or an oral dose of almitrine there was a significant increase in arterial oxygen tension which ranged from 7.5 to 18.4 mmHg, with an average improvement of 10.6 mmHg (Table 13.1). This was associated with an increased oxygen saturation of haemoglobin. This improvement in oxygenation of the blood was not balanced by a similar fall in arterial carbon dioxide tension which ranged from 1.0 to 11.1 mmHg, with a mean of 5.4 mmHg (Table 13.1), approximately half of the value for the increase in oxygen tension. The reason for this small change in carbon dioxide tension can be deduced from the figures for the increase in minute ventilation which ranged from zero to 3.4 l/min, with a mean value of only 1.5 l/min (Table 13.1). The improved oxygenation of the blood after almitrine therapy cannot be explained entirely by an increase in minute ventilation (Sergysels et al. 1978; Naeije et al. 1981; Simonneau et al. 1981; Connaughton et al. 1983; Melot et al. 1983; Powles et al. 1983; Stradling et al. 1984; Maxwell et al. 1985).

The clinical investigations summarised in Table 13.1 involved patients with chronic bronchitis and emphysema in whom the blood gases had been stable for several weeks prior to study. Nevertheless, almitrine has beneficial effects on patients with severe hypoxaemia and hypercarbia during exacerbations of their disease due to concurrent pulmonary infections (Lambropoulos et al. 1986). In a study of 16 such patients with acute respiratory failure who were in a state of pre-coma or coma, almitrine at a dose of 100 mg/day had improved arterial oxygen tension from 45 to 75 mmHg by 4 days. Placebo tablets were associated with a much smaller increase from 49 to 56 mmHg (Lambropoulos et al. 1986). Almitrine has also been shown to be effective in improving

Table 13.1. Influence of short-term almitrine therapy on blood gases and ventilation in patients with chronic obstructive airways disease

No. of patients	Dose	Pa_{O_2} (mmHg)		Pa_{CO_2} (mmHg)		$\dot{V}E$ (l/min)		References
		Before	After	Before	After	Before	After	
36	400 mg/day orally for 15 days	47.8	66.2	57.4	46.3	—	—	Neukirch et al. 1974
12	1 mg/kg i.v. for 30 min	60.1	70.3	50.0	42.5	10.0	12.3	Sergysels et al. 1978
10	0.5 mg/kg i.v. for 30 min	63.7	77.0	38.3	32.9	8.8	9.9	Schrijen and Romero Colomer 1978
6	200 mg single dose orally	59	73	46	39	no change		Simonneau et al. 1981
12	0.25 mg/kg i.v. for 30 min	43.7	53.9	55.0	50.0	10.4	12.9	Naeije et al. 1981
14	0.5 mg/kg i.v. for 60 min	51.9	61.9	52.8	45.7	—	—	Weitzenblum et al. 1982
11	0.5 mg/kg i.v. for 2 h	48	56	54	48	8.2	11.6	Powles et al. 1983
12	100 mg single dose orally	54.4	62.5	43.4	39.0	13.2	14.7	Stanley et al. 1983c
6	100 mg single dose orally	59.3	67.5	49.0	44.7	6.8	7.5	Stradling et al. 1983
8	100 mg single dose orally	57.0	69.0	45.7	42.0	12.6	12.7	Yernault et al. 1983
9	100 mg/day orally for 14 days	52	60	48	44	—	—	Connaughton et al. 1985
2	100 mg single dose orally	77	88	35	34	10.9	12.7	de Backer et al. 1985
12	100 mg/day orally for 4 weeks	54.0	61.5	43.5	39.0	11.3	12.2	Stanley 1986
28	200 mg single dose orally	67.7	76.8	37.0	33.0	12.6	14.2	Magnussen et al. 1987

$\dot{V}E$, minute ventilation.

arterial oxygen tension in a 5-year-old boy with severe bronchopulmonary dysplasia (Gaultier et al. 1983). However, not all forms of pulmonary disease which are associated with hypoxaemia respond to almitrine therapy. Magnussen et al. (1987) compared the effects of the drug on patients with chronic bronchitis and emphysema, bronchial asthma and interstitial pulmonary fibrosis. Almitrine improved ventilation in all three groups but with pulmonary fibrosis and asthma the increase in arterial oxygen tension was respectively 2.1 and 5.7 mmHg, which was statistically insignificant. Only in chronic bronchitis and emphysema was a significant improvement of 9.1 mmHg obtained. Thus, it is likely that almitrine will find its main therapeutic application in the treatment of chronic bronchitis and emphysema.

The ventilatory response to hypoxia in patients with chronic obstructive airways disease has been tested using a rebreathing apparatus. Such studies have demonstrated that almitrine greatly enhances the ventilatory response to hypoxia, suggesting that it stimulates the carotid bodies even in patients who are already hypoxic (Stanley et al. 1983c; Maxwell et al. 1985; Stanley 1986). A small increase in the ventilatory response to hypercarbia also occurs, and may be mediated by the carotid bodies (Stanley et al. 1983c). Exercise has an effect upon ventilation similar to that of hypoxia, and here too almitrine produces an exaggerated change in minute ventilation and tensions of oxygen and carbon dioxide when results are compared with exercising patients given placebo only (Simonneau et al. 1981).

As suggested above, the small, often insignificant, increases in minute ventilation cannot explain the large improvements in oxygenation of the blood which occur with almitrine therapy. In studies in which patients with chronic obstructive airways disease are asked to hyperventilate in a controlled manner, improvements in oxygen saturation fall far short of those induced by almitrine (Tweney 1987). Furthermore, in patients who are mechanically ventilated so that minute ventilation is held constant, almitrine still causes a significant increase in arterial oxygen tension. Such considerations have led respiratory physiologists to suggest that almitrine improves ventilation–perfusion ratios in the lungs. Support for this idea comes from the observations that the alveolar–arterial oxygen gradient is decreased (Naeije et al. 1981; Simonneau et al. 1981; Stradling et al. 1984) as is the mean venous admixture (Melot et al. 1983; Powles et al. 1983). Both of these entities accompany improvement in matching of ventilation and perfusion.

Increases in pulmonary arterial pressure have also been recorded following acute administration of almitrine (Naeije et al. 1981; Weitzenblum et al. 1982; Dull et al. 1983; Melot et al. 1983). In one study the pulmonary artery mean pressure rose from 32 to 45 mmHg 30 minutes after intravenous infusion of almitrine (Dull et al. 1983). This was associated with an increase in pulmonary vascular resistance of 64%. Baseline values for pulmonary arterial pressure are higher than normal because of pulmonary vasoconstriction secondary to alveolar hypoxia in these emphysematous patients. Thus almitrine is apparently capable of augmenting pulmonary vasoconstriction in addition to a pre-existing hypoxic pressor response. In this way it might divert blood from hypoxic areas of the lung to better ventilated ones, thereby improving matching of ventilation and perfusion.

Long-Term Clinical Trials

Chronic bronchitis and emphysema associated with hypoxaemia, pulmonary arterial hypertension, right ventricular hypertrophy and episodes of systemic oedema has a poor prognosis. Approximately three-quarters of patients with this combination die within five to six years of the first episode of ankle oedema (Tweney 1987). Long-term domiciliary oxygen therapy is said to improve the quality of life for them but almitrine might be thought to offer a chance to prevent the development of the blue-and-bloated syndrome. Several long-term clinical trials of almitrine were organised, including the Vectarion international multicentre study (VIMS) in 1983. Included in this double-blind, placebo-controlled study were 701 patients from 70 centres in 12 countries (Voisin et al. 1987).

The results of long-term studies on the efficacy of almitrine to improve the tensions of respiratory gases in the blood are summarised in Table 13.2. All involved oral administration of the drug, most

with one 50 mg tablet taken twice a day, but in some the daily dose was increased to 150 or 200 mg/day in those patients who showed a poor response. In all of the trials there was a significant improvement in the arterial oxygen tension in each group given almitrine as contrasted with patients given placebo. The magnitude of the improvement ranged from 5.2 to 12.0 mmHg, with a mean for all the trials of 7.3 mmHg. There is a trend towards a greater improvement in those patients with a low initial level of oxygen saturation (Bourgouin-Karauni et al. 1986). Prolonged administration of almitrine also resulted in a slight reduction in arterial carbon dioxide tension, which averaged 2.3 mmHg but in some trials it was absent altogether. This was reflected by insignificant increases in minute ventilation in most patients.

Some patients in these trials were also treated with long-term domiciliary oxygen therapy. In them an increase in arterial oxygen tension was induced by almitrine in addition to that created by long-term oxygen therapy alone, although there was no significant effect on hypercarbia (Evans et al. 1986, 1990; Voisin et al. 1987) When the rate of delivery of oxygen was increased, almitrine still maintained an increased arterial oxygen tension until normal levels of oxygenation were approached, when its effect diminished (Evans et al. 1990). Thus almitrine apparently improves oxygenation of the blood in addition to oxygen therapy alone and may be a useful adjunct to it.

As the clinical trials progressed it became clear that some patients respond well to almitrine, with increases in arterial oxygen tension of over

Table 13.2. Changes in blood gases in patients with chronic obstructive airways disease after long-term therapy with almitrine

No. of patients	Dose (expressed in mg/day)	Duration of therapy in months	ΔPa_{O_2} (mmHg)	ΔPa_{CO_2} (mmHg)	References
128	100	2	+ 7.6	−4.1	Grassi et al. 1986
44	100	2	+10.0	−2.0	Vulterini et al. 1986
26	100	2	+ 6.0	0	Johanson et al. 1986
25	200	2	+12.0	−4.0	Johanson et al. 1986
19	100	3	+ 9.4	0	Evans et al. 1986
28	100	3	+ 5.3	0	Evans et al. 1990
84	200 falling to 100	6	+ 5.2	−3.9	Arnaud et al. 1983
10	100	6	+ 6.4	−3.7	Paramelle et al. 1983
81	200 falling to 100	6	+ 5.2	−3.9	Bourgouin-Karauni et al. 1986
415	100–150	6	+ 7.0	−2.2	Marsac 1986
10	100	12	+ 7.8	+0.4	Paramelle et al. 1983
86	100	12	+ 6.0	−2.7	Bourgouin-Karauni et al. 1986
344	100–200	12	+ 6.3	−2.3	Voisin et al. 1987 (VIMS study)
13	100	12	+ 9.5	−3.4	Watanabe et al. 1989
43	100–200	24	+ 5.3	−2.2	Howard et al. 1989a

5 mmHg reaching in some cases 17.7 mmHg (Suggett et al. 1986). On the other hand, approximately 25% of patients are poor responders, with an increased arterial oxygen tension of less than 5 mmHg (Arnaud et al. 1983; Bourgouin-Karauni et al. 1986; Marsac 1986; Voisin et al. 1987). This differing response has been noted during acute studies with almitrine (Feist 1986), but with long-term trials the distinction between the two groups is more clear-cut. It was found that some poor responders have a low level of almitrine in their plasma. Increasing the dose in such patients may improve blood gases but in others the plasma almitrine is high and increasing the dose is without any beneficial effect (Ansquer 1986; Voisin et al. 1987). It is not clear why some patients should show such a poor response to prolonged almitrine therapy. Retrospective examination of their clinical histories and pulmonary function tests show that it is impossible to predict in advance which patients might respond and which might not. The differing response takes some 3 months of almitrine therapy to manifest itself in any case (Marsac 1986). In a detailed investigation to try to identify factors by which a response to almitrine may be predicted, nine good responders and nine poor responders were examined. It was found that poor responders could not be recognised by their age, weight, baseline blood gases, spirometry measurements, ventilation, lung volume, level of dyspnoea or concentration of almitrine in the blood (Suggett et al. 1986). There was a difference between the two groups of patients in their response to exercise with good responders being more breathless at the end of a 6 minute walking test. This suggests that almitrine may be of greater efficacy in patients with severe hypoxaemia and hypercarbia ("blue bloaters") whilst those with relatively mild hypoxaemia and no hypercarbia ("pink puffers") benefit only when the levels of their blood gases deteriorate (Suggett et al. 1986).

Those patients who respond to almitrine with a large increase in arterial oxygen tension show other clinical improvements. Most of them experience a reduction in the degree of their dyspnoea (Arnaud et al. 1983; Marsac 1986; Voisin et al. 1987). There is also a reduction in the proportion of patients admitted to hospital for acute exacerbations of their disease (Arnaud et al. 1983; Bourgouin-Karauni et al. 1986; Voisin et al. 1987; Howard et al. 1989a). In one study the hospitalisation rate was only 18% in patients taking almitrine, as contrasted with 33% in those taking a placebo for 6 months (Bourgouin-Karauni et al. 1986). Effort tolerance is also increased and appears to improve progressively with the dur-ation of almitrine therapy (Marsac 1986; Grassi et al. 1987). The increased arterial oxygen tension is often associated with a reduction in the level of polycythaemia (Arnaud et al. 1983; Voisin et al. 1987; Howard et al. 1989a). Finally, the patients may feel subjectively that there is an improvement in their symptoms (Marsac 1986). Despite these improvements in patients taking almitrine there is no apparent difference in the death rate between them and those taking placebo (Voisin et al. 1987). A similar conclusion was reached when 89 patients from the VIMS trial were studied for a further year, despite the fact that improved blood gases and clinical improvements were maintained (Howard et al. 1989a). It may be that much longer studies will be required before any differences in survival rate due to almitrine therapy will become manifest.

Hypoventilation Syndromes

Temporary hypoventilation may complicate recovery from anaesthesia and almitrine has been shown to improve ventilation and blood gases in such patients (Tweney 1987). Blue-and-bloated patients with chronic obstructive airways disease are both hypoxic and severely hypercarbic and often suffer from episodes of hypoventilation and increased hypoxaemia during sleep. An improvement in oxygen saturation of the blood during sleep in these patients has been noted after oral administration of almitrine (Connaughton et al. 1983; Marrone et al. 1986; Racineux et al. 1987). Although this treatment for nocturnal hypoxaemia increases the mean oxygen saturation of the blood, the drop in arterial oxygen tension during sleep does not change. This is because patients taking almitrine start the night with a higher level of arterial oxygen tension which places their oxygen saturation closer to the horizontal region of the oxygen–haemoglobin dissociation curve (Connaughton et al. 1985). The number of nocturnal hypoxaemic episodes is also reported to be significantly reduced after almitrine therapy by some workers (Connaughton et al. 1983, 1985) but not by others (Marrone et al. 1986; Racineux et al. 1987). Almitrine appears to be more effective in improving sleep hypoxaemia than the centrally acting respiratory stimulant medroxyprogesterone. Thus, when the two drugs were tested in the same patients, almitrine treatment resulted in a higher level of oxygenation of the blood, with fewer arousals during sleep (Daskalopoulou et al. 1990).

The efficacy of almitrine has been studied in children with the rare congenital central hypoventilation syndrome (Ondine's curse). Almitrine therapy reduces the duration of apnoea and increases minute ventilation and frequency of breathing with an associated improvement in blood gases (Naeije et al. 1981; Fleming et al. 1983; Shannon et al. 1983). Two children with this syndrome were described; they required ventilation day and night, but with almitrine therapy they could be weaned off the respirator during the day (Shannon et al. 1983).

Effects on the Pulmonary Vasculature

Since it has been suggested that almitrine improves matching of ventilation and perfusion by stimulating pulmonary vasoconstriction, numerous experiments have been performed to try to demonstrate this pressor response. The results of such studies are often contradictory. In a normoxic environment almitrine induces significant pulmonary vasoconstriction in dogs, cats, rats and ferrets (Bee et al. 1981, 1983; Barer et al. 1983; Hughes et al. 1983). When these species are ventilated with hypoxic gas mixtures, almitrine produces a variety of responses including vasoconstriction (Romaldini et al. 1983), vasoconstriction followed by prolonged vasodilatation (Bee et al. 1981, 1983; Barer et al. 1983), vasodilatation (Hughes et al. 1983) or no change (Schmoller et al. 1983) Even when rats are subjected to prolonged hypoxia, almitrine produces a smaller increase in pulmonary arterial pressure than it does in normoxic controls (Bee et al. 1983). Such findings do not exclude the possibility that almitrine has a local vasoconstrictor effect on the pulmonary vasculature. In an attempt to mimic the mismatching of ventilation that occurs in pulmonary emphysema, animal experiments were performed in which one lobe of lung was ventilated with hypoxic gas whereas the remainder of the lung was normoxic or hyperoxic. Such experiments failed to demonstrate any diversion of blood away from the poorly oxygenated lobe (Barer et al. 1983; Hughes et al. 1983, 1986; Schmoller et al. 1983; Wach et al. 1986). On the contrary, almitrine increases the pulmonary vascular resistance in the well-oxygenated lung but reduces it slightly in the hypoxic lobe (Hughes et al. 1986), so that the blood flow through the lobe actually increases (Barer et al. 1983; Wach et al. 1986).

In one of the above experiments it was noted that almitrine continued to cause vasoconstriction in normally ventilated lungs when the common carotid arteries were clamped, suggesting that it might act directly on the pulmonary vasculature (Hughes et al. 1983). In preparations of isolated lung of rats, cats and dogs almitrine partially inhibits the pulmonary vasoconstriction induced by hypoxia (Mazmanian and Lockhart 1982; Barer et al. 1983). In ferrets almitrine causes vasodilatation in the majority of isolated lungs but in a few constriction occurs (Bee et al. 1981). No clear-cut conclusions can be drawn from these studies on isolated lungs, since in chemodenervated and vagotomised rats the increase in pulmonary arterial pressure normally induced by either hypoxia or almitrine is almost totally abolished, suggesting that carotid and aortic bodies are a vital component of the vascular reflexes (Lagneaux 1987).

In humans a direct action of almitrine on the pulmonary vasculature is unlikely since, following carotid glomectomy, improvements in ventilation–perfusion ratios seen before operation are completely abolished (de Backer et al. 1985). Nevertheless, as noted earlier in this chapter, short-term studies on the effects of almitrine in patients with chronic obstructive airways disease reveal an increase in pulmonary arterial presssure and resistance, but this is short lived (Weitzenblum et al. 1982) and may be more pronounced in patients with severe hypercarbia (Dull et al. 1983). When almitrine is administered for 6 months to 1 year no difference in resting pulmonary arterial pressure is found compared with patients taking placebo tablets (Paramelle et al. 1983; Bourgouin-Karauni et al. 1986). The apparent lack of a pressor response of almitrine on the pulmonary vasculature after prolonged therapy would not preclude an influence of it on ventilation–perfusion ratios if almitrine were selectively to constrict vessels in underventilated areas of lung and dilate them in well-oxygenated regions. Improved matching of ventilation and perfusion could then occur in the absence of any overall increase in vascular resistance. In this context it has been noted that almitrine diverts blood from the lung bases to the better ventilated apices (Rigaud et al. 1981). A diversion of blood from hypoxic units of the lung to better oxygenated ones has also been demonstrated by using the inert gas technique (Castaing et al. 1983; Melot et al. 1983). In one of these studies (Castaing et al. 1983) the patients were mechanically ventilated so that changes in ventilation could not affect the observed improvement in distribution of perfusion.

Effects on the Pattern of Breathing

Some workers argue that almitrine does not improve the distribution of perfusion in the lungs but that it changes the volume and pattern of breathing to improve the distribution of ventilation (Stradling et al. 1984). The small increases in minute ventilation noted in patients with chronic obstructive airways disease treated with almitrine are due largely to an increase in tidal volume (Stradling et al. 1983; Grassi et al. 1986; Watanabe et al. 1989). However, changes in alveolar ventilation can occur in the absence of any change in external ventilation and are difficult to detect (Tweney 1987). It has been noted that almitrine increases the inspiratory flow with only small changes in minute ventilation or ventilatory rate (Maxwell et al. 1985; Grassi et al. 1986). Thus the effect of almitrine in these patients is to create a shorter, deeper and faster inspiration followed by a longer expiration thus keeping the duration of each breath unaltered (Stradling et al. 1983). This change in breathing pattern may increase the ventilation of poorly oxygenated regions of lung. In rats there is also an increase in functional residual capacity which widens the alveoli thereby increasing the area for gaseous diffusion (Marmo 1986). Such an effect has not been demonstrated in patients.

Although there is strong evidence that almitrine improves ventilation–perfusion ratios there is insufficient evidence at present to determine with certainty whether this is achieved by ventilatory or pulmonary haemodynamic changes. Animal models do not support the concept of selective pulmonary vasoconstriction but such experiments cannot reproduce the complex inhomogeneity of ventilation–perfusion ratios found in patients with chronic obstructive airways disease. This particular problem will need to be resolved by further studies.

Side Effects and Dosage

Side effects have been reported as occurring after prolonged treatment with almitrine (Table 13.3). Most of these symptoms are mild in nature and do not require discontinuation of the therapy. Although almitrine treatment usually reduces dyspnoea and increases exercise tolerance, a small minority of patients exhibit increased dyspnoea and awareness of breathing. These symptoms may become sufficiently distressing to force the patient to discontinue the drug.

Table 13.3. Side effects associated with prolonged almitrine therapy

Symptoms	References
Respiratory (increased dyspnoea, awareness of breathing)	Arnaud et al. 1983; Johanson et al. 1986; Voisin et al. 1987; Watanabe et al. 1989
Tachycardia	Arnaud et al. 1983; Grassi et al. 1986
Digestive upsets (nausea, gastric pain, diarrhoea, constipation)	Arnaud et al. 1983; Bourgouin-Karauni et al. 1986; Grassi et al. 1986; Voisin et al. 1987
Loss of weight	Howard et al. 1989a; Watanabe et al. 1989
Central nervous disturbances (dizziness, fatigue, insomnia, headache, irritability)	Arnaud et al. 1983; Voisin et al. 1987
Peripheral neuropathy (peripheral paraesthesiae, numbness, leg-pain cramps, abnormal ground perception)	Chedru et al. 1985; Lerebours et al. 1987; Voisin et al. 1987; Howard et al. 1989a; Watanabe et al. 1989

A side effect which has given rise to considerable concern is peripheral neuropathy (Table 13.3). In the VIMS trial, five times as many patients taking almitrine complained of peripheral paraesthesiae, especially of the lower limbs, than amongst those taking placebo tablets (Voisin et al. 1987). This was observed after approximately 7 months of treatment and led to 4% of patients withdrawing from the trial. Peripheral neuropathy can occur in cases of chronic obstructive airways disease with severe hypoxaemia, in the absence of almitrine administration (Howard 1985). This realisation has led to investigations into the incidence and electrophysiology of peripheral neuropathy occurring in association with chronic bronchitis and emphysema. It was found that between 60% and 75% of such patients exhibited electrophysiological abnormalities of peripheral nerve conduction suggestive of partial denervation, and reduced action potentials (Gasnault et al. 1987). Histological examination of neuromuscular biopsies showed decreased nerve fibre density, demyelination and axonal degeneration (Chedru et al. 1985; Paramelle et al. 1986). These were usually subclinical but tended to become clinically overt as hypoxaemia worsened, especially in heavy cigarette-smokers (Paramelle et al. 1986). In two studies electrophysiological parameters were measured in patients before and after taking almitrine (Lerebours et al. 1986, 1987). It was found that 66% of patients had at least one abnormal parameter before treatment and this proportion did not change after 3 months (Lerebours et al. 1986) or 1 year of almitrine administration (Lerebours et al. 1987). Furthermore, there was no deterioration in conduction velocity in those patients who started with low conduction values before treatment. Thus, peripheral neurological disturbances are frequent in patients with chronic bronchitis and emphysema and their genesis should not necessarily be attributed to a side effect of almitrine. However, almitrine does appear to exaggerate the symptoms of such disturbances in patients with existing electrophysiological abnormalities (Voisin et al. 1987). It is not understood why almitrine should do this but clinicians should be alert to the possibility of symptoms of peripheral neuropathy developing in patients being treated with almitrine long-term.

The optimum therapeutic dose of almitrine is under constant review. For short-term treatment the drug is usually given by intravenous infusion at a dose recommended by the manufacturers ranging from 1–3 mg/kg per day. For prolonged treatment, as in the therapy of chronic bronchitis and emphysema, Vectarion® has been provided in the form of tablets each containing 50 mg of almitrine. Until recently two of these tablets per day was recommended and this was the dose used in the majority of clinical trials (Table 13.2). However, since the drug accumulates in the body to give a half-life of nearly 1 month, plasma levels may become too high with this simple regime leading to some of the side effects discussed above. An alternative regime has been tested with success (Howard et al. 1989b). In this, tablets of almitrine, each containing 25 mg, and given twice daily at a dose of 50–100 mg/day for the first 3 months are followed by a 1 month wash-out period. Almitrine at the same dose is then given for 2-month periods alternating with intervals of 1 month in which it is withheld. This regime of reduced, intermittent dosing has been shown to maintain the efficacy of the drug whilst reducing the incidence of side effects over a 6 month period (Howard et al. 1989b).

Mechanism of Action

It is not known how almitrine triggers a chemosensory response in the carotid body but at first sight it appears to mimic closely the action of hypoxia in that ventilation is increased in association with an increased rate of discharge in the carotid sinus nerve. Furthermore, in rats both acute administration of almitrine and acute hypoxia cause depletion of dopamine due to its release from the chief cells (Pallot and Al Neamy 1983). An increase in the levels of dopamine is associated with hypercarbia but, if almitrine is administered at the same time, depletion of the amine occurs in the same way as it does with acute hypoxia. When exogenous dopamine is infused intravenously into rats it depresses the ventilatory response to both hypoxia and almitrine, an effect which is abolished by the dopamine antagonist domperidone (Wach et al. 1989). An interpretation of these experiments is that almitrine acts on chief cells in a manner similar to that of hypoxia by depleting them of dopamine, which, due to its inhibitory influence on chemoreception, leads to stimulation of the carotid bodies (Pallot and Al Neamy 1983).

If these short-term studies are extended to 15 days a different interpretation emerges. Prolonged hypoxia elicits a great increase in dopamine content associated with an enlargement of dense-core vesicles. Almitrine given for a similar period of time has a reverse effect, causing a depletion of dopamine (Pequignot et al. 1987). Further experiments with cats also suggest that almitrine, unlike hypoxia, does not act through catecholaminergic or cholinergic mechanisms (O'Regan et al. 1987).

Further evidence to suggest that almitrine acts differently to hypoxia comes from the few pathological studies to be made of its effect on experimental animals. In rats given almitrine at a dose of 4 mg/kg per day no subjective differences are to be found in the glomic tissue compared with control animals. Morphometric analysis of electron micrographs from these carotid bodies reveals a small reduction in the size of the nuclei and volume of cytoplasm of chief cells (Dhillon et al. 1984). These minor changes are in sharp contrast to the vascular dilatation and hyperplasia of chief cells which characterise a response to hypoxia of similar duration (see Chapter 9). In the spontaneously hypertensive rat almitrine administered orally for 13 weeks does not cause the two- to three-fold enlargement of the carotid bodies seen after prolonged hypoxia (Habeck et al. 1988). If rabbits are given almitrine in ascending doses of 1–9 mg/day (approximately 0.5–4.5 mg/kg per day) over a period of 5 weeks there is no difference between test and control rabbits with respect to the volume of their carotid bodies, the number of chief cells per unit area or in the relative proportions of light, dark and sustentacular cells (Smith et al. 1986). Furthermore, the qualitative appearances of the carotid bodies from the two groups are identical. The fact that so far no specific pathological change attributable to almitrine administration has been described, plus the evidence from physiological studies discussed above, suggest that almitrine and hypoxia stimulate the carotid bodies by different mechanisms. This is, however, still far from certain and further research is required into the pharmacological action of this drug. Also, pathological studies of the carotid bodies of patients who die during treatment may provide valuable clues as to the mechanism of action of the drug and might explain why some people respond well to it whereas others do not.

References

Ansquer JC (1986) A one-year double-blind placebo-controlled study of the efficacy and safety of almitrine bismesylate in hypoxic COLD patients. Monitoring progress status on 490 patients entered into the VIMS study before January 1st 1984. Eur J Respir Dis 69[suppl 146]:703–712

Arnaud F, Bertrand A, Charpin J et al. (1983) Long term almitrine bismesylate treatment in patients with chronic bronchitis and emphysema: a multicentre double-blind placebo controlled study. Eur J Respir Dis 64[suppl 126]:323–330

Barer GR, Bee D, Wach RA, Gill GW, Dhillon DP, Suggett T, Evans W (1983) Does almitrine bismesylate improve V̇/Q̇ matching? An animal study. Eur J Respir Dis 64[suppl 126]:209–214

Bee D, Emery CJ, Barer GR, Gill GW (1981) Pulmonary vascular actions of the respiratory stimulant S2620. Clin Sci 61:32P

Bee D, Gill GW, Emery CJ, Salmon GL, Evans TW, Barer GR (1983) Action of almitrine on the pulmonary vasculature in ferrets and rats. Bull Eur Physiopathol Respir 19:539–545

Bisgard GE (1981) The response of few-fiber carotid chemoreceptor preparations to almitrine in the dog. Can J Physiol Pharmacol 59:396–401

Bourgouin-Karauni D, Mercier J, Prefaut CH (1986) Long term studies on almitrine bismesylate in COPD patients. Eur J Respir Dis 69[suppl 146]:695–701

Campbell DB, Gordon B, Taylor A, Taylor D, Williams J (1983) The biodisposition of almitrine bismesylate in man: a review. Eur J Respir Dis 64[suppl 126]:337–348

Castaing Y, Manier G, Guénard H (1983) V̇A/Q̇ ratios distribution and oral almitrine bismesylate in COAD patients under mechanical ventilation: preliminary results. Eur J Respir Dis 64[suppl 126]:243–247

Chedru F, Nodzenski R, Dunand JF et al. (1985) Peripheral neuropathy during treatment with almitrine. Br Med J 290:896–897

Connaughton JJ, Morgan AD, Douglas NJ, Shapiro CN, Pauly N, Flenley DC (1983) Almitrine reduces the frequency of recurrent transient nocturnal hypoxaemia in chronic bronchitis and emphysema. Thorax 38:717 (abstract)

Connaughton JJ, Douglas NJ, Morgan AD et al. (1985) Almitrine improves oxygenation when both awake and asleep in patients with hypoxia and carbon dioxide retention caused by chronic bronchitis and emphysema. Am Rev Respir Dis 132:206–210

Daskalopoulou E, Patakas D, Tsara V, Zoglopitis F, Maniki E (1990) Comparison of almitrine bismesylate and medroxyprogesterone acetate on oxygenation during wakefulness and sleep in patients with chronic obstructive lung disease. Thorax 45:666–669

de Backer W, Vermeire P, Bogaert E, Janssens E, Van Maele R (1985) Almitrine has no effect on gas exchange after bilateral carotid body resection in severe chronic airflow obstruction. Bull Eur Physiopathol Respir 21:427–432

Dhillon DP, Barer GR (1982) Respiratory stimulation by almitrine during acute or chronic hypoxia/hypercapnia in rats. Bull Eur Physiopathol Respir 18:751–764

Dhillon DP, Barer GR, Walsh M (1984) Morphometry of the rat carotid body chronically stimulated by hypoxia +/− hypercapnia or the peripheral chemoreceptor drug almitrine: a preliminary report. In: Pallot DJ (ed) The peripheral arterial chemoreceptors. Croom Helm, Beckenham, Kent, pp 269–275

Dull WL, Polu JM, Sadoul P (1983) The pulmonary heamodynamic effects of almitrine infusion in men with chronic hypercapnia. Clin Sci 64:25–31

Evans TW, Tweney J, Waterhouse JC, Suggett AJ, Nicholl A, Howard P (1986) The interaction of long term domiciliary oxygen therapy and almitrine bismesylate in hypoxic cor pulmonale. Eur J Respir Dis 69[suppl 146]:665–669

Evans TW, Tweney J, Waterhouse JC, Nichol J, Suggett AJ, Howard P (1990) Almitrine bismesylate and oxygen therapy in hypoxic cor pulmonale. Thorax 45:16–21

Feist H (1986) The effects of almitrine at rest and on exercise in COPD patients. Eur J Respir Dis 69[suppl 146]:651–655

Fleming PJ, Levine MR, Lewis GTR, Pauly N (1983) Almitrine bismesylate in congenital central hypoventilation. Eur J Respir Dis 64[suppl 126]:307–312

Gasnault J, Moore N, Arnaud F, Rondot P (1987) Peripheral neuropathies during hypoxaemic chronic obstructive airways disease. Bull Eur Physiopathol Respir 23[suppl 11]:199–202

Gaultier C, Boule M, Mercier JC, Beaufils F (1983) Lung function after almitrine bismesylate administration in a five year old boy with chronic airways obstruction. Eur J Respir Dis 64[suppl 126]:319–321

Gautier H, Bonora M (1982) Effects of hypoxia and respiratory stimulants in conscious intact and carotid denervated cats. Bull Eur Physiopathol Respir 18:565–582

Gordon BH, Pallot DJ, Mir A, Ings RMJ, Evrard Y, Campbell DB (1987) Kinetics of almitrine bismesylate and its metabolites in the carotid body and other tissues of the rat. In: Ribeiro JA, Pallot DJ (eds) Chemoreceptors in respiratory control. Croom Helm, Beckenham, Kent, pp 394–407

Grassi C, Cerveri I, Rampulla C, Tantucci C, Grassi V (1987) Influence of the carotid body on gas exchange during exercise. Bull Eur Physiopathol Respir 23[suppl 11]:191–194

Grassi V, Tantucci C, Sorbini CA et al. (1986) A review of clinical studies carried out in Italy with almitrine bismesylate. Eur J Respir Dis 69[suppl 146]:671–676

Guillerm R, Radziszewski E (1974) Effets ventilatoires chez l'homme sain d'un nouvel analeptique respiratoire, le S2620. Bull Physiopathol Respir 10:775–791

Habeck J-O, Huckstorf C, Mewes H (1988) The carotid bodies of spontaneously hypertensive rats (SHR) after long term stimulation by almitrine bismesylate. Biomed Biochim Acta 47:543–545

Howard P (1985) Peripheral neuropathy during treatment with almitrine. Br Med J 290:1288 (letter)

Howard P, de Backer W, Vermeire P et al. (1989a) Two year follow-up of patients with hypoxic chronic obstructive airways disease (COAD) randomised between almitrine bismesylate and placebo. Am Rev Respir Dis 139 [suppl 4]:12 (abstract)

Howard P, Empey D, Harrison B et al. (1989b) Lower dose almitrine bismesylate with intermittent scheduling in hypoxaemic COAD. Am Rev Respir Dis 139[suppl 4]:10 (abstract)

Hughes JMB, Allison DJ, Goatcher A, Tripathi A (1983) Action of almitrine bismesylate on pulmonary vasculature in the dog: preliminary report. Eur J Respir Dis 64[suppl 126]:215–224

Hughes JMB, Allison DJ, Goatcher A, Tripathi A (1986) Influence of alveolar hypoxia on pulmonary vasomotor response to almitrine in the dog. Clin Sci 70:555–564

Johanson WG, Mullins RC, Bell RC, West LG, Bachand RT (1986) The efficacy of almitrine in improvement of hypoxaemia in patients with COPD. Eur J Respir Dis 69[suppl 146]:663

Lagneaux D (1987) Peripheral chemoreceptor stimulation and pulmonary circulation in the rat. In: Ribeiro JA, Pallot DJ (eds) Chemoreceptors in respiratory control. Croom Helm, Beckenham, Kent, pp 351–360

Lambropoulos S, Chatzipappas A, Tsekos G, Tsantoulis K (1986) The role of almitrine bismesylate in acute respiratory failure. Eur J Respir Dis 69[suppl 146]:657–661

Laubie M, Diot F (1972) Étude pharmacologique de l'action stimulante respiratoire du S2620. Rôle des chémorecepteurs carotidiens et aortiques. J Pharmacol (Paris) 3:363–374

Laubie M, Schmitt H (1980) Long-lasting hyperventilation induced by almitrine: evidence for a specific effect on carotid and thoracic chemoreceptors. Eur J Pharmacol 61:125–136

Lerebours G, Ozenne G, Senant J, Moore N, David PH, Nouvet G (1986) Abnormalities in the electrophysiological examination of patients with chronic obstructive lung disease: a preliminary study of the effects of almitrine. Eur J Respir Dis 69[suppl 146]:713–714

Lerebours G, Senant J, Moore N, et al. (1987) Evolution of peripheral nerve function in hypoxaemic COPD patients taking almitrine bismesylate: a prospective long-term study. Bull Eur Physiopathol Respir 23[suppl 11]:203–206

Magnussen H, Radenbach D, Kiwull-Schöne H (1987) The acute effect of a single oral dose of 200 mg almitrine on gas exchange in patients with chronic obstructive bronchitis and emphysema, bronchial asthma and lung fibrosis. Bull Eur Physiopathol Respir 23[suppl 11]:211–216

Marmo E (1986) The pharmacological properties of almitrine bismesylate. Eur J Respir Dis 69[supp 146]:619–632

Marrone O, Milone F, Coppola P et al. (1986) Effects of almitrine bismesylate on nocturnal hypoxaemia in patients with chronic bronchitis and obesity. Eur J Respir. Dis 69[suppl 146]:641–648

Marsac J (1986) The assessment of almitrine bismesylate in the long-term treatment of chronic obstructive bronchitis. Eur J Respir Dis 69[suppl 146]:685–693

Maxwell DL, Cover D, Hughes JMB (1985) Almitrine increases the steady-state hypoxic ventilatory response in hypoxic chronic air-flow obstruction. Am Rev Respir Dis 132:1233–1237

Mazmanian G, Lockhart A (1982) Almitrine reduces hypoxic pulmonary vasoconstriction in isolated rat lungs. Am Rev Respir Dis 125:270 (abstract)

Melot C, Naeije R, Rothchild T, Mertens B, Mols P, Hallimans R (1983) Improvement in ventilation–perfusion matching by almitrine in COPD. Chest 83:528–533

Naeije R, Melot C, Mols P et al. (1981) Effects of almitrine on decompensated chronic respiratory insufficiency. Bull Eur Physiopathol Respir 17:153–161

Neukirch F, Castillon du Perron M, Verdier F et al. (1974) Action d'un stimulante ventilatoire (S2620), administré oralement, dans les bronchopneumopathies obstructives (Étude statistique). Bull Physiopathol Respir 10:793–800

O'Regan RG, Kennedy M, Przybyszewski AW (1987) Carotid body responses to administration of almitrine bismesylate. In: Ribeiro JA, Pallot DJ (eds) Chemoreceptors in respiratory control. Croom Helm, Beckenham, Kent, pp 386–393

Pallot DJ, Al Neamy KW (1983) The effect of hypoxia, hypercapnia and almitrine bismesylate on carotid body catecholamines. Eur J Respir Dis 64[suppl 126]:203–207

Paramelle B, Levy P, Pirotte C (1983) Long term follow-up of pulmonary arterial pressure evolution in COPD patients treated by almitrine bismesylate. Eur J Respir Dis 64[suppl 126]:333–336

Paramelle B, Vila A, Stoebner P et al. (1986) Peripheral neuropathies and chronic hypoxaemia in chronic obstructive lung disease. Eur J Respir Dis 69[suppl 146]:715

Pequignot JM, Tavitian E, Boudet C, Peyrin L (1987) Reduction in dopaminergic activity in the rat carotid body after acute or chronic almitrine. In: Ribeiro JA, Pallot DJ (eds) Chemoreceptors in respiratory control. Croom Helm, Beckenham, Kent, pp 378–385

Powles ACP, Tuxen DV, Mahood CB, Pugsley SO, Campbell EJM (1983) The effect of intravenously administered almitrine, a peripheral chemoreceptor agonist, on patients with chronic air-flow obstruction. Am Rev Respir Dis 127:284–289

Racineux JL, Meslier N, Aubert P (1987) The effect of long-term almitrine therapy on sleep hypoxaemia in patients with chronic airway obstruction. Bull Eur Physiopathol Respir 23[suppl 11]:183–184

Rigaud D, Dubois F, Godart J, Paramelle B (1981) Changes in pulmonary perfusion induced by almitrine in COPD patients. Am Rev Respir Dis 123:112 (abstract)

Romaldini H, Rodriguez-Roisin R, Wagner PD, West JB (1983) Enhancement of hypoxic pulmonary vasoconstriction by almitrine in the dog. Am Rev Respir Dis 128:228–293

Roumy M, Leitner LM (1981) Stimulant effect of almitrine (S2620) on the rabbit carotid chemoreceptor afferent activity. Bull Eur Physiopathol Respir 17:255–259

Schmoller T, Schumacker PT, Wagner PD, West JB (1983) Effects of almitrine on the distribution of pulmonary blood flow in dogs with hypoxic and hyperoxic lobes. Am Rev Respir Dis 127:245 (abstract)

Schrijen F, Romero Colomer P (1978) Effets hémodynamiques d'un stimulant ventilatoire (Almitrine) chez des pulmonaires chroniques. Bull Eur Physiopathol Respir 14:775–784

Sergysels R, Etien J-L, Schandevyl W et al. (1978) Functional evaluation of a new analeptic drug (Vectarion) in patients with chronic hypoxaemia and hypercapnia (preliminary study). Acta Tuberc Pneumol Belg 69:237–245

Shannon DC, Sullivan K, Perret L, Kelly DH (1983) Use of almitrine bismesylate to stimulate ventilation in congenital central hypoventilation. Eur J Respir Dis 64[suppl 126]:295–301

Simonneau G, Denjean A, Raffestin B et al. (1981) Improved pulmonary gas exchange by almitrine in chronic airway obstruction; stimulation of ventilation vs changes in $\dot{V}A/\dot{Q}$ ratios. Am Rev Respir Dis 123:88 (abstract)

Smith P, Heath D, Fitch R, Hurst G, Moore D, Weitzenblum E (1986) Effects on the rabbit carotid body of stimulation by almitrine, natural high altitude, and experimental normobaric hypoxia. J Pathol 149:143–153

Stanley NN (1981) Almitrine a promising new drug for sleep hypoxia. Lancet 14:1105–1106

Stanley NN (1986) Effect of almitrine bismesylate 100 mg daily for four weeks on respiratory regulation in patients with chronic bronchitis. Eur J Respir Dis 69[suppl 146]:635–639

Stanley NN, Galloway JM, Gordon B, Pauly N (1983a) Increased respiratory chemosensitivity induced by infusing almitrine intravenously in healthy man. Thorax 38:200–204

Stanley NN, Galloway JM, Flint KC, Campbell DB (1983b) Increased respiratory chemosensitivity induced by oral almitrine in healthy man. Br J Dis Chest 77:136–146

Stanley NN, Pieczora JA, Pauly N (1983c) Effects of almitrine bismesylate on chemosensitivity in patients with chronic airways obstruction. Eur J Respir Dis 64[suppl 126]:233–237

Stradling JR, Barnes P, Pride NB (1982) The effects of almitrine on the ventilatory response to hypoxia and hypercapnia in normal subjects. Clin Sci 63:401–404

Stradling JR, Nicholl CG, Cover D, Davies EE, Hughes JMB, Pride NB (1983) Pattern of breathing and gas exchange following oral almitrine bismesylate in patients with chronic obstructive pulmonary disease. Eur J Respir Dis 64[suppl 126]:255–264

Stradling JR, Nicholl CG, Cover D, Davies EE, Hughes JMB, Pride NB (1984) The effects of oral almitrine on pattern of breathing and gas exchange in patients with chronic obstructive pulmonary disease. Clin Sci 66:435–442

Suggett AJ, Proctor A, Smyllie H, Peake MD, Cayton RM, Howard P (1986) Prediction of which hypoxaemic patients will respond to almitrine. Am Rev Respir Dis 133:339 (abstract)

Tweney J (1987) Almitrine bismesylate: current status. Bull Eur Physiopathol Respir 23[suppl 11]:153–163

Voisin C, Howard P, Ansquer JC (1987) Almitrine bismesylate: a long-term placebo-controlled double-blind study in COAD – Vectarion international multicentre study group. Bull Eur Physiopathol Respir 23[suppl 11]:169–182

Vulterini S, Dardes N, Moreo GC, Chianese R, Maiorano V, Camimeo N (1986) Almitrine bismesylate and hypoxaemia in COPD patients (results of an Italian multicentric trial). Eur J Respir Dis 69[suppl 146]:677–683

Wach RA, Gill GW, Suggett AJ, Bee D, Barer GR (1986) Action of almitrine bismesylate on ventilation–perfusion matching in cats and dogs with part of the lung hypoventilated. Clin Exp Pharmacol Physiol 13:453–467

Wach RA, Bee D, Barer GR (1989) Dopamine and ventilatory effects of hypoxia and almitrine in chronically hypoxic rats. J Appl Physiol 67:186–192

Watanabe S, Kanner RE, Cutillo AG et al. (1989) Long-term effect of almitrine bismesylate in patients with hypoxemic chronic obstructive pulmonary disease. Am Rev Respir Dis 140:1269–1273

Weitzenblum E, Ehrhart M, Schneider JC, Hirth C, Roegel E (1982) Effets hémodynamiques pulmonaires de l'almitrine intraveineuse chez les bronchiteux chroniques insuffisants respiratoires. Bull Eur Physiopathol Respir 18:765–774

Yernault JC, Van Muylem A, Noseda A, Ravez P, Paiva M (1983) Effect of almitrine bismesylate on the distribution of ventilation and mechanics of breathing in patients with COLD. Eur J Respir Dis 64[suppl 126]:265–270

14   The Carotid Bodies in Systemic Hypertension

In this chapter the interesting association between systemic hypertension and carotid body hyperplasia is explored.

Increase in Weight

The first intimation that an elevated blood pressure in the systemic circulation may affect the carotid body in humans was signalled when a relation was found between left ventricular weight and the weight of the carotid bodies in 40 successive subjects coming to necropsy (Heath et al. 1970). Several of these cases were patients who had been hypertensive in life. There proved to be a positive statistical correlation between the weights of the isolated left ventricle and of the carotid bodies. Subsequently a detailed study was made of 15 cases of systemic hypertension (Edwards et al. 1971) in all of which a left ventricular weight exceeding 200 g was found by the method of Fulton et al. (1952) and indicated unequivocal left ventricular hypertrophy. The combined weights of the carotid bodies ranged from 17.6 to 80.8 mg, with a mean of 38.1 mg. Ten cases had a combined weight in excess of 30 mg, which we regard as the upper limit of normal (see Chapter 2). In a series of 100 consecutive necropsies studied some years later (Smith et al. 1982), a highly significant statistical correlation was found between left ventricular weight and combined carotid body weight. Thirty-six of these cases had a history of systemic hypertension and all but two had left ventricular weights exceeding 200 g. The mean combined carotid body weight of the 36 hypertensives was significantly greater than that of the remaining 64 cases but there was a considerable overlap of the 95% probability limits between the two groups. The increase in weight of the glomus appears to be related directly to the elevated systemic blood pressure and not to increased left ventricular weight per se because in cases of hypertrophic cardiomyopathy or aortic stenosis, where left ventricular hypertrophy occurs with a normal systemic blood pressure, the carotid bodies are not enlarged (Smith et al. 1982).

When systemic hypertension and pulmonary emphysema are present together it might be anticipated that there would be an additive effect of these stimuli on the size of the carotid bodies. However, it is not established that this does in fact occur, since in five cases with this combination the mean combined carotid body weight was 51.8 mg, which was similar to that in cases with pulmonary emphysema alone (Edwards et al. 1971). In two further examples the carotid body weights were similar to those with isolated systemic hypertension (Smith et al. 1982). On the other hand, Habeck (1986) found that the carotid bodies of three of five cases of severe lung disease coexisting with essential hypertension were significantly bigger than those in patients with either of these diseases alone.

Histology

The histological appearance of the carotid bodies in systemic hypertension is indistinguishable from that found in emphysema. The glomic lobules are enlarged, with a reduction in the quantity of stromal connective tissue, suggesting hyperplasia of glomic tissue. The lobules typically show sustentacular cell hyperplasia (see Chapter 10) in which there are numerous elongated sustentacular and Schwann cells forming broad whorls around diminished cores of chief cells. The sustentacular cell count averages 54%, and when it is plotted against lobule diameter there is a strongly positive correlation (Smith et al. 1982). The numerous elongated cells are associated with a profusion of nerve axons (Fitch et al. 1985).

A somewhat different account of the histology of the carotid bodies in 27 cases of systemic hypertension, 17 being essential and 10 renal, was provided by Habeck (1986). He found only insignificant differences in the histology of carotid bodies from hypertensives and controls. However, in 7 of his 27 cases he found a proliferation of what he termed "spindle-shaped cells", which was defined by him as including both sustentacular and Schwann cells. This proliferation tended to be more pronounced in elderly patients with essential hypertension. Hyperplasia of the carotid bodies involving sustentacular cells was found in only two of ten cases with renal hypertension. These appearances were in sharp contrast to those in nine patients with "chronic lung disease" comprising four with severe chronic obstructive lung disease, one with cystic fibrosis, three with interstitial pulmonary fibrosis, and one with anthracosilicosis. Eight of these subjects showed extensive proliferation of sustentacular cells in the carotid body, the exception being a boy of 15 years with cystic fibrosis (Habeck 1986). There is no doubt that the hypertensive subjects studied by Habeck (1986) had significant elevations of intravascular pressure since systolic and diastolic blood pressures as high as 220/150 mmHg are quoted. However, there is no way of assessing the duration of the hypertension in these cases as left ventricular weights are not provided. In our experience a clinical history of systemic hypertension alone does not necessarily mean that the carotid bodies will be hyperplastic in that case. Evidence of sustained systemic hypertension in the form of a

left ventricle weighing more than 200 g is needed to signify unequivocal left ventricular hypertrophy (Smith et al. 1982).

Coarctation of the Aorta

In this congenital anomaly the carotid bodies are exposed to an elevated systemic blood pressure. In the case of a man of 61 years (Heath et al. 1986), there was severe left ventricular hypertrophy (361 g) secondary to a coarctation distal to the origin of the left subclavian artery where the diameter of the lumen was reduced to only 0.4 cm. Both carotid bodies were enlarged, with a combined weight of 68.6 mg. The maximum area of cross-section of both was enlarged when compared with eight pairs of controls and the area occupied by glomic tissue was strikingly increased from 1.02 mm^2 to 2.91 mm^2. The area of the largest lobule was also considerably greater than

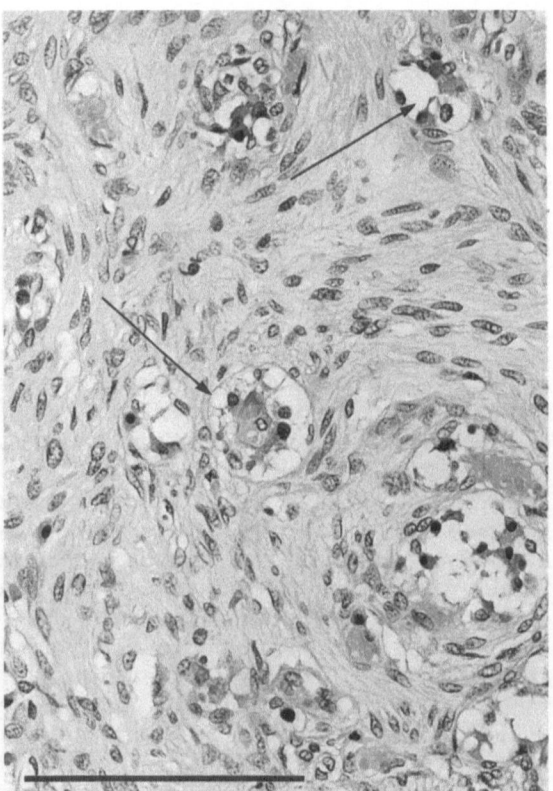

Fig. 14.1. Carotid body of a 61-year-old man with coarctation of the aorta. Cell clusters are widely separated by a profusion of elongated cells forming palisaded sheets. Cell clusters (*arrows*) are atrophic and contain chief cells with highly vacuolated cytoplasm and dark nuclei. (Haematoxylin–eosin (HE)) Scale line = 100μm.

in any of the control cases and was due to a profuse proliferation of elongated cells (Fig. 14.1). These were in dense masses which not only produced concentric whorls around the cores of chief cells but also formed large sheets of palisaded cells (Fig. 14.1). The elongated cells comprising the proliferation contained large, plump nuclei with a pale, lightly stippled or vesicular chromatin pattern and an eosinophilic indistinct cytoplasm and it is likely that many of them were Schwann cells (Fig. 14.2). A differential cell count of these elongated cells revealed that they accounted for 73% of the total. They were associated with large numbers of axons which took an undulating path amongst them. Some of these ended in contact with a sustentacular cell (Fig. 14.3).

Chief cells appeared to have been compressed by the proliferating sustentacular and Schwann cells, for the size of the central cores in clusters was much reduced, sometimes to isolated groups of three or four cells (Fig. 14.1). Many of them had darkly staining nuclei with compact chromatin. The cytoplasm was also often darkly stained and highly vacuolated but this appearance may have been, at least partly, the consequence of autolysis (Heath et al. 1986). Thus, although the glomic tissue in this case showed features of sustentacular cell hyperplasia, this was more extreme than in cases of systemic hypertension, with a greater degree of atrophy of chief cells. This may be a reflection of the greatly elevated blood pressure in the carotid and glomic arteries, which would be anticipated in coarctation of the aorta.

Origin of Carotid Body Hyperplasia in Systemic Hypertension

It is difficult to explain why systemic hypertension should promote hyperplasia of the carotid bodies, since they are generally considered to respond to the stimulus of hypoxaemia rather than systemic blood pressure. A possible explanation might lie in the accentuated development of atherosclerosis in carotid arteries, the formation of occlusive intimal fibrosis in glomic arteries, and arterioles and ischaemia of the tissues of the carotid body. Such ischaemia might stimulate the glomic tissue to undergo hyperplasia just as it responds to the hypobaric hypoxia of high altitude or the alveolar hypoxia of pulmonary emphysema. We have found extensive intimal fibroelastosis in the interlobular glomic arteries in systemic hypertension and similar lesions have been referred to by Habeck et al. (1983). They studied 13 cases of raised systemic blood pressure and found thicken-

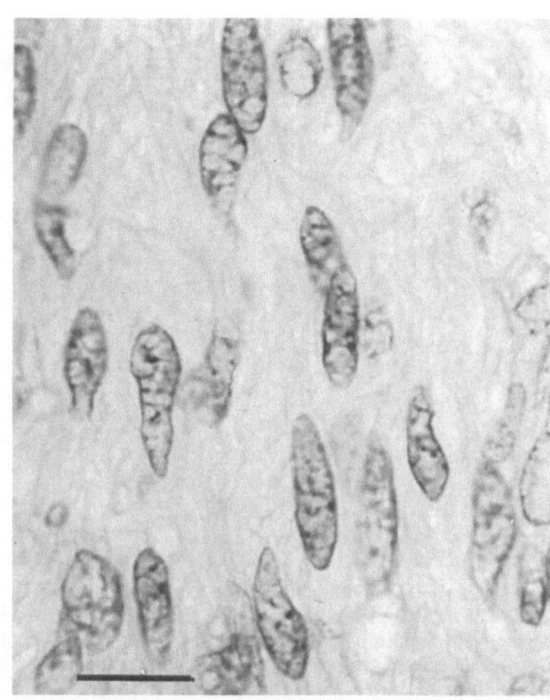

Fig. 14.2. Detail of the elongated cells from the same case as Fig. 14.1. Their nuclei are plump with a stippled or vesicular chromatin pattern. Many of them are probably Schwann cells. (HE) Scale line = 10 μm.

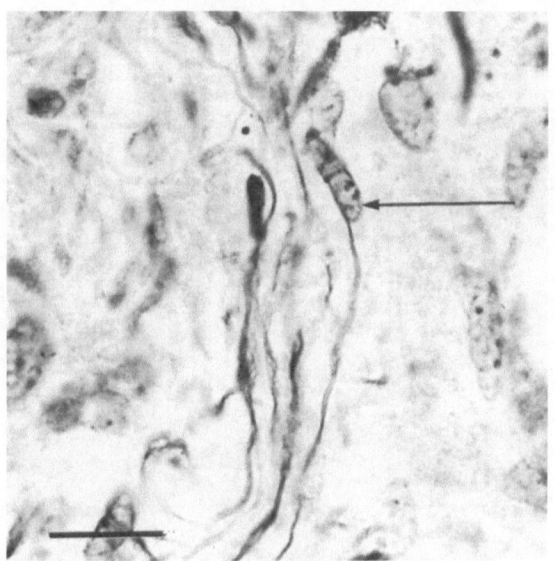

Fig. 14.3. Section of carotid body from a man of 61 years with coarctation of the aorta. Nerve axons run through the mass of proliferated sustentacular cells and one of them ends in contact with a sustentacular cell (*arrow*). (Bodian silver) Scale line = 10 μm.

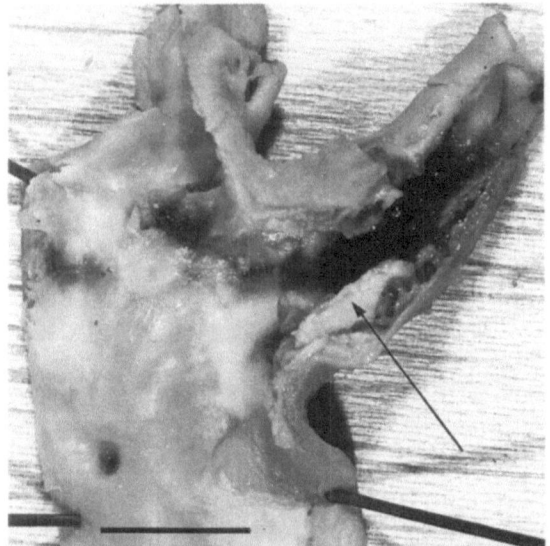

Fig. 14.4. Left carotid bifurcation from a man of 97 years. The vessel has been opened to show an atheromatous plaque (*arrow*) with superimposed thrombosis occluding the origin of the internal carotid artery. The thrombus blocks the entrance to the left glomic artery. Scale line = 1 cm.

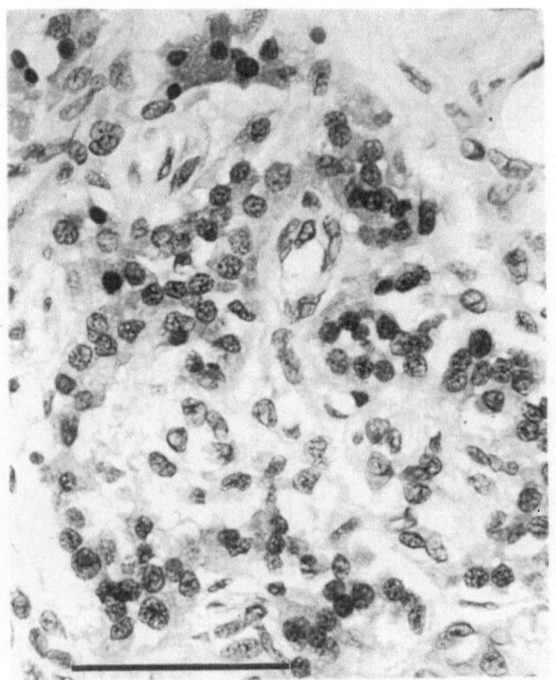

Fig. 14.5. Glomic lobule from the left carotid body from the case in Fig. 14.4. There is a prominence of the dark variant of chief cell which in an elderly patient suggests an early cellular response to hypoxaemia. (HE) Scale line = 50 μm.

ing of the media by eosinophilic hyaline material and partial or complete occlusion of the lumen by concentric intimal fibro-elastosis in the smaller interlobular glomic arteries and intralobular glomic arterioles. In three cases there was fibrinoid necrosis. However, not all cases of carotid body hyperplasia associated with systemic hypertension show such vascular lesions. In the case of coarctation of the aorta referred to above, therefore, the glomic arteries showed but slight intimal fibrosis and were widely patent and hardly likely to induce ischaemia (Heath et al. 1986). Vascular obstruction may occur more proximally in the main glomic artery or in the major carotid vessel giving rise to it. The carotid sinus in particular is commonly the site of severe atherosclerosis (see Chapter 20) and in some cases this could obstruct the flow of blood into the glomic artery, leading to ischaemia of the carotid body. We have encountered such a situation in a man of 97 years in whom there was no systemic hypertension but almost total occlusion of the left carotid bifurcation and external carotid artery by atherosclerosis with superimposed thrombosis (Fig. 14.4). The origin of the left glomic artery was obstructed but that on the right side was patent. The left carotid body was 22.1 mg in weight but the right weighed only 11.3 mg. Histologically, the left carotid body contained more lobules than the right; they were small, with indistinct borders, and their clusters contained foci of dark cell proliferation (Fig. 14.5). These features are suggestive of an early response of the carotid body to hypoxaemia (see Chapter 9).

The concept that ischaemia of glomic tissue causes hyperplasia of the carotid bodies in systemic hypertension is attractive, since it would allow all causes of carotid body enlargement to be united by the same mechanism of hypoxaemia. It is, nevertheless, not proven and the possibility that other factors are responsible has to be considered. For example there is experimental evidence that the carotid bodies are involved in sodium and water regulation. The tendency to retain sodium and water in patients with pulmonary emphysema showing the blue-and-bloated syndrome and in systemic hypertension may lead to an increased need for a natriuretic factor thought to be derived from the carotid bodies, perhaps causing them to become hyperplastic (see Chapter 15).

Experimental Systemic Hypertension

The relationship between carotid body hyperplasia and systemic hypertension has been studied

extensively in animals. Most of these investigations have used genetically hypertensive rats
such as the Okamoto-Aoki, New Zealand, Kyoto
or Milan hypertensive strains. These animals all
develop a severe elevation of systemic blood
pressure when they are a few weeks old and may
ultimately die from left ventricular failure or
stroke. For example, in Okamoto-Aoki rats an
elevation of systolic blood pressure is detectable
by 6 weeks of age and reaches a maximum of
between 180 and 220 mmHg by 10 weeks (Wong et
al. 1982), and mean blood pressures as high as
160 mmHg have been recorded (Habeck et al.
1985). This high systemic blood pressure is associated with hypertrophy of the left ventricle and an
increased medial thickness of the aorta (Smith et
al. 1984).

The carotid bodies of Okamoto-Aoki rats are
enlarged in comparison to those of normotensive
Wistar rats (Habeck et al. 1981, 1984a; Smith et al.
1984). This difference in size is detectable by 5–
6 weeks of age and becomes increasingly great up
to an age of 30–40 weeks (Habeck et al. 1984a).
When the volume of the carotid body is expressed
in terms of body weight, its size in Okamoto-Aoki
rats is consistently three times that of the Wistar
strain (Smith et al. 1984). Although the carotid
body is much bigger, its histological appearance is
similar to that of controls, consisting of round or
oval clusters of chief cells surrounded by thin rims
of sustentacular cells (Fig. 14.6). There is no
vascular engorgement such as typifies enlargement
of the carotid body in experimental hypoxia (see
Chapter 9), neither is there a proliferation of
sustentacular cells such as occurs in humans.
Differential cell counts reveal that the relative
proportions of chief and sustentacular cells remain
the same as normal (Smith et al. 1984). It has been
estimated that there are on average 18 000 chief
cells in the carotid bodies of Okamoto rats compared with only 8000 in controls, suggesting extensive hyperplasia in the former (Smith et al. 1984).
Chief cells in the rat are not characterised by three
distinct variants as in humans, but comprise a
homogeneous population of cells with slightly
ovoid, moderately haematoxyphilic, nuclei and
palely eosinophilic, finely vacuolated, cytoplasm
(Fig. 14.6; see Chapter 21). These differences
between rats and man in the reaction of the carotid
body to systemic hypertension suggest that any
comparisons of physiological responses between
the two species must be made with caution.

The glomic arteries of spontaneously hypertensive rats show striking changes which are progressive with age. As early as 13 weeks of age the small
muscular glomic arterioles show an intimal proli-

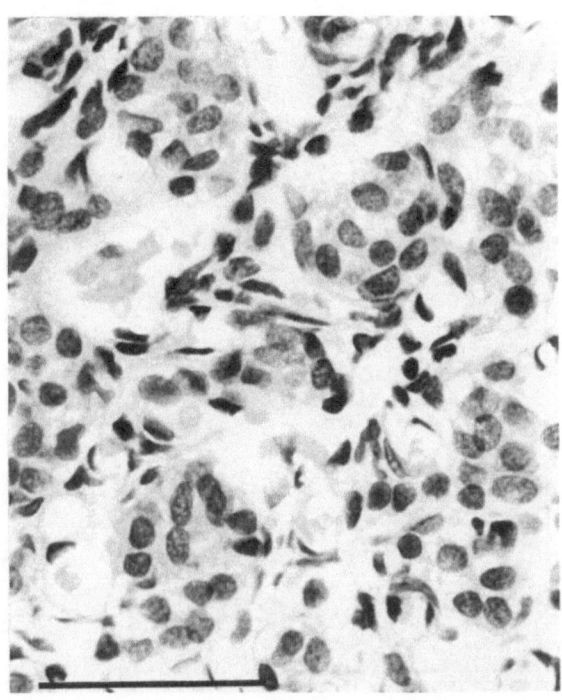

Fig. 14.6. Carotid body from a spontaneously hypertensive rat
of the Okamoto-Aoki strain. Chief cells have slightly oval
nuclei of uniform appearance. They are arranged in clusters
which are associated with smaller numbers of elongated sustentacular cells. Note that there is neither vascular engorgement
nor sustentacular cell hyperplasia. (HE) Scale line = 50 μm.

feration consisting of dark, elongated cells embedded within a loose matrix. These changes become
increasingly extensive as the rats age (Fig. 14.7).
Initially these lesions were interpreted as consisting of large cells and copious light cytoplasm
(Habeck et al. 1981; Honig et al. 1981), but it was
later recognised that what appears to be copious
cytoplasm is actually mucopolysaccharide ground
substance and that the dark cells are derived from
smooth muscle (Smith et al. 1984; Habeck and
Holzhausen 1985; Habeck et al. 1985). Electron
microscopy shows that these cells contain rough
endoplasmic reticulum resembling that of fibroblasts with peripheral myofilaments inserted into
dense attachment points similar to smooth muscle.
They are therefore probably myofibroblasts which
orginate from smooth muscle cells in the media of
the glomic arteries (Smith et al. 1984; Habeck and
Holzhausen 1985). They are surrounded by a
loose matrix of ground substance resembling multiple basal laminae arranged in parallel rows or as
whorled structures which are presumably secreted
by the myofibroblasts. Endothelial cells are prominent with convoluted inner margins and contain
an abundance of free ribosomes and mitochon-

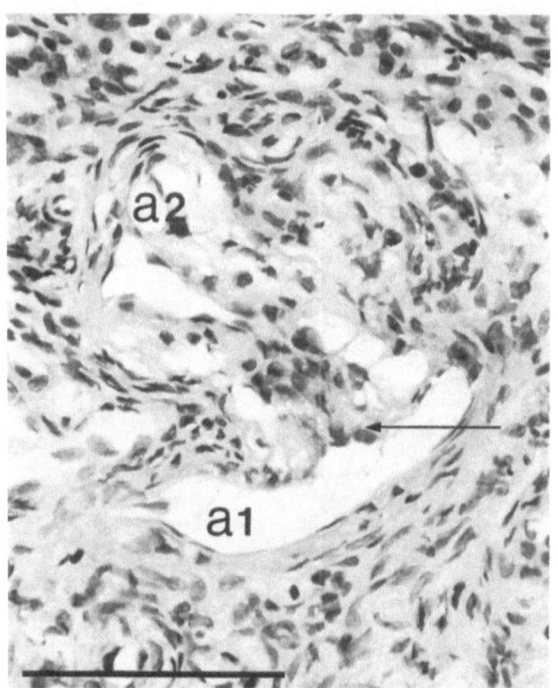

Fig. 14.7. Intimal proliferation in the glomic arteries of an Okamoto-Aoki rat. A first-order branch of the main glomic artery (*a1*) gives rise to a second-order branch (*a2*) which is dilated with an attenuated media. At the origin of the branch is an enlarged sphincter (*arrow*) from which a mixture of irregular, elongated cells and mucopolysaccharide ground substance stream into the branch, virtually obliterating its lumen. (HE) Scale line = 100 μm.

dria. The intimal proliferation typically adopts the form of broad cushions at the origin of second-order branches of the glomic artery, where they reduce the lumen to a narrow slit or may even occlude it (Fig. 14.7). They appear to originate from small sphincters which are situated at the orifices of branches of the glomic arterioles. With increasing age intimal cushions involve larger vessels so that by 41 weeks of age many first-order branches of the glomic artery are involved and in a few animals even the main glomic artery itself (Smith et al. 1984). There are usually associated changes in the media of the affected vessels with fraying of elastic laminae and dilatation of a segment of artery, usually near to its origin (Fig. 14.7). It is assumed that the raised systemic pressure dilates and damages glomic arteries and arterioles and the intimal proliferation develops in response to this damage (Smith et al. 1984; Habeck et al. 1985). The effects of systemic hypertension on the glomic vasculature may be exaggerated in Okamoto-Aoki rats since, in this strain, protective sphincters at the origin of the main glomic artery from the external carotid

artery are reduced in size (Habeck et al. 1981, 1984b; Smith et al. 1984).

The occlusion of glomic arterioles in spontaneously hypertensive rats may be analogous to the intimal fibrosis which occludes the glomic vasculature in human systemic hypertension. Indeed it is frequently cited as evidence to support the theory that vascular occlusion leads to carotid body hyperplasia. Such comparisons should not be accepted uncritically, since spontaneously hypertensive rats have been bred specially from normotensive stock and as such may show genetic differences in carotid body size. Such a possibility was originally discounted on the grounds that an elevation of blood pressure preceded, and hence caused, enlargement of the carotid body (Pfeiffer et al. 1984). However, when different strains of spontaneously hypertensive rats are compared with normotensive individuals of the same genetic stock, carotid body volumes are similar, despite intimal thickening of glomic arteries in the hypertensive animals (Barer et al. 1987). Also, when hypertension is induced in normal rats by constricting the renal artery, massive changes occur in the glomic arteries including fibrinoid necrosis and perivascular granulomas, but their carotid body volume is, nevertheless, normal (Habeck et al. 1987a). The possible genetic nature of the enlarged carotid bodies was elegantly demonstrated by Habeck et al. (1987b) by cross-breeding spontaneously hypertensive rats with a normotensive strain. In the first generation of offspring the carotid bodies were small with few arterial thickenings and yet the systemic blood pressure was high. When these animals were interbred to produce an F_2 generation the carotid bodies were twice the normal size with extensive vascular intimal thickening, and yet the blood pressure was lower than in the F_1 generation. It was concluded that the carotid body enlargement may not be related to blood pressure or vascular occlusion but may be genetically determined. This conclusion was supported by the observation that, although administration of the antihypertensive drug propranolol to Okamoto-Aoki rats reduces both the blood pressure and the number of vascular intimal cushions, it does not influence the volume of the carotid bodies (Habeck and Huckstorf 1987).

Regardless of whether the large carotid bodies of spontaneously hypertensive rats are genetic or acquired, they are apparently more active than in normotensive strains. Thus, Okamoto-Aoki rats have a raised partial pressure of oxygen and a lowered partial pressure of carbon dioxide in their systemic arterial blood secondary to alveolar hyperventilation (Przybylski 1978). A respiratory alkalosis is consistently found in spontaneously hypertensive rats (Honig et al. 1981; Pfeiffer et al. 1984). This increase in minute ventilation is apparently dependent more on the level of systemic blood pressure than on the size of the carotid bodies (Barer et al. 1987). Thus systemic hypertension increases the sensitivity of the carotid bodies in the rat at least, although this effect appears to be minimal in the hypertensive rabbit (Angell-James et al. 1982).

Chemoreceptor Activity in Humans

There is evidence that the chemoreceptor drive to ventilation is increased in systemic hypertension in humans (Trzebski et al. 1982; Quies et al. 1983). The studies on which this conclusion is based were performed on young male students with mild, early systemic hypertension. In the first investigation, 20 normotensive students were compared with 20 hypertensive volunteers between the ages of 20 and 26 years. Mean systolic and diastolic blood pressures in the normotensive control group were 115/76 mmHg, whereas these values were 141/98 mmHg in the group with early hypertension (Trzebski et al. 1982). The effects of isocarbic hypoxia and hyperoxic hypercarbia were studied using a rebreathing apparatus. Progressive hypoxia was associated with an increase in minute ventilation in both groups, but the hypertensive subjects showed an exaggerated response. It was noted that the controls were able to rebreathe for longer and to attain a greater degree of hypoxia than were the hypertensive subjects, who felt discomfort earlier, compelling them to remove the mouthpiece. Only three hypertensive subjects were able to reach the same level of hypoxia as that achieved by the controls. This augmentation of the hypoxic ventilatory response resulted mainly in an increase in tidal volume. The force generated by the respiratory muscles was also measured by occluding the mouthpiece for 0.2 second during inspiration. The occlusion pressure thus produced in the hypertensive group was nearly twice that of the controls, demonstrating a heightened ventilatory drive. It was estimated that in hypertensive subjects the hypoxic ventilatory drive was increased four-fold and that this could not be due simply to hyperreactivity of respiratory muscles as hypercarbia induced a negligible effect.

Similar studies were performed on 32 male medical students, 12 of whom showed a mild elevation of systemic blood pressure (Quies et al. 1983). Systolic and diastolic blood pressures in the control group were 115/66 mmHg and in the hypertensive group 142/78 mmHg. Even whilst breathing normal air, the hypertensive students exhibited a significantly higher minute ventilation of 8.3 l/min as compared with 7.0 l/min in the controls. When breathing a hypoxic gas mixture of 11% oxygen, both groups showed an increase in minute ventilation but, unlike the findings of Trzebski et al. (1982), this was less pronounced in the hypertensive group so that the difference between the two diminished. On the other hand, pure oxygen led to a depression of ventilation which was more pronounced in hypertensives so that the difference between the two groups was again reduced. In other words the greatest difference in ventilatory drive between hypertensive and normotensive subjects occurs when they are breathing normal air. In order to explain these findings Quies et al. (1983) assumed that the cause of the stimulation of the arterial chemoreceptors in systemic hypertension is ischaemia secondary to occlusion of the glomic vasculature. They postulated that even when normal air is inspired, the carotid bodies are stimulated by this ischaemia leading to an increased resting ventilatory drive. However, ischaemia of the glomic tissue has little influence upon the activity of the chemoreceptors when patients are breathing 100% or 11% oxygen, since the chemoreceptors can not be more active than maximally nor less active than minimally. In effect, the response curve of chemoreceptor activity is displaced to the right in systemic hypertension. Quies et al. acknowledged that the above scheme was speculative and that it depended entirely upon a state of ischaemia being produced in the glomic tissue. It is, nevertheless, questionable whether intimal fibrosis of glomic arteries is likely to have developed in youthful individuals with such a mild elevation of systemic blood pressure. It may be that the carotid bodies respond to changes in intravascular pressure directly through the intermediary of baroreceptors in the dilated portions of the glomic arteries (see Chapter 16).

It is not known whether the increased ventilatory drive in young subjects with early systemic hypertension is associated with structural changes in their carotid bodies. Considering the youth of the patients one might predict that, if hyperplasia of their carotid bodies were present, it would involve the chief cells, but it seems unlikely that any significant change in the glomic arteries or parenchyma could have developed as a consequence of such mild elevations of systemic arterial blood pressure. Flenley (1986) has shown a sevenfold variability in the sensitivity of the hypoxic drive, believed to be mediated by the carotid bodies in healthy subjects. It may be that the differences in ventilatory drive in the experiments described above are simply a reflection of this wide variation and that individuals with moderately elevated systemic blood pressure have a drive in the upper range of normal. Physiological data on ventilatory drive are lacking from elderly patients with established, severe systemic hypertension and left ventricular hypertrophy. It is known that their carotid bodies are often enlarged and show sustentacular cell hyperplasia. By analogy with patients with chronic obstructive lung disease, whose carotid bodies are identical in appearance, a blunted hypoxic ventilatory drive might be predicted.

References

Angell-James JE, Clarke JA, de Burgh Daly M, Taton A (1982) Carotid body chemoreceptor function and structure in experimental renal hypertensive rabbits. J Physiol (Lond) 326:30P

Barer G, Bee D, Pallot D, Jones S (1987) Size of the carotid body and ventilatory responses to hypoxia in genetically hypertensive rats. Biomed Biochim Acta 46:919–923

Edwards C, Heath D, Harris P (1971) The carotid body in emphysema and left ventricular hypertrophy. J Pathol 104:1–13

Fitch R, Smith P, Heath D (1985) A quantitative study of nerve axons in carotid body hyperplasia. Arch Pathol Lab Med 109:234–237

Flenley D (1986) Long-term oxygen therapy and the pulmonary circulation. In: Heath D (ed). Aspects of hypoxia. Liverpool University Press, Liverpool, pp 45–59

Fulton RM, Hutchinson EC, Jones AM (1952) Ventricular weight in cardiac hypertrophy. Br Heart J 14:413–420

Habeck J-O (1986) Morphological findings at the carotid bodies of humans suffering from different types of systemic hypertension or severe lung diseases. Anat Anz 162:17–27

Habeck J-O, Holzhausen HJ (1985) An ultrastructural study of the vascular alterations within the carotid bodies of spontaneously hypertensive rats (SHR). Exp Pathol 27:195–200

Habeck J-O, Huckstorf C (1987) The carotid bodies of spontaneously hypertensive rats after long-term antihypertensive treatment with propranolol. Biomed Biochim Acta 46:915–917

Habeck J-O, Honig A, Pfeiffer C, Schmidt M (1981) The carotid bodies in spontaneously hypertensive (SHR) and normotensive rats – a study concerning size, location and blood supply. Anat Anz 150:374–384

Habeck J-O, Waller H, Protze J (1983) Pathological alterations of the arterial vessels of the carotid bodies in hypertensive humans. Dtsch Gesundheitswes 38:1970–1972

Habeck J-O, Huckstorf C, Honig A (1984a) Influence of age on position, shape and size of the carotid bodies in spontaneously hypertensive (SHR) and normotensive (NCR) rats. Anat Anz 157:351–363

Habeck J-O, Huckstorf C, Honig A (1984b) Influence of age on the carotid bodies of spontaneously hypertensive (SHR) and normotensive rats. I. Arterial blood supply. Exp Pathol 26:195–203

Habeck J-O, Huckstorf C, Honig A (1985) Influence of age on the carotid bodies of spontaneously hypertensive (SHR) and normotensive rats. II. Alterations of the vascular wall. Exp Pathol 27:79–89

Habeck J-O, Kreher C, Huckstorf C, Behm R (1987a) The carotid bodies of renal hypertensive rats. Anat Anz 163:49–55

Habeck J-O, Tafil-Klawe M, Klawe J (1987b) The carotid bodies of hypertensive/normotensive hybrid rats. Biomed Biochim Acta 46:903–906

Heath D, Edwards C, Harris P (1970) Post-mortem size and structure of the human carotid body. Its relation to pulmonary disease and cardiac hypertrophy. Thorax 25:129–140

Heath D, Smith P, Hurst G (1986) The carotid bodies in coarctation of the aorta. Br J Dis Chest 80:122–130

Honig A, Habeck J-O, Pfeiffer C, Schmidt M, Huckstorf C, Rotter H, Eckermann P (1981) The carotid bodies of spontaneously hypertensive rats (SHR): a functional and morphologic study. Acta Biol Med Ger 40:1021–1030

Pfeiffer C, Habeck J-O, Rotter H, Behm R, Schmidt M, Honig A (1984) Influence of age on carotid body size and arterial chemoreceptor reflex effects in spontaneously hypertensive (SHR) and normotensive rats. Biomed Biochim Acta 43:205–213

Przybylski J (1978) Alveolar hyperventilation in young spontaneously hypertensive rats. IRCS Med Sci 6:315

Quies von W, Claus T, Honig A (1983) Die Hyperventilation der Hypertoniker, – eine Pilotstudie zum Verhalten der Reflexe der arteriellen Chemorezeptoren bei hypertensiven Erkrankungen. Dtsch Gesundheitswes 16:612–617

Smith P, Jago R, Heath D (1982) Anatomical variation and quantitative histology of the normal and enlarged carotid body. J Pathol 137:287–304

Smith P, Jago R, Heath D (1984) Glomic cells and blood vessels in the hyperplastic carotid bodies of spontaneously hypertensive rats. Cardiovasc Res 18:471–482

Trzebski A, Tafil M, Zoltowski M, Przybylski J (1982) Increased sensitivity of the arterial chemoreceptor drive in young men with mild hypertension. Cardiovasc Res 16:163–172

Wong GWK, Melax H, Singh DNP, Jeria MJ (1982) Ultrastructural myocardial degeneration in spontaneously hypertensive rats under chronic saline intake. IRCS Med Sci 10:155–156

15   The Carotid Bodies and Sodium Metabolism

Mild hypoxia induces polyuria and, in those low-landers who tolerate rapid ascent to high altitude well, there may be a diuresis that lasts for days. This brings about a contraction of extracellular and plasma volume producing the "haemocon-centration" characteristic of early exposure to high altitude (Asmussen and Nielsen 1945; Berger et al. 1949; Heath and Williams 1989, pp 182–183). Since the osmotic pressure of extracellular fluid is precisely controlled, the reduced volume is achieved by a negative sodium balance. Many subjects, on the other hand, who ascend mountains become oliguric during the first few hours of their exposure to the hypobaric hypoxia. At the same time there is a redistribution of water in the body, with a shift of water into the extracellular compartment, leading to acute mountain sickness and predisposing to high-altitude pulmonary and cerebral oedema (Heath and Williams 1989, pp 196–201). This regulation of sodium and water excretion is brought about by the interaction of several hormones listed below and particularly by aldosterone and antidiuretic hormone. Honig (1989) has suggested that the natriuretic response to the hypobaric hypoxia of high altitude may be influenced in addition by the carotid bodies. It has been appreciated for many years that the haemo-concentration of early exposure to the mountain environment may play a part in acclimatisation to high altitude, for it enables the blood to carry more oxygen per unit volume. Thus a natriuretic response to hypoxia not only reduces the risk of pulmonary or cerebral oedema, but also increases oxygenation of the tissues. The possible involve-ment of the carotid bodies in the regulation of sodium and water metabolism in hypoxia is dis-cussed in this chapter. The reduction in plasma volume which accompanies acclimatisation to hypoxia can be attained by two mechanisms: a voluntary reduction in dietary intake of salt and water and their increased excretion by the kidneys (Honig 1989).

Decreased Sodium Intake

Experiments with spontaneously hypertensive rats suggest that the voluntary intake of salt is reduced by hypoxia. Rats of this strain are useful for such experiments because they have a naturally high appetite for salt, thus rendering changes in their sodium intake to be detected readily. In one such experiment (Behm et al. 1984) the rats were fed on a normal diet but were provided with two drinking bottles, one containing tap water and the other hypertonic saline. The animals which were exposed to hypobaric hypoxia at a pressure of 460 mmHg for 20 days showed a reduced intake of saline which was greater in degree than that of hypoxic, normotensive rats. Conceivably stimu-lation of the carotid chemoreceptors by hypoxia might influence this. Certainly, administration of almitrine bismesylate, a stimulant of the carotid body (see Chapter 13), has a similar effect (Honig 1989). In contrast a reduction in activity of the carotid bodies generated by inhalation of hyper-oxic gas mixtures increases voluntary salt intake (Behm et al. 1987a). Since the effects of both hypoxia and almitrine are attenuated by section of the carotid sinus nerves it would appear that the carotid bodies are responsible for the reduced salt ingestion.

Increased Sodium Excretion

An increased urinary excretion of sodium chloride which was possibly mediated by the carotid body was first demonstrated in cats (Honig 1983; Honig and Schmidt 1980). In these experiments the carotid bodies of anaesthetised cats were isolated from the general circulation and then perfused with venous blood. This hypoxaemic stimulation caused an increased ventilatory rate, a mild ele-vation of systemic arterial pressure, and a signifi-cantly elevated urinary excretion of sodium, which increased steadily with time. This reaction was independent of whether the vagus nerve was intact and was not related to the transient increase in blood pressure, since it still occurred when the renal blood flow was maintained at a constant level (Honig 1983). Furthermore, artificial ventilation to prevent respiratory alkalosis did not abolish the natriuresis. The one factor which prevented the loss of sodium was denervation of the carotid bodies (Honig 1989), indicating that they con-stitute an important component of the natriuretic response to hypoxia. In further experiments all neural connections to one kidney were severed and the carotid bodies perfused with hypoxic blood as before (Honig and Schmidt 1980). The innervated kidney showed vasoconstriction, whereas the denervated one showed little pressor effect, thus demonstrating that it was isolated from the nervous system. Nevertheless, there was still an increased excretion of sodium from the denervated kidney. These experiments suggest a reflex control of sodium excretion which is medi-ated by the carotid bodies and which may be effected by a circulating substance or hormone.

Chemoreceptors and the Kidney

The natriuresis stimulated by hypoxia is the result of an inhibition of absorption of sodium by the renal tubules (Honig 1989). Hence the process may be influenced by the renal vasculature as well as by the carotid body. In an attempt to distinguish between these two factors Honig et al. (1985) clamped the common carotid arteries of anaesthe-tised cats. In this species the occipital arteries provide an adequate perfusion of the brain, thus avoiding the complication of cerebral hypoxia and its effects on the cardiovascular system. Following clamping the reduced pressure in the carotid sinuses led to a reflex increase in systemic arterial pressure, whereas hypoxaemia of the carotid bodies led to hyperventilation. There was a pro-nounced natriuresis due to inhibition of renal tubular sodium absorption. Inactivation of the carotid bodies with acetic acid attenuated, but did not abolish, the natriuresis and so did lowering renal arterial pressure to normal levels. Simulta-neous implementation of both procedures stopped the natriuresis. It was concluded that both the increased perfusion pressure in the kidney and the activity of the carotid chemoreceptors were involved in causing the natriuresis.

In experimental conditions where the blood pressure in the carotid sinuses is not altered, stimulation of the isolated carotid bodies by hypoxaemic blood leads to constriction of the renal arteries (Parker et al. 1975; Schmidt et al. 1985). When the renal nerves are sectioned this vasoconstriction does not occur, suggesting that the carotid body affects renal vascular tone via the nervous system (Schmidt et al. 1985). Increased sodium excretion occurs irrespective of whether the nervous supply to the kidneys is intact or not, suggesting a humoral stimulus from the carotid body. These two influences of the carotid bodies induce conflicting effects in the kidney, since vasoconstriction derived from the neural pathway would tend to decrease urinary sodium and water excretion. The kidney appears to overcome this apparent conflict by increasing both the flow resistance in the renal vascular bed and the filtration fraction (Honig 1989).

Almitrine permits less invasive techniques to be applied to the question of interactions between carotid body and kidney. This drug has the advan-tage of increasing chemoreceptor activity in the absence of hypoxia without there being recourse to complex surgery. A single injection of almitrine into rats causes diuresis and natriuresis, an effect which is abolished by section of the carotid sinus nerves (Bardsley and Suggett 1987). In chronically hypoxic rats with enlarged carotid bodies the natriuretic response is attenuated suggesting that this, as well as the hypoxic ventilatory drive, may be blunted after prolonged residence at high altitude. When almitrine is administered orally to spontaneously hypertensive rats, natriuresis ensues but in a biphasic manner. After 4 days there is a peak of sodium excretion followed by another peak 2 days later (Behm et al. 1987b). Denervation of both the carotid and aortic bodies abolishes only the second response, suggesting that the first peak is independent of the chemo-receptors (Behm et al. 1987b). In healthy humans oral almitrine also induces increased sodium excretion associated with a brief initial renal vaso-dilatation followed by prolonged constriction (Ledderhos et al. 1987). Here the pattern of the response more closely resembles that of hypoxia.

The above experiments suggest that the carotid bodies may exert an influence on renal function. This is mediated in part neurally and in part hormonally, but as yet a specific natriuretic hor-mone synthesised by the carotid bodies has not been identified. For example, radioimmunoassay has failed to demonstrate the presence of atrial natriuretic factor in human carotid bodies (Heath et al. 1988). There are, however, many polypeptide hormones within the chief cells (see Chapter 8) whose function is largely unknown. It is conceivable that one of these substances might be released after stimulation by hypoxaemia and inhibit renal tubular absorption of sodium. In this context, Poston et al. (1981) have suggested that in patients with systemic hypertension there is a circulating sodium transport inhibitor which is continuously correcting a tendency for sodium retention by the kidney.

Chemoreception and the Endocrine System

The regulation of salt and water is brought about by the interaction of several hormones including renin, angiotensin, aldosterone, antidiuretic hor-mone (ADH) and atrial natriuretic factor (ANF). The carotid body may be involved in their release. Adrenal cortical glucocorticoids and aldosterone both influence sodium absorption in the renal tubules, the former tending to decrease it and the latter to increase it (Honig 1989). A sustained increase in release of cortisol into the adrenal vein of anaesthetised dogs occurs when they breathe 10% oxygen, and denervation of either the carotid or aortic bodies reduces this release by 30%–40% (Marotta 1972). Denervation of both chemo-receptors abolishes cortisol release completely. Thus nerve impulses from both glomera appear to affect adrenocortical activity, probably via the hypothalamus and the pituitary (Marotta 1972). Furthermore, a release of adrenocorticotrophic hormone (ACTH) by the pituitary has been demonstrated in hypoxic dogs, and is attenuated

by denervation of the carotid bodies (Raff et al. 1984). In contrast there is a reduced secretion of aldosterone in response to hypoxia (Honig 1989). The influence of hypoxia upon both of these adrenocortical hormones, therefore, has the effect of reducing renal tubular absorption of sodium with consequent natriuresis. The carotid chemo-receptors may play a significant rôle in this process by affecting secretion of pituitary hormones.

There is scanty and conflicting information on the part played by the carotid body in regulating other hormones involved in salt and water meta-bolism. For example, it is not known whether the carotid body influences the release of renin from the juxtaglomerular apparatus (Honig 1989). Neither is it certain whether it exerts any influence over the conversion of angiotensin, although an increase in the activity of angiotensin-converting enzyme occurs in hypoxic lungs apparently inde-pendently of the chemoreceptors. Some reports suggest that ADH release may be altered by exposure to hypoxia. In humans stimulation of the carotid bodies by almitrine produces a slight decrease in the level of vasopressin in the blood, which could augment any diuresis produced through other mechanisms (Honig et al. 1987). In general, however, the effects of the carotid bodies on this hormone are unknown.

There is some evidence that exposure of animals to acute or chronic hypoxia increases the level of ANF in the blood (Honig 1989). In humans stimulation of the carotid bodies with almitrine produces a trivial increase in plasma ANF (Koller et al. 1989). It is not known whether there is any specific reflex effect of the chemoreceptors upon release of ANF, particularly as direct stimulation of cardiac stretch and pressure receptors by the cardiovascular changes occurring in hypoxia could well account for its observed release.

Clinical Implications

The natriuretic response to hypoxia differs accord-ing to age. While young healthy people respond to acute hypobaric hypoxia at high altitude by a decrease in plasma volume and haemocon-centration, the response in the elderly is greatly reduced or even reversed, so that they may show an increase in plasma volume with haemodilution (Jung et al. 1971). The elderly have attenuated ventilatory and cardiovascular responses to hypoxia (Kronenberg and Drage 1973). Thus it seems likely that the ventilatory and natriuretic effects of the carotid body are linked and that both decline with age. This is in keeping with the histological changes which develop in the carotid body in the elderly, in which there is atrophy and fibrosis of the glomic tissue (see Chapter 4), which would be consistent with a reduction in activity of the organ. Patients with pulmonary emphysema also tend to have a reduced hypoxic drive (Flenley et al. 1970) and it is likely that they too show a reduced natriuretic response to hypoxia. Emphysematous patients who show the blue-and-bloated pattern have the most pronounced attenuation of their hypoxic drive, and their systemic oedema may be the result of blunted natriuretic activity of their carotid bodies, with consequent rentention of salt and water.

The carotid bodies exert some influence over sodium metabolism even in healthy individuals at sea level. This is presumably due to the activity of the carotid bodies which persists in the absence of hypoxia, the so-called "resting drive". In anaesthetised normoxic cats, inactivation of the carotid chemoreceptors is followed by a significant decrease in sodium excretion (Honig 1989). If one kidney is denervated, carotid body inactivation causes a decrease in sodium excretion from the innervated kidney only. Thus the resting drive of the carotid bodies in the cat inhibits tubular sodium reabsorption in normoxia. Since this effect requires intact renal nerves, it is possible that the carotid bodies normally antagonise the antinatriuretic effects of the efferent renal nerves, and when the chemosensory resting drive is abolished salt and water are retained (Honig 1989). If this occurs also in humans, the carotid bodies could be involved in the aetiology of systemic hypertension. Honig (1989) suggested that any reduction in activity of the carotid bodies could lead to an enhanced voluntary sodium intake as well as a reduced ability of the kidneys to excrete sodium and water. This increase in plasma volume could then lead to primary systemic hypertension. In support of this hypothesis, he quoted work performed in his own laboratory in which it has been demonstrated that patients with systemic hypertension have a blunted natriuretic response when their carotid bodies are stimulated by almitrine (Honig 1989). On the other hand, a stimulus which causes persistent hyperactivity of the carotid bodies such as chronic hypoxia might be expected to prevent systemic hypertension. Thus, chronic exposure to hypobaric hypoxia inhibits the development of systemic hypertension in spontaneously hypertensive rats. Furthermore, Caucasians living permanently at high altitude have a lower systemic arterial pressure than those of similar age living at sea level. Such a proposed link between the activity of the carotid bodies and systemic arterial pressure is highly speculative. Indeed, the evidence that the carotid bodies are involved in sodium metabolism at all is circumstantial. Definite proof requires elucidation of the precise pathway by which they affect sodium metabolism. Conceivably, this may involve the discovery of a natriuretic hormone released by the carotid bodies.

References

Asmussen E, Nielsen M (1945) Studies on the initial increase in O$_2$-capacity of the blood at low O$_2$-pressure. Acta Physiol Scand 9:75–87

Bardsley PA, Suggett AJ (1987) The carotid body and natriuresis: effects of almitrine bismesylate. Biomed Biochim Acta 46:1017–1022

Behm R, Honig A, Griethe M, Schmidt M, Schneider P (1984) Sustained suppression of voluntary sodium intake of spontaneously hypertensive rats (SHR) in hypobaric hypoxia. Biomed Biochim Acta 43:975–985

Behm R, Franz U, Sitarek U (1987a) The link between chemoreceptor activity and voluntary salt intake in spontaneously hypertensive rats: a hypothesis. Biomed Biochim Acta 46:987–991

Behm R, Gerber B, Habeck J-O, Huckstorf C (1987b) Effect of almitrine on renal sodium excretion in chemoreceptor denervated spontaneously hypertensive rats. Biomed Biochim Acta 46:1011–1015

Berger EY, Galdston M, Horwitz SA (1949) The effect of anoxic anoxia on the human kidney. J Clin Invest 28:648–652

Flenley D, Franklin D, Millar J (1970) The hypoxic drive to breathing in chronic bronchitis and emphysema. Clin Sci 38:503–518

Heath D, Williams DR (1989) High altitude medicine and pathology. Butterworths, London

Heath D, Quinzanini M, Rodella A, Albertini A, Ferrari R, Harris P (1988) Immunoreactivity to various peptides in the human carotid body. Res Commun Chem Pathol Pharmacol 62:289–293

Honig A (1983) Role of the arterial chemoreceptors in the reflex control of renal function and body fluid volumes in acute arterial hypoxia. In: Acker H, O'Regan RG (eds) Physiology of the peripheral arterial chemoreceptors. Elsevier, Amsterdam, pp 395–429

Honig A (1989) Peripheral arterial chemoreceptors and reflex control of sodium and water homeostasis. Am J Physiol 257:R1282–R1302

Honig A, Schmidt M (1980) Kidney function during carotid chemoreceptor stimulation: influence of unilateral renal nerve section. In: Lichardus B, Schrier RW, Ponec J (eds) Hormonal regulation of sodium excretion. Elsevier, Amsterdam, pp 93–98

Honig A, Schmidt M, Arndt H, Hanus U, Kranz G, Rogoll I (1985) Kidney function during common carotid artery occlusion in anaesthetised cats: influence of vagotomy, constant ventilation, blood pressure stabilization, and carotid body chemoreceptor inactivation. Biomed Biochim Acta 44:261–273

Honig A, Landgraf R, Ledderhos C, Quies W (1987) Plasma vasopressin levels in healthy young men in response to stimulation of the peripheral arterial chemoreceptors by almitrine bismesylate. Biomed Biochim Acta 46:1043–1049

Jung RC, Dill DB, Horton R, Horvath SM (1971) Effects of age on plasma aldosterone levels and haemoconcentration at altitude. J Appl Physiol 31:593–597

Koller EA, Schopen M, Keller M, Lang RE, Valloton MB (1989) Ventilatory, circulatory, endocrine, and renal effects of almitrine infusion in man: a contribution to high altitude physiology. Eur J Appl Physiol 58:419–425

Kronenberg RS, Drage CW (1973) Alteration of the ventilatory and heart rate responses to hypoxia and hypercapnia with ageing in normal man. J Clin Invest 52:1812–1819

Ledderhos C, Quies W, Schuster R, Peters R (1987) Renal hemodynamics and excretory function of healthy young men during stimulation of their peripheral arterial chemoreceptors by almitrine bismesylate. Biomed Biochim Acta 46:1035–1042

Marotta SF (1972) Roles of aortic and carotid chemoreceptors in activating the hypothalamo-hypophyseal-adrenocortical system during hypoxia. Proc Soc Exp Biol Med 141:915–922

Parker PE, Dabney JM, Scott JB, Haddy FJ (1975) Reflex vascular reponses in kidney, ileum and forelimb to carotid body stimulation. Am J Physiol 228:46–51

Poston L, Sewell RB, Wilkinson SP et al. (1981) Evidence for a circulating sodium transport inhibitor in essential hypertension. Br Med J 282:847–849

Raff H, Shinsako J, Dallman MF (1984) Renin and ACTH responses to hypercapnia and hypoxia after chronic carotid chemodenervation. Am J Physiol 247:R412–R417

Schmidt M, Ledderhos C, Honig A (1985) Kidney function during arterial chemoreceptor stimulation. I. Influence of unilateral renal nerve section, bilateral cervical vagotomy, constant artificial ventilation and carotid body chemoreceptor inactivation. Biomed Biochim Acta 44:695–709

16 The Glomic Vasculature

The carotid body receives its blood supply through one, two or three glomic arteries which arise from the carotid bifurcation in 88% of cases, from the posterior aspect of the external carotid artery in 5%, from the anterior aspect of the carotid sinus in 5% and from the ascending pharyngeal artery in 2% (Smith et al. 1982; Khan et al. 1988; see Chapter 2). Exceptionally the glomic arteries may arise from the common carotid artery.

The Main Glomic Artery

This vessel, passing from the carotid arterial tree into the carotid body, is ensheathed by the fibro-elastic tissue of Mayer's ligament (see Chapter 1), which also contains the sinus nerve. It is a short vessel which was estimated as being 1–2 mm long in one study (Heath and Edwards 1971) and 3.5–4 mm in another (Jago et al. 1982). Its course to the carotid body may be tortuous. Along its length it may divide into main branches which continue into the carotid body before subdividing again (Jago et al. 1982). The ostium of the glomic artery occurs abruptly in the wall of its parent vessel and has no muscular sphincter. The lumen at its origin is instead encircled by closely packed bands of thick elastic fibres derived from the parent vessel. The elastic fibres of the media of the glomic artery proper are finer than those of the parent vessel.

The luminal diameter of the first free portion of the glomic artery is in the region of 200 μm. Despite its small diameter the main glomic artery retains the structure of an elastic vessel, which is more typical of large systemic arteries such as the common carotid artery (Jago et al. 1982). Thus, soon after its origin the media of the glomic artery becomes thin and consists of closely adjacent elastic fibres. Those of the inner media are small (0.5 μm) and longitudinally oriented to form a continuous band surrounding the lumen. The elastic fibres of the outer media are thicker (1–2 μm) and are arranged circularly. Between the elastic fibres are a small number of smooth muscle cells, fibrocytes and collagen fibrils. Non-myelinated nerve axons occur throughout the media of the glomic artery at this level and show neurotubules and neurofilaments. It is not possible to identify these on morphological grounds as either autonomic or somatic fibres. However, their simple structure and lack of relation to the few smooth muscle cells present suggest that they have an afferent function.

Such histological appearances, which are also seen in first-order branches of the main glomic artery (Fig. 16.1), are reminiscent of the thinned area of the carotid sinus (see Chapter 20), raising the possibility that the glomic artery may subserve a similar baroreceptor function. There is also a similarity in the ultrastructure of the free portion of the glomic artery and its first-order and inter-lobular branches (Fig. 16.2) to that of the thinned area of the carotid sinus (Jago et al. 1982). Thus, the media of both carotid sinus and glomic arteries includes plentiful elastic fibres woven into skeins forming a series of layers encircling the vessel

(Fig. 16.2). The spaces between successive elastic fibres are narrow and contain smooth muscle cells. In the main glomic artery these are scanty and attenuated but in the interlobular arteries they become more plentiful (Fig. 16.2). In addition there are ground substance and fibrocytes and also collagen fibres. Electron microscopy confirms that non-myelinated nerve axons are commonly encountered throughout the media of both vessels but, in the case of the main glomic artery in humans, the nature and connections of these nerves have not yet been established. We return to this question of baroreception by glomic arteries below.

After leaving its parent vessel, the glomic artery becomes surrounded by loose connective tissue. Its wall is from 25 to 60 μm thick and consists of circularly oriented elastic fibres pressed closely together with a few intervening attenuated fibrocytes. The lumen, 200 μm in diameter, is lined by a single layer of flat endothelial cells. In older subjects the intima shows minimal fibrosis, which represents a normal age-related change. As the glomic artery moves away from the parent vessel it becomes thinner, with a medial thickness of between 12 μm and 40 μm.

Shortly before reaching the carotid body the glomic artery usually divides into two or three first-order branches which enter the substance of

Fig. 16.2. Electron micrograph of part of a transverse section through an interlobular artery. Abundant elastic tissue is present in an irregular anastomosing network. Between the elastic fibrils are small groups of smooth muscle cells (m). Scale line = 10 μm.

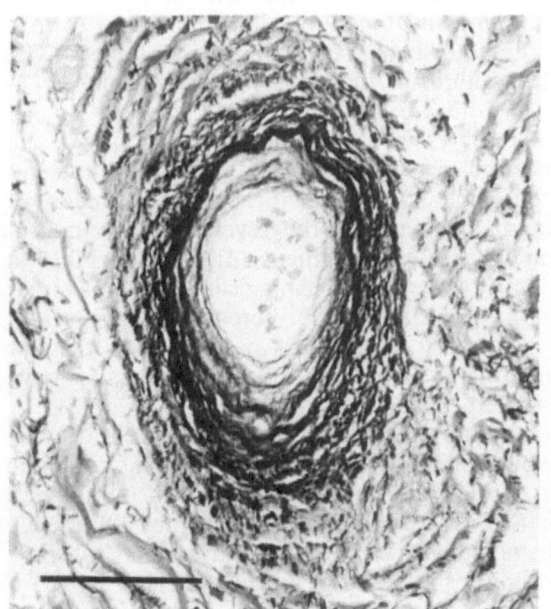

Fig. 16.1. A first-order branch of the main glomic artery in a woman of 80 years who died from ischaemic heart disease. The media consists almost entirely of concentric rings of elastic tissue of variable thickness. There is also moderate fibrosis of the intima associated with ageing. (elastic-Van Gieson (EVG)) Scale line = 100 μm.

the organ. The media of these vessels still retains an elastic structure but is considerably thicker than the media of the free portion. Their elastic laminae vary greatly in thickness from coarse, well-defined fibres to short wisps of elastin (Fig. 16.1). These fibres are more widely separated than in the free portion of the vessel, with sparsely distributed smooth muscle cells between them. Nevertheless, the histological appearance remains very much a predominantly elastic structure (Fig. 16.1).

When the glomic parenchyma is first encountered, a greater concentration of smooth muscle cells is seen in the outer part of the media. Branches now become overtly muscular and represent the transition into the microcirculation of the carotid body.

Pattern of Branching

The manner in which the glomic arteries divide and ramify throughout the carotid body can be studied by corrosion or latex casts of the vasculature followed by scanning electron microscopy (McDonald and Larue 1983; Seidl 1975) or by reconstruction of serial sections (McDonald and Larue 1983; Heath et al. 1983).

The human carotid body is sufficiently large to examine latex casts with a dissecting microscope. Such a specimen is shown in Fig. 16.3. In the cast illustrated, in addition to the main glomic artery there is an accessory vessel arising from a separate site on the external carotid artery. Occasionally there is a third, accessory glomic artery. Both the main and supernumerary vessels continue for a very short distance of about 1 mm towards the carotid body without branching but, just before entering its substance, each divides into an average of three. These first-order branches diverge to the peripheral aspects of the organ where they give rise to numerous lateral branches (Fig. 16.3). Finally, the branches of the glomic artery drain

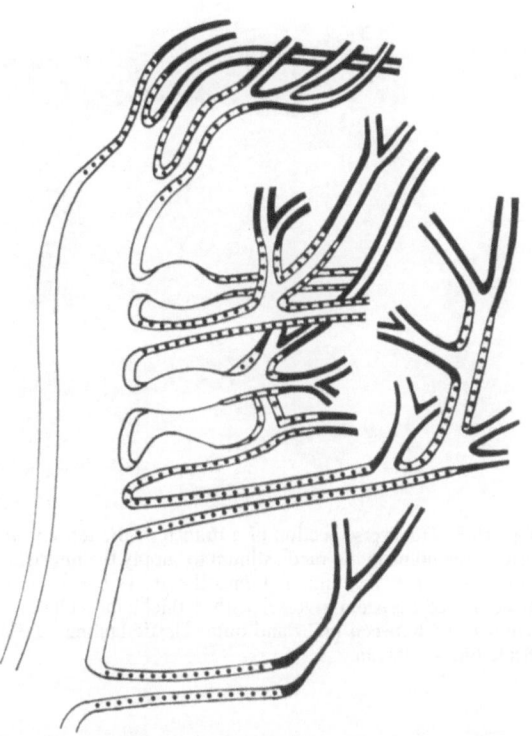

Fig. 16.4. Diagrammatic, two-dimensional representation of the glomic vasculature of the left carotid body of a man of 23 years who died from a cerebellar haemorrhage. The diagram was constructed from serial histological sections stained by the elastic-Van Gieson technique. A first-order branch of the main glomic artery is unshaded; interlobular arteries are stippled; intralobular arterioles are black. Note that the origins of some of the interlobular arteries are dilated and thin-walled.

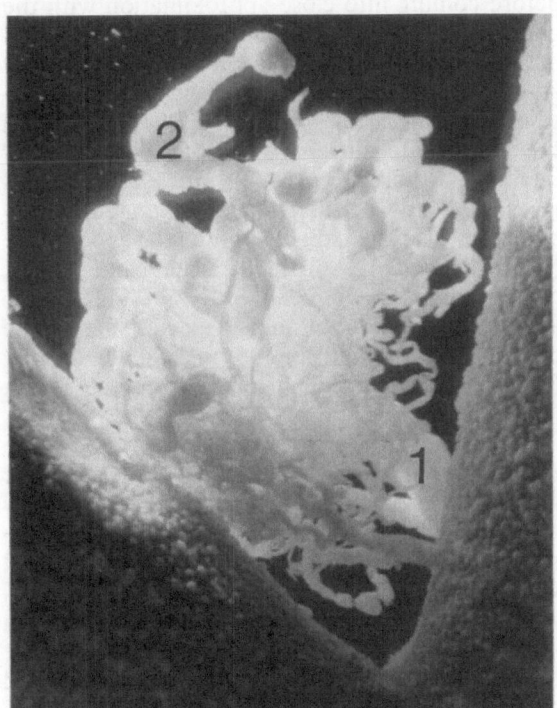

Fig. 16.3. Latex cast of the glomic vasculature in a normal carotid body. The main glomic artery (*1*) arises from the external carotid artery. A smaller, subsidiary glomic artery arises from a separate location in front of it. Both vessels divide into three branches. The large convoluted vessels at the upper pole are glomic veins (*2*).

into glomic veins which are identified in latex casts as large convoluted vessels which drain from the upper pole of the carotid body (Fig. 16.3).

Further division of the lateral arterial branches is best studied by reconstruction of serial sections of the carotid body. The results of such a study in a man of 23 years are illustrated diagrammatically in Fig. 16.4. This diagram shows a first-order branch of the main glomic artery passing through the connective tissue in the periphery of the carotid body. From its inner aspect it gives rise to several branches which arise at right angles and traverse the connective tissue septa between lobules. The origins of some of these tributaries are dilated and thin-walled. These interlobular arteries may themselves divide or terminate directly as muscular intralobular arterioles which penetrate the lobules of glomic tissue. Some arterioles are short, whereas others continue for some considerable distance before ending as precapillaries. Towards the upper pole of the carotid body the first-order branches of the glomic artery narrow until they

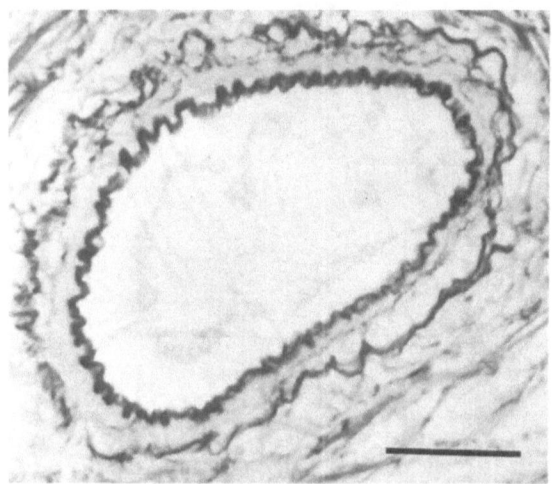

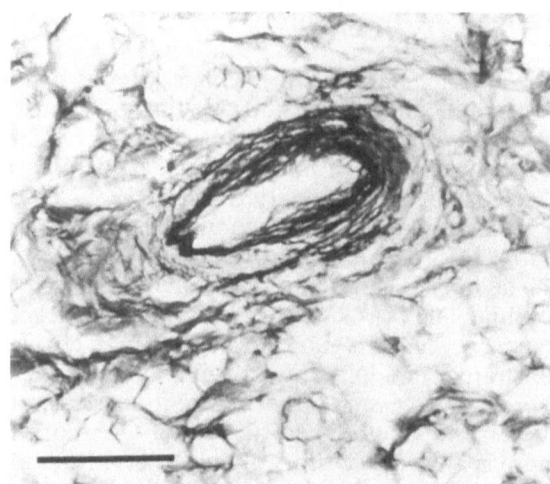

Fig. 16.5. Transverse section of a branch of the left coronary artery ascending in the mediastinum to supply the intertruncal glomera. Before reaching a glomus the artery has the typical structure of a systemic vessel with a thick, muscular media sandwiched between inner and outer elastic laminae. (EVG) Scale line = 100 μm.

Fig. 16.6. A further section from the artery shown in Fig. 16.5 at a point where it comes into proximity with a glomus. Here it undergoes an abrupt transition into an elastic artery so that part of the wall has the structure of a typical muscular systemic artery while the remainder is highly elastic. (EVG) Scale line = 100 μm.

adopt the structure of interlobular arteries. They then leave their peripheral position and enter the interlobular septa, giving rise to the several intralobular arterioles. Thus all the branches of the main glomic artery terminate eventually as precapillaries within the glomic tissue.

Baroreception and the Glomic Arteries

The carotid sinus subserves baroreception (see Chapter 20) and shows histological modifications which appear to be associated with this function. These include thinning of the outer part of the media, which assumes a highly elastic structure with considerable loss of smooth muscle cells between the elastic laminae. The junction of media and adventitia has a rich nerve supply (see Chapter 20). As we have just seen, the main glomic arteries and their first-order branches show the same highly elastic structure, with focal dilatations and medial thinning, quite inappropriate for an artery of such small dimensions. Such considerations raise the possibility that the carotid body is itself directly sensitive to elevation of intravascular pressure through the agency of the glomic arteries and their branches. There is close interlinking of the nerve network supplying the carotid body and that servicing the carotid sinus.

It is of interest to note that the arteries supplying glomic tissue elsewhere in the body also develop the same elastic structure immediately before coming into close approximation with the glomus. Thus, the vessels supplying the mediastinal glomera are branches of the coronary artery and have the typical histological structure of muscular systemic arteries (Fig. 16.5). However, in the immediate vicinity of glomic tissue they undergo an abrupt transition into elastic vessels (Figs. 16.6, 16.7). Having passed the glomus they revert back to the expected muscular appearance (Edwards and Heath 1970). The same elastic characteristics have been noted in the glomic arteries of a variety of animal species such as the cat and dog (Addison 1944), the hedgehog (Adams 1958, pp 153 and 161) and the rabbit (Becker 1966).

The paucity of smooth muscle cells and close relation of nerve endings to elastic fibres in glomic arteries suggests that the nerve endings carry afferent signals and that they do so in response to some change in the physical status of the vessel wall. Jago (1986) compares such elastic vessels with a simple weighing machine containing a spring. In this analogy, intravascular pressure represents the load, elastic fibres the spring, and nerve axons a means of recording the load. Elastic fibres are the only components of the glomic artery which appear to be capable of responding to changes in intravascular pressure in a predictable manner. It is possible to isolate the intact skeleton of elastin from rings of aorta by digestion in formic

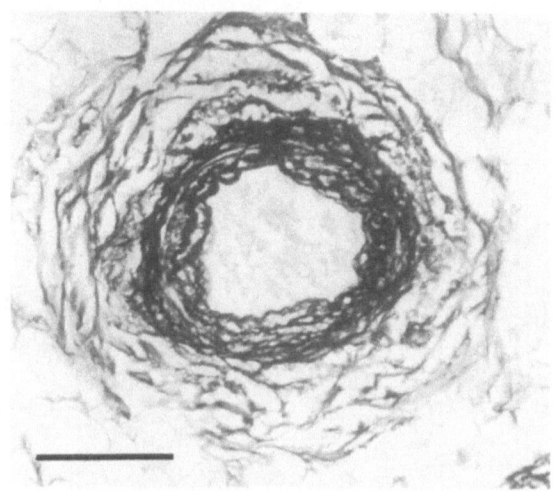

Fig. 16.7. A further section from the artery shown in Fig. 16.5 where the artery is directly supplying blood to the glomus. The vessel now has a highly elastic structure in spite of its diminutive size, a histological feature consistent with its having assumed a baroreceptor function. (EVG) Scale line = 100 μm.

acid (Hass 1942a) and to demonstrate, by applying different loads to the loops of elastic fibres and measuring the extensions, that elastin obeys Hooke's law (Hass 1942b). Since elastic fibres have the ability to undergo considerable alteration in length in response to tension, they are likely to be sensitive to small fluctuations in intravascular pressure. Their circular arrangement around the lumen in the media of glomic arteries and their loose, crenated appearance suggest that the wall of the vessel is capable of considerable distension. Furthermore, ultrastructural studies have shown that individual elastic fibres have the general property of being organised in a helical fashion (Gross 1949), suggesting an additional capacity for elastic fibres to increase in length. Because of their great extensibility, elastic fibres have a low tensile strength but they are reinforced in the arterial wall by other structural components of high tensile strength such as the collagen fibrils of the outer media and adventitia.

The tension in the elastic fibres of the media of glomic arteries seems to be almost entirely due to intravascular pressure because there are only a few smooth muscle cells present and any contribution made by them to transmural tension can probably be neglected. At the proximal end of this elastic system the pressure wave-form will be pulsatile, reflecting systolic and diastolic pressure, but distally it will be dampened and the pressure of blood delivered to glomic tissue constant (Jago 1986). By application of Laplace's law the tension

in the wall of the glomic arteries is a product of its radius and the pressure in its lumen. It is this simple mechanical relation which allows the elastic fibres in the media of the glomic artery to respond in a predictable manner to changes in systemic blood pressure (Jago 1986).

It is apparent that, if the elastic tissue in the glomic arteries allows the carotid body to be sensitive to the level of intravascular pressure, then in some way the messages passing through afferent nerve axons between the elastic fibres probably bring about both hormonal changes and a cellular response in the carotid body. The cellular response is hyperplasia of sustentacular cells and it is identical with that following the stimulus of hypoxaemia (see Chapter 14). At present the way in which raised intravascular pressure leads to the same proliferation of sustentacular cells produced by hypoxaemia is obscure. However, in cases of hyperplasia of sustentacular cells complicating systemic hypertension and coarctation of the aorta (Heath et al. 1986), nerve fibres have been demonstrated in close proximity to the proliferating sustentacular cells.

Interlobular Arteries

The main glomic artery and its first-order branches give rise to arterial branches which extend between the lobules of glomic tissue (Fig. 16.4). Because of their location they are best termed interlobular arteries (Heath et al. 1983). They range in external diameter from 75 to 150 μm. These interlobular arteries, like their parent vessel, retain a highly elastic structure. In the media the elastic fibres present an irregular anastomosing arrangement, the individual fibres being of variable thickness and presenting a ragged appearance. Between the elastic fibres are numerous smooth muscle cells which are oriented circumferentially (Fig. 16.2). There is no clearly defined external lamina to these vessels, so that the elastic fibres of the media tend to merge with those in the connective tissue of the adventitia. In this way the histological appearance differs markedly from the media of solid smooth muscle that would be anticipated in the typical systemic artery of this diameter.

Their ultrastructure is irregular and distinct from that of most other systemic arteries. Thus, the prominent elastic laminae are arranged in the form of a net which divides the media into compartments (Heath et al. 1983). Within the individual compartments are isolated smooth muscle cells, oriented circularly in the main, and a hap-

hazard arrangement of collagen fibres (Fig. 16.2).
Nerve axons are plentiful, usually embedded within the bundles of elastic and collagen fibres (Heath
et al. 1983). Isolated smooth muscle cells may also
be found within the intima and adventitia, so that
the limits of the media become indistinct.

The medial thickness of interlobular glomic
arteries varies widely over short distances. In
some section it may occupy more than 20% of the
external diameter (Heath et al. 1983), whereas in a
nearby section it may be extremely thin. Thin
segments of interlobular arteries are most commonly found close to their origin, and may involve
their whole circumference or only part of it (Fig.
16.4). The media of these thin portions consists
merely of a few beaded elastic laminae compressed tightly with no discernible smooth muscle
between them. The diameter of the vessel is often
greater here than elsewhere so that there may be a
resemblance to a miniature carotid sinus. It is
possible that these thin regions are focal baroreceptors.

Glomic arteries commonly show an irregular
intimal proliferation of collagen with fine elastic
fibrils which is often eccentrically situated and of
variable thickness along the vessel. It appears to
become commoner and more pronounced with
advancing age, in common with the other arteries
throughout the body. It seems to be exaggerated
in the presence of systemic hypertension.

Intralobular Arterioles

As the interlobular arteries pass between lobules
of glomic tissue, arteriolar branches arise which
are predominantly muscular in nature. The terminations of the interlobular arteries undergo the
same transition to a muscular vessel without com-
municating with a vein or venule. These intralobular arterioles are very small, having an external
diameter ranging from 15 to 75 μm (Heath et al.
1983). They closely resemble systemic arterioles
but there are certain histological differences which
are described below. Shortly after undergoing
muscularisation, these arterioles enter the lobules
of the carotid body. Each has a well-defined media
of circularly oriented smooth muscle sandwiched
between inner and outer elastic laminae. In some
instances the appearance of the elastic fibres is
suggestive of a helical arrangement of the smooth
muscle cells that lie between them. Many of the
glomic arterioles show ovoid areas of pallor in
their media, 10 to 40 μm in diameter, distorting
them and causing variations in their thickness
(Fig. 16.8). Ultrastructurally, the smaller pallid

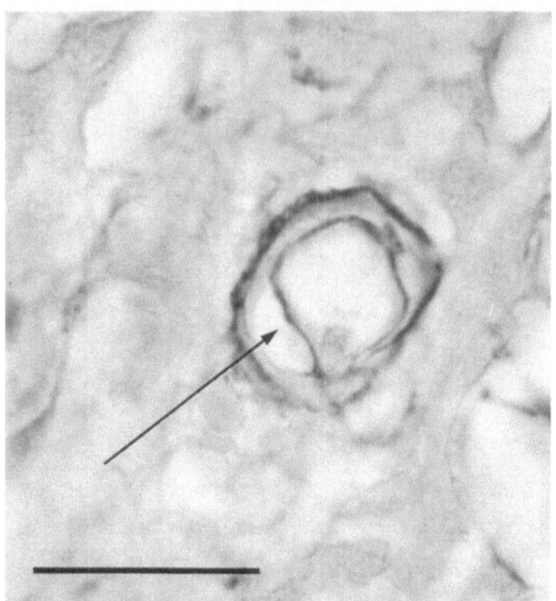

Fig. 16.8. Transverse section of an intralobular arteriole
showing a large area of pallor (*arrow*) within the media. The
inner elastic lamina appears distorted in this region. (EVG)
Scale line = 20 μm.

areas correspond to enlarged smooth muscle cells,
with distension of the mitochondria attributable to
autolysis, while larger ones correspond to
increased amounts of fluid separating smooth
muscle from the internal elastic lamina (Jago
1986). Such focal areas of pallor may, therefore,
be artifactual but it is not known why they should
occur in the carotid body and not in other organs.

Even in the glomic arterioles the elastic tissue is
deposited irregularly, often showing discontinui-
ties, fraying and branching. There is considerable
variation in the quantity of elastic tissue present in
arterioles from different carotid bodies, so that in
some the arterioles are clearly defined whereas in
others they may require close examination to
identify them.

At the point where the interlobular arteries give
rise to intralobular arterioles there are occasionally small muscular protuberances into the lumen.
These correspond to "blocking mechanisms"
(arterial rings) as described in the fetal and adult,
human and mammalian, systemic and pulmonary
circulations (Wagenvoort et al. 1964, pp 21, 46,
63–66, 138).

The histological features of intralobular
arterioles suggest that they have a different
function from those of the interlobular arteries.
Their muscular composition and small diameter
suggest that they may constrict to reduce blood

flow to the glomic tissue and the plump nuclei of the endothelium may play a part in occluding the lumen. This potential for constriction is supported by the "blocking mechanisms".

Glomic Precapillaries and the Capillary Bed

As the intralobular arteriole moves nearer glomic tissue, its muscle coat is gradually lost and it undergoes a transition into precapillaries in which only the outer elastic lamina remains. These vessels lead into the capillary bed. Glomic capillaries are lined by plump endothelial cells linked by tight junctions and are surrounded by prominent elongated pericytes arranged in a helical fashion (Fig. 16.9). A surrounding zone of ground substance separates the capillary from the nearby sustentacular and chief cells of the glomic parenchyma (Fig. 16.9). The distance from capillary lumen to the cytoplasm of the chief cells is of the order of 6 μm. These small vessels are lined by endothelial cells and hence are true glomic capillaries and not sinusoids. Glomic capillaries follow a tortuous path through the cell clusters which is

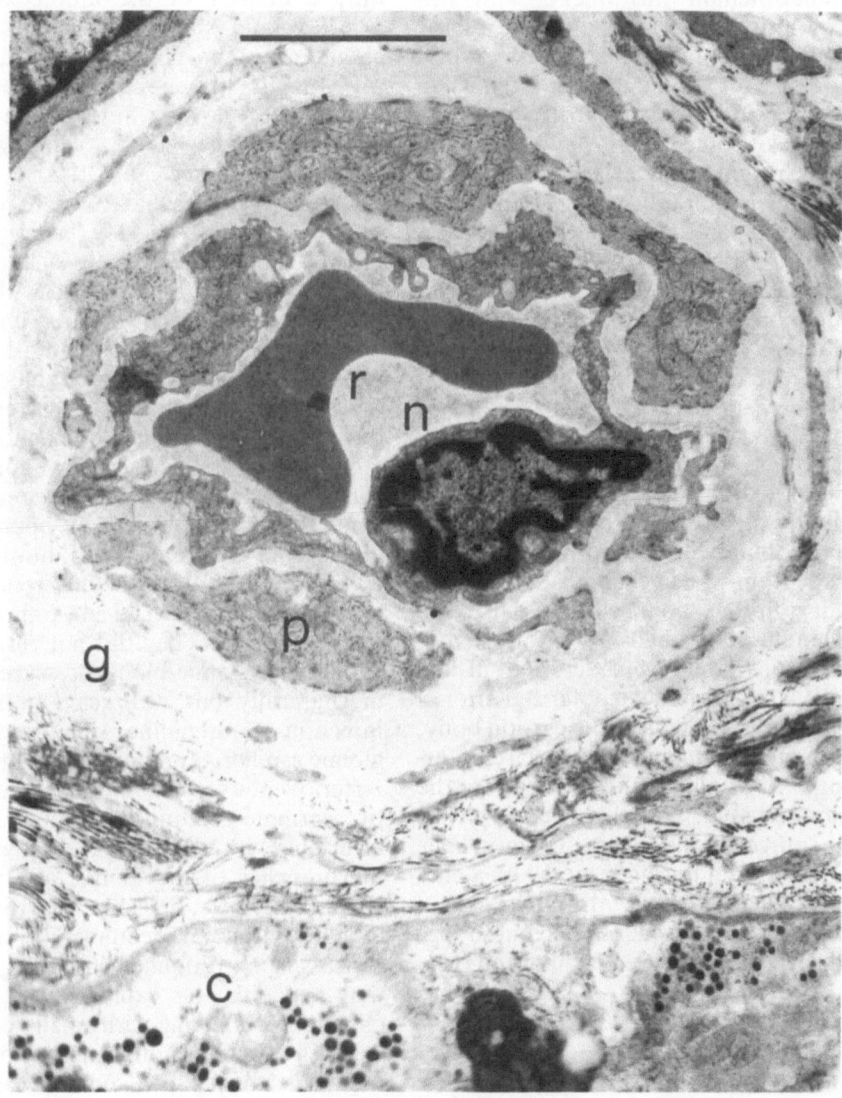

Fig. 16.9. Electron micrograph of human glomic capillary from a male infant aged 3 years killed in a road traffic accident. Its lumen contains a red blood cell (*r*) and is lined by an endothelial cell whose prominent nucleus (*n*) protrudes internally. There are surrounding pericytes (*p*). Around the capillary is a zone of ground substance (*g*) separating it from chief cells (*c*). Scale line = 3 μm.

impossible to trace using conventional histological sections. It requires serial reconstructions of sections 1 μm thick to determine their pathway. Using this technique on rats McDonald and Larue (1983) were able to identify two types of glomic capillary. Type I capillaries were the most numerous with a large diameter (8 to over 20 μm), a thin endothelium and a tortuous course amongst the chief cells. Type II capillaries were narrower in diameter (6 μm) and thus resembled capillaries of striated muscle. They were found to take a more direct route from precapillaries to venules and were less intimately associated with chief cells, often skirting around the perimeter of cell clusters. Their endothelium was thicker and was surrounded by more pericytes than were type I capillaries. Human glomic capillaries correspond to the type I vessels of rats and it is not established whether type II occur in human carotid bodies.

Glomic capillaries come into close juxtaposition with chief and sustentacular cells over approximately half their circumference. The remainder contacts stromal elements. This unit comprising a capillary with its small group of closely adjacent cells is assumed to be the smallest functional unit in the carotid body. In the rabbit it has been termed a "glomoid" (Seidl 1975).

Glomic Venules and Veins

Glomic capillaries drain into venules within the lobules of glomic tissue which themselves empty into larger interlobular veins that form a series of tortuous, elastic, thin-walled channels surrounded by interlobular fibrous connective tissue. The interlobular venules run radially to form the tributaries of numerous small veins carried in the fibrous capsule. Thus, unlike the polarised arterial supply entering the lower pole of the carotid body, the venous drainage occupies the entire circumference of the organ, forming a plexus enclosing the lobules of glomic tissue. It is this venous plexus which is the source of the severe haemorrhage commonly encountered during removal of a chemodectoma (see Chapter 19). At the upper pole of the carotid body the venous plexus becomes more concentrated as larger veins are encountered (Fig. 16.3).

Glomic Microcirculation and Arteriovenous Anastomoses

It has been demonstrated in experimental animals that the blood flow to the carotid body is very high.

In the cat the venous outflow from the carotid body is 40 μl/min which would correspond to a value of 2000 ml/100 g per min if the weight of the carotid body in this species is taken to be 2 mg (Daly et al. 1954). This is an extremely high rate, being roughly five times the blood flow in the kidney or four times that in the thyroid (Joels and Neil 1963). However, such estimates have to be viewed with circumspection, for, if they are equated with flow through glomic capillaries, they would represent such a high flow as not to expose the chemoreceptor cells to sufficient hypoxia to excite the chemosensory nerve endings (Joels and Neil 1963). Indeed these authors calculated that only a quarter of the blood from the glomic arteries could pass into the capillaries to allow for chemoreception. Such considerations have led physiologists to believe that a sizeable proportion of the blood flowing through the glomic arteries must reach the glomic veins without flowing through the glomic capillaries.

Several workers have sought to establish the microanatomical basis for "arteriovenous anastomosis". The carotid bodies of the cat have been observed during ventilation with different gas mixtures (de Castro 1951). Hypoxia caused a decrease in their size, with an associated reddening due to increased surface blood flow. With a high oxygen content in the inspired air the carotid body became pale and swollen and this was associated with a reduction in venous outflow. de Castro concluded that, in cats, peripheral arteriovenous shunts are normally operating, diverting blood from the glomus cells, but in hypoxia there is an increase in this shunt, rendering the cells even more hypoxic. The anastomoses also opened in hypercarbia. They did not respond to an elevation of systemic blood pressure brought about mechanically but *contracted* in response to an injection of adrenaline, squeezing blood into the glomic capillaries and distending the carotid body. Arteriovenous anastomoses were demonstrated by carmine-gelatin injections linking glomic arterioles with veins. In the dog, wide-bore terminal branches of the glomic artery end in thick-walled arterioles, but some pass directly into veins in the periglomic connective tissue (Serafini-Fracassini and Volpin 1966). The glomic arterioles may end either as capillaries or as anastomotic channels connecting with small veins. References to such arteriovenous anastomoses have been made in many papers on the carotid bodies of animals but it is worthy of comment that in a recent study involving 390 murine carotid bodies, such anastomoses were found in only two instances (Habeck et al. 1984). Despite this rarity, great weight is commonly attached by physiologists to the haemodynamic significance of the elusive arteriovenous anastomosis. There is no convincing evidence that arteriovenous anastomoses exist in the human carotid body. We have examined serial sections of the glomus obtained from humans at necropsy and we have failed to reveal any direct communication between glomic arteries or arterioles and glomic veins (Heath et al. 1983). We found in our cases that all the first-order branches of the glomic artery terminated as intralobular arterioles which then became translated into precapillaries.

Glomic Vasculature of the Rat

The carotid body of the rat is a poor model for the study of diseases of human glomic tissue. In particular it does not show the variants of chief cells in the parenchyma which are of importance in human disease (see Chapter 21). The same comment may be applied to the glomic vasculature. In the rat the media of the main glomic artery is composed of smooth muscle cells without the predominance of the elastic tissue which characterises the corresponding human vessels. This suggest that in this species the main artery acts simply as a muscular tube transporting blood from the carotid arterial tree into the glomic parenchyma. Its structure does not suggest that it has a rôle in baroreception (Jago 1986). The origin of the main artery to the murine carotid body is guarded by a flattened sphincter-like cushion, unlike that in humans (Jago 1986).

References

Adams WE (1958) The comparative morphology of the carotid body and carotid sinus. Charles C. Thomas, Springfield, IL

Addison WHF (1944) The extent of the carotid pressoreceptor area in the cat as indicated by its special elastic-tissue wall. Anat Rec 88:418–419 (abstract of a paper which in the event was not presented due to the exigencies of World War II)

Becker AE (1966) The glomera in the region of the heart and great vessels. A microscopic-anatomical and histochemical study. MD thesis, Laboratory of Pathological Anatomy, University of Amsterdam

Daly M De B, Lambertsen CJ, Schweitzer A (1954) Observations on the volume of blood flow and oxygen utilization of the carotid body in the cat. J Physiol (Lond) 125:67–89

de Castro F (1951) Sur la structure de la synapse dans les chemorecepteurs: leur mécanisme d'excitation et rôle dans la circulation sanguine locale. Acta Physiol Scand 22:14–43

Edwards C, Heath D (1970) Site and blood supply of the intertruncal glomera. Cardiovasc Res 4:502–508

Gross J (1949) The structure of elastic tissue as studied with the electron microscope. J Exp Med 89:699–708

Habeck J-O, Honig A, Huckstorf C, Pfeiffer C (1984) Arteriovenous anastomoses at the carotid bodies of rats. Anat Anz 156:209–215

Hass GM (1942a) Elastic tissue I, Description of a method for the isolation of elastic tissue. Arch Pathol Lab Med 34: 807–819

Hass GM (1942b) Elastic tissue. II. A study of the elasticity and tensile strength of elastic tissue isolated from the human aorta. Arch Pathol Lab Med 34:971–981

Heath D, Edwards C (1971) The glomic arteries. Cardiovasc Res 5:303–312

Heath D, Jago R, Smith P (1983) The vasculature of the carotid body. Cardiovasc Res 17:33–42

Heath D, Hurst G, Smith P (1986) The carotid body in coarctation of the aorta. Br J Dis Chest 80:122–130

Jago R (1986) The glomic vasculature. In: Heath D (ed) Aspects of hypoxia. Liverpool University Press, Liverpool, pp 97–112

Jago R, Heath D, Smith P (1982) Structure of the glomic arteries. J Pathol 138:205–218

Joels N, Neil E (1963) The excitation mechanism of the carotid body. Br Med Bull 19:21–24

Khan Q, Heath D, Smith P (1988) Anatomical variations in human carotid bodies. J Clin Pathol 41:1196–1199

McDonald DM, Larue DT (1983) The ultrastructure and connections of blood vessels supplying the rat carotid body and carotid sinus. J Neurocytol 12:117–153

Seidl E (1975) On the morphology of the vascular system of the carotid body of cat and rabbit and its relations to type I cells. In: Purves MJ (ed) The peripheral arterial chemoreceptors. Cambridge University Press, London and New York, pp 239–299

Serafini-Fracassini A, Volpin D (1966) Some features of the vascularization of the carotid body in the dog. Acta Anat (Basel) 63:571–579

Smith P, Jago R, Heath D (1982) Anatomical variation and quantitative histology of the normal and enlarged carotid body. J Pathol 137:287–304

Wagenvoort CA, Heath D, Edwards JE (1964) Blocking mechanisms. In: The pathology of the pulmonary vasculature. Charles C Thomas, Springfield, IL, pp 21, 46, 63–66 and 138

17 Normal Ultrastructure

Most electron microscopical studies of the carotid body have been performed on animals. Thus, the normal organ has been examined in several species at sea level as well as in animals exposed to the natural hypoxia of high altitude or the induced hypoxia of simulated altitude in hypobaric chambers. By contrast, with the exception of the chemodectoma, there is scant information on the ultrastructure of either the normal or abnormal human carotid body. This imbalance in our knowledge has come about in part because most physiological studies of the organ have involved animals and it is convenient to compare structure with function in the same species. Another factor is the difficulty in obtaining suitable material for electron microscopy from humans. Grimley and Glenner (1968) obtained fresh specimens of carotid body resected during therapeutic glomectomy in relief of bronchial asthma but most studies must rely on tissue obtained from post-mortem examinations, with all the problems of autolysis and artifact that they bring. Nevertheless, provided that the interval between death and necropsy is less than 12 hours, preservation of cellular detail is sufficiently good to derive much information about the ultrastructure of the carotid body. In this chapter these appearances in the normal human glomus are described, with occasional reference to studies on animals to illustrate similarities and differences between species.

Chief Cells

A quick inspection of glomic chief cells under the electron microscope conveys the impression that they are all symmetrically rounded but closer scrutiny reveals that they have a complex border. Blunt, cytoplasmic extensions interdigitate and, in some cells, prolongations of the cytoplasm may extend for several cell diameters overlapping with those of their neighbours. In some cells the entire cytoplasm is elongated to form streamers which may encircle adjacent cells (Grimley and Glenner 1968). The cytoplasmic prolongations usually contain fewer organelles than does the main body of the cells and appear as small, pale islands of cytoplasm when sectioned transversely. They may contain microfilaments and microtubules (Grimley and Glenner 1968).

The three variants of chief cell, which characterise the carotid body at the light microscopical level, are readily identified under the electron microscope. Light chief cells are typically polygonal with an average diameter of 15 μm. Their nucleus is round or oval and contains chromatin consisting of a few, small aggregates dispersed within an almost transparent background. Larger clumps of heterochromatin may be attached to the nuclear membrane (Fig. 17.1). A small nucleolus is often seen. The cytoplasm of light cells is distinctively pale and contains the most characteristic organelle of chief cells, the dense-core vesicle (Fig. 17.2). These structures in all species typically consist of a central, dark core surrounded by an apparently empty, clear zone or "halo", which in turn is encapsulated by a single, limiting membrane. In animals such as the rat or cat the peripheral halo of the dense-core vesicles is broad and conspicuous (Fig. 17.3) but in humans it is considerably narrower in proportion to the diameter of the vesicles (Fig. 17.2). Also, comparison

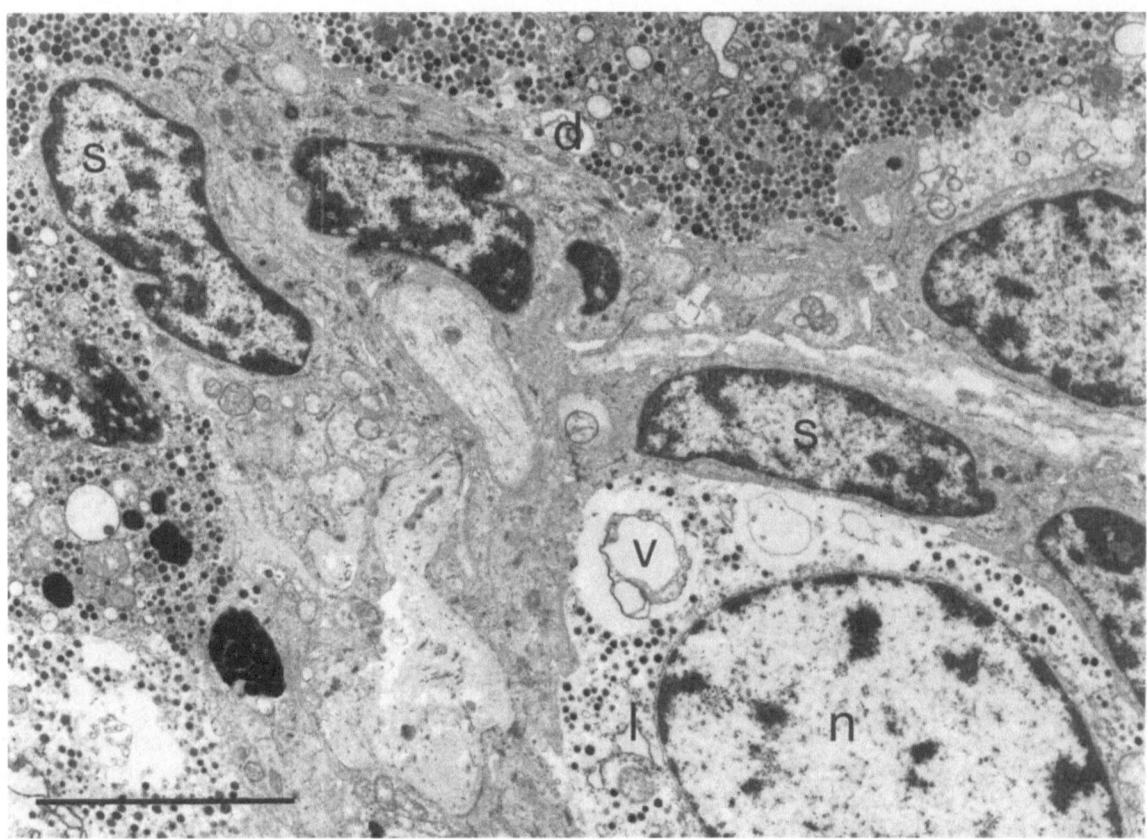

Fig. 17.1. Low power electron micrograph of the junction between two cell clusters. A light cell (*l*) has pale cytoplasm in which there are scanty dense-core vesicles appearing as black dots. The nucleus (*n*) is also pale and rounded, with a few clumps of heterochromatin predominantly at the periphery. The cell also contains a vacuole (*v*) in which there is a myelin figure derived from cellular membranes. Part of a dark cell (*d*) from the adjacent cluster is shown in which there are numerous dense-core vesicles within a dark cytoplasm. A row of sustentacular cells (*s*) runs between the clusters. These cells have elongated, moderately heterochromatic nuclei and attenuated cytoplasm in which there are few organelles. All figures in this chapter except Fig. 17.3 are of the human carotid body. Scale line = 5 μm.

of Figs. 17.2 and 17.3 demonstrates that human dense-core vesicles are considerably bigger than those of rats. Grimley and Glenner (1968) estimated their diameter to range between 100 and 200 nm. This is in contrast to the cat in which their diameter is said to range from 60 to 80 nm (Ross 1959) or from 50 to 150 nm (Lever et al. 1959), or in the rat where the diameter measures between 50 and 170 nm (McDonald and Mitchell 1975). There is considerable variation in intensity of staining of the central cores in the granules (Fig. 17.2).

The cytoplasm of light cells contains numerous vacuoles (Fig. 17.1) which range in diameter from little more than that of a dense-core vesicle to that of the nucleus. They may show no limiting membrane and contain myelin figures derived from breakdown of cytoplasmic membranes (Fig. 17.1). They are, therefore, most likely to be areas of autolytic degradation of the cytoplasm. They

probably correspond to the large, empty vacuoles seen in light chief cells by conventional light microscopy which are thus likely to be artifactual. Other cytoplasmic organelles in light cells include numerous round mitochondria measuring about 1 μm in diameter, isolated membranes of rough endoplasmic reticulum, free ribosomes and a Golgi apparatus. The last of these is often large in light cells and, in fortuitous sections, the matrix of the dense-core vesicles can be see forming within the membrane stacks (Grimley and Glenner 1968). Large, dense, irregular inclusions associated with paler lipid droplets are a frequent feature of light chief cells but not of the other two variants. These are granules of lipofuscin pigment such as one sees in a variety of cells as a consequence of ageing.

The dark variant contrasts sharply with the light cell because both its nucleus and cytoplasm are

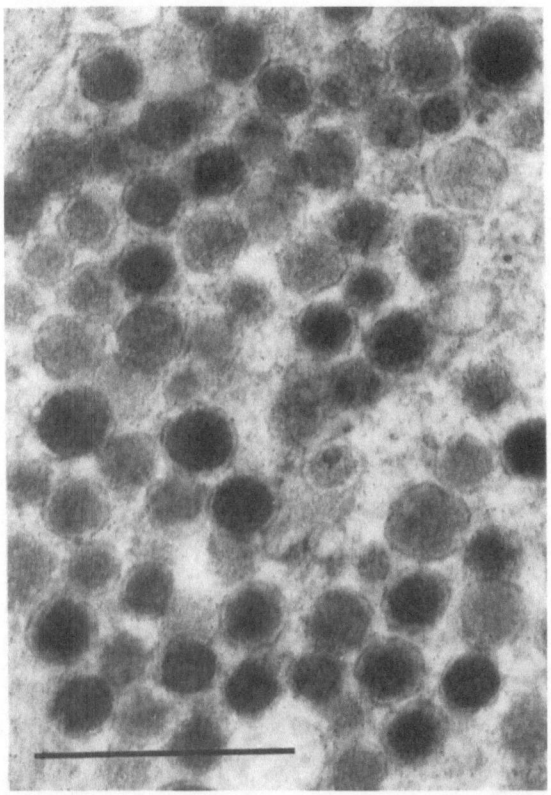

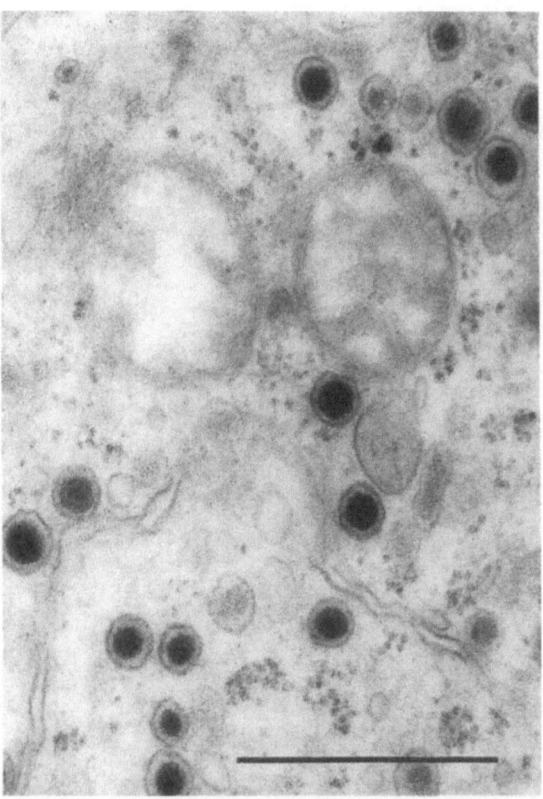

Fig. 17.2. Detail of dense-core vesicles in a dark cell. They are numerous and each consists of a large, dark, central core surrounded by a narrow clear zone and a single limiting membrane. There is considerable variation in staining density of the central cores. Scale line = 0.5 μm.

Fig. 17.3. Detail of dense-core vesicles in the carotid body of an Okamoto-Aoki rat for comparison with Fig. 17.2. Note that they are more sparsely distributed and smaller in diameter than those in humans. Also the peripheral "haloes" are more conspicuous. Scale line = 0.5 μm.

denser. The nucleus is slightly smaller than that of light cells and contains large aggregates of hetero-chromatin scattered throughout its interior, with a broad, irregular ring around its perimeter (Fig. 17.4; Jago et al. 1984). The dense cytoplasm contains an abundance of dense-core vesicles which are conspicuously more numerous than in light cells (Fig. 17.4). Many dark cells appear to be packed with vesicles almost to the exclusion of other organelles such as mitochondria, rough endoplasmic reticulum and the Golgi apparatus. There are no apparent differences in configuration or size of dense-core vesicles between light and dark cells when assessed subjectively (Grimley and Glenner 1968) but no quantitative studies have been performed to confirm this. A few vesicles in dark cells may be pale and enlarged to form microvacuoles up to 400 nm in diameter (Jago et al. 1984).

Chief cells with dense cytoplasm, which have been designated dark cells, have been described in other species including dog, cat, horse, rabbit, mouse and seal (Lever and Boyd 1957; Lever et al. 1959; Höglund 1967; Morita et al. 1969, 1970). In the rabbit, dark cells are readily identified by their abundant, intensely dark, pleomorphic granules, which contrast with the sparser, smaller, circular profiles of vesicles in the light variant (Lever et al. 1959; Smith 1986; Smith et al. 1986; see Chapter 21). In the rat, although light and dark variants of chief cell cannot be recognised, two populations can be identified by the size of their dense-core vesicles. These have been designated type A cells, containing many vesicles with a mean diameter of 116 nm, and type B cells, which have fewer vesicles with a mean diameter of 90 nm (McDonald and Mitchell 1975). In the macaque monkey four types of chief cell can be recognised on the basis of size and configuration of their dense-core vesicles (Hansen 1985).

The progenitor cell has not been described in animals and was not referred to by Grimley and Glenner (1968) in their studies of human carotid bodies. The most distinctive feature of this cell is

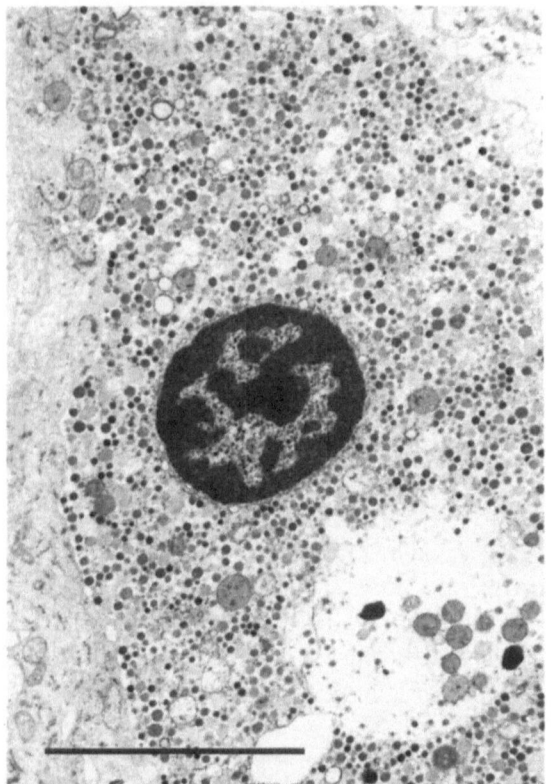

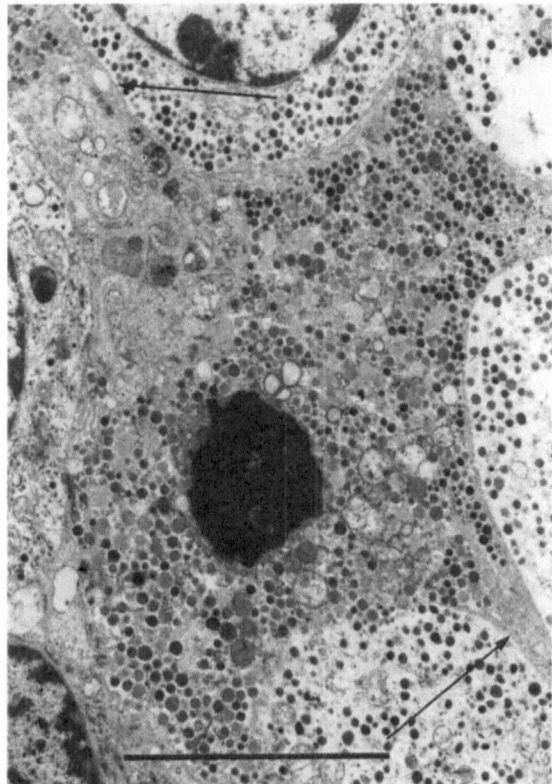

Fig. 17.4. A dark chief cell showing the round nucleus with large clumps of heterochromatin forming an irregular ring around its periphery. The cytoplasm contains an abundance of dense-core vesicles. Scale line = 5 μm.

Fig. 17.5. A progenitor cell with a small, totally heterochromatic nucleus. The cytoplasm is dense and contains numerous dense-core vesicles similar to the dark cell. Its border is thrown into several cytoplasmic extensions (*arrows*). Scale line = 5 μm.

its nucleus which is small and round or slightly indented. It consists almost entirely of highly condensed chromatin with a particularly dense band, some 1–2 μm thick, adjacent to the nuclear envelope. Two or three large aggregates of heterochromatin are present in the interior, surrounded by dense nucleoplasm. Sometimes the entire nucleus consists of dense heterochromatin (Fig. 17.5). The cytoplasm is similar to that of dark cells with numerous dense-core vesicles and sparsely distributed, round, intact mitochondria. There is no evidence of cytoplasmic degeneration, emphasising the fact that, despite their small, dark nuclei resembling those of pyknosis, these cells are far from being effete (Jago et al. 1984).

The Nature of Dense-Core Vesicles

The dense-core vesicles within chief cells have attracted much attention, not only because they are conspicuous, but also because they resemble

the densely cored variety of vesicle in synapses or the secretory granules in the adrenal medulla. Electron microscopic studies employing immuno-gold techniques (Kobayashi et al. 1983) or potassium dichromate cytochemistry (Christie and Hansen 1983) have shown that catecholamines are contained within the dense-core vesicles. Foremost amongst these is dopamine but there are also significant amounts of 5-hydroxytryptamine, noradrenaline and adrenaline. In a series of 76 pairs of carotid bodies removed at necropsy, the content of these four biogenic amines was, respectively, 64%, 18.8%, 14.8% and 2.5% (Steele and Hinterberger 1972). The concentration of dopamine in the rabbit carotid body is 20–40 μg/g, while that of noradrenaline is only 1.5 μg/g (Dearnaley et al. 1968). In the rat at least there is evidence that dopamine and noradrenaline may be stored separately in the chief cells (Christie and Hansen (1983). Electron microscopic immunocytochemistry has demonstrated that dense-core vesicles also contain methionine-enkephalin (see

Chapter 8) and furthermore that this peptide is contained in the same vesicles that store biogenic amines (Kobayashi et al. 1983). Although similar experiments have not been performed for leucine-enkephalin, it is likely that this peptide too resides in the vesicles. Thus, the characteristic vesicles of chief cells contain several products and it is tempting to suggest that peptides reside in the cores, since they stain darkly, whilst the more soluble amines are in the peripheral haloes.

Support for such an idea comes from experiments in which reserpine was used to stimulate degranulation of biogenic amines in cats and hamsters (Duncan and Yates 1967; Chen et al. 1969). Although this drug induces striking vacuolation of granules in sympathetic nerve endings and the adrenal medulla, it fails to induce discharge of granules in the carotid body. Interestingly, the nature of the fixative used for electron microscopy influences the observed effect of reserpine. Thus, fixation in osmium tetroxide alone reveals a striking reduction in the density of dense-core vesicles, whereas after fixation in glutaraldehyde, followed by osmium tetroxide, no change is observed (Duncan and Yates 1967; Chen et al. 1969). Since the latter method is a better fixative for protein it may be that the dense cores are largely proteinaceous in nature and thus show no response to reserpine. Christie and Hansen (1983) also used reserpine to deplete the carotid bodies of catecholamines in rats. They found that both dopamine and noradrenaline were reduced to 20% of their normal levels. This was associated with 89% of chief cells showing an absence of staining of their dense-core vesicles. In this study potassium dichromate was used to label the catecholamines, rather than the uranyl acetate and lead citrate as used in conventional electron microscopy. This technique does not stain the proteinaceous cores, which thus adopt an empty appearance once the catecholamines have been discharged.

Böck (1973) considered that the configuration of dense-core vesicles was an artifact of fixation. He pointed out that a mixture of noradrenaline and glutaraldehyde immediately forms a precipitate which stains darkly with osmium. Dopamine and glutaraldehyde behave similarly but the precipitate takes longer to form. Thus, during the interval between death and fixation of the carotid bodies, an uncontrolled quantity of dopamine diffuses out of the vesicles, accounting for the variability in staining intensity of the dense cores. The peripheral halo may even be an artifact due to shrinkage of the residual dopamine during fixation. Such studies ignore the fact that peptides

have been localised to the dense-core vesicles so that a significant proportion of the staining intensity of the cores can be accounted for by the presence of protein. It is not known whether catecholamines also share the cores or are confined to the clear periphery.

The function of dense-core vesicles is not known. Although structurally they resemble synaptic dense-core vesicles, they are considerably larger and do not appear to congregate in the vicinity of synaptic junctions. Indeed, chief cells of the rat contain small, clear, synaptic vesicles which are close to the cell membrane where nerve endings contact it, as discussed below. Dense-core vesicles may, therefore, be endocrine in nature, discharging their contents at the surface of the cell. The problem with this theory is that exocytosis of the vesicles is never encountered in electron micrographs of the carotid body even when it has been stimulated by acute hypoxia (see Chapter 18). If the carotid body of the rat is fixed by perfusion with glutaraldehyde containing a high concentration of potassium, exocytotic profiles of vesicles are seen in the intercellular spaces with some caught in the act of extrusion from the cell surface (Grönblad 1983). The explanation for this is that potassium ions cause depolarisation of the chief cell membranes, which triggers exocytosis. This process is then fixed in situ almost immediately afterwards by the glutaraldehyde. However, although the technique indisputably demonstrates that chief cells are capable of discharging their vesicles, it is uncertain to what extent it represents a normal physiological process.

Sustentacular Cells

Sustentacular cells do not contain distinctive organelles as do chief cells and their main distinguishing feature is the possession of long, fine cytoplasmic extensions which make close contact with chief cells. Each sustentacular cell gives rise to several of these processes, some of which encircle the perimeter of the cell clusters, enclosing them in a kind of basket. Others penetrate deep into the clusters where they partially enfold the chief cells (Fig. 17.6). The cytoplasmic processes taper gradually from the cell body so that their terminations, several cell diameters away, are extremely thin, measuring less than 50 nm (Jago et al. 1984). Where processes from neighbouring cells meet they overlap one another but only occasionally do they form cell junctions (Grimley and Glenner 1968). The cytoplasmic

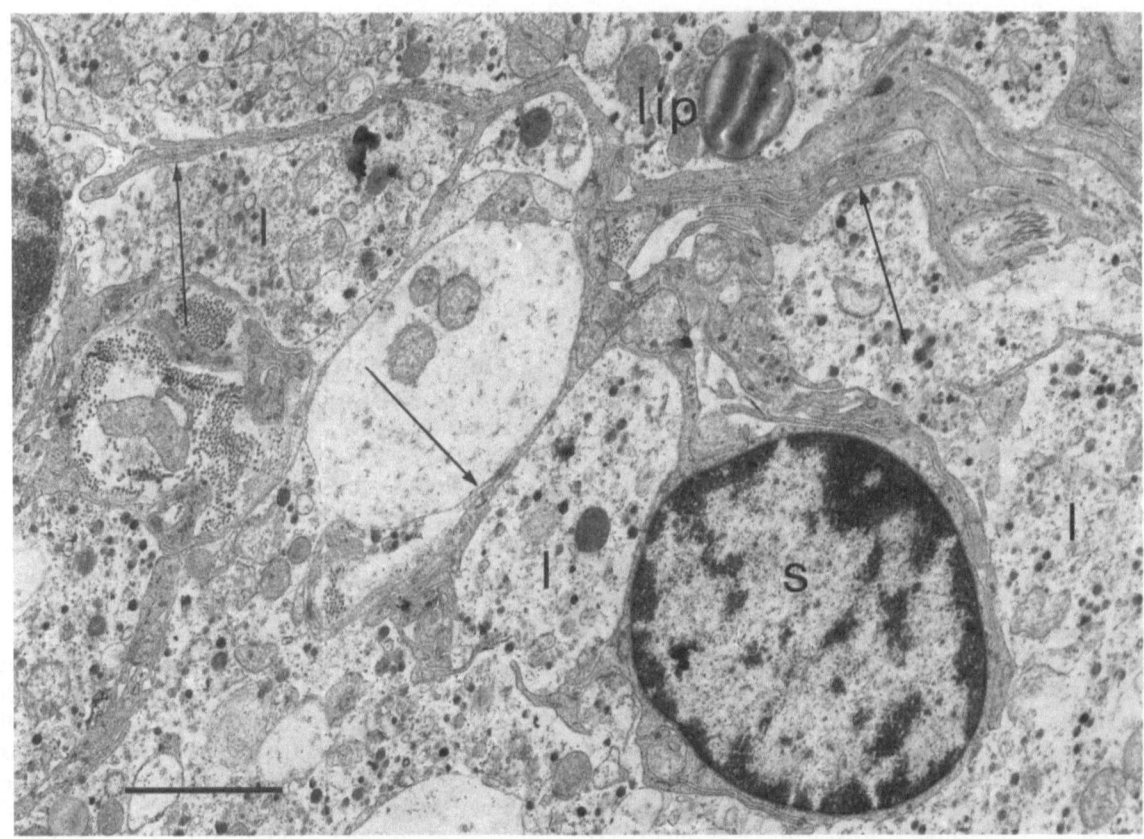

Fig. 17.6. A sustentacular cell (s) is buried deep within a core of light chief cells (l) and sends out numerous, fine, cytoplasmic extensions (arrows) which penetrate between, and partially encircle, the light cells. The nucleus of the sustentacular cell appears round because it has been sectioned transversely. One of the light cells contains a droplet of lipofuscin (lip). Scale line = 2 μm.

extensions of sustentacular cells contain nerve axons which they convey into intimate contact with the chief cells in the region of the synaptic nerve endings. In the normal human carotid body such axons are sparse and careful examination of electron micrographs may be required to detect them.

The main bodies of sustentacular cells are confined largely to the periphery of cell clusters (Fig. 17.1) but may also be found surrounded by chief cells within them (Fig. 17.6). Their nuclei are oval in shape and measure 3 μm × 9 μm (Jago et al. 1984), sometimes with an undulating profile (Fig. 17.1). The chromatin consists of numerous small aggregates within a grey nucleoplasm, with most of the heterochromatin being condensed at the periphery (Figs. 17.1, 17.6). The cytoplasm contains few organelles which are concentrated within the body of the cell. These comprise small segments of rough and smooth endoplasmic reticulum and occasional mitochondria. There is a randomly oriented network of intermediate (10 nm) filaments but these do not terminate in

peripheral attachment points as in smooth muscle. The processes of sustentacular cells always come between the chief cells and capillary endothelium. The capillaries are in turn surrounded by a basement membrane and pericytes and so there may be several layers of cytoplasm and ground substance separating chief cells from the blood to which they are believed to respond (Fig. 17.7).

Schwann Cells

Since the carotid body is a richly innervated organ, there are Schwann cells in association with its neurons. They are especially numerous in the interlobular septa where they are associated with large bundles of axons, most of which are myelinated. Nerve fibres between the cell clusters are thinner and only a few myelinated. The majority are contained within deep depressions of the cytoplasm of Schwann cells with which they exhibit a simple mesaxonal relationship. As the axons approach the edge of the cell clusters they lose the

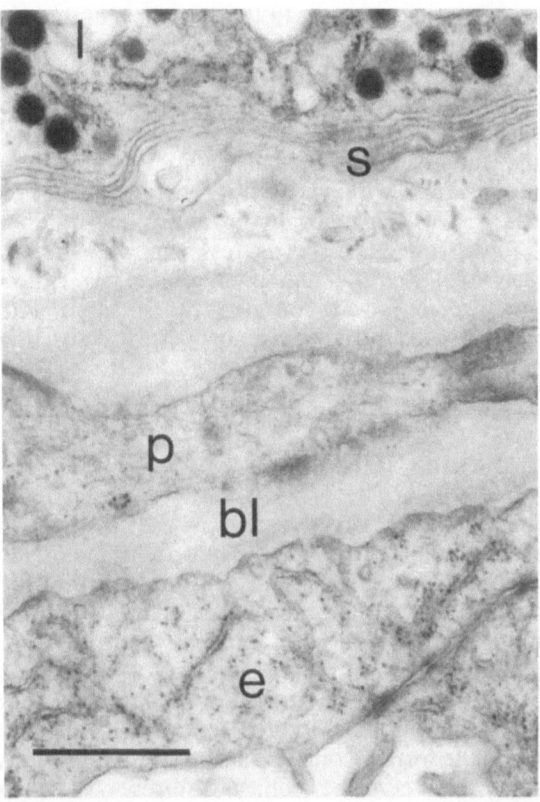

Fig. 17.7. An electron micrograph to show the various structures which may separate chief cells from the blood supply to the carotid body. A light cell (*l*) is enclosed by the fine, overlapping processes of sustentacular cells (*s*). These are separated from the capillary endothelium (*e*) by a pericyte (*p*) embedded within a thick basal lamina (*bl*). Scale line = 0.5 μm.

protection of Schwann cells but are taken up instead by sustentacular cells as described above.

The ultrastructure of Schwann cells is similar to that of the sustentacular cells. Their nuclei may be slightly larger, plumper and paler and their cytoplasm is not usually so finely attenuated. In other respects the two types of cell are identical and can be distinguished only be their location. Thus, Schwann cells are situated remotely from cell clusters and are associated either with myelin sheaths or numerous nerve axons in their cytoplasm. Sustentacular cells, on the other hand, penetrate the clusters and give rise to processes which enfold both axons and chief cells.

Axons and Nerve Endings

Nerve axons rapidly undergo autolysis and it is, therefore, rare that they can be demonstrated

clearly in human carotid bodies obtained post mortem. Grimley and Glenner (1968) examined carotid bodies removed surgically in which details of axoplasmic organelles were well preserved.

The fine bundles of nerves which approach and penetrate cell clusters contain regularly aligned neurotubules 18–20 nm in diameter, together with neurofilaments. They are enclosed within the cytoplasm of sustentacular cells until the zone of synaptic contact is reached. In this region the axon is often dilated to form a terminal bulb which contains a collection of small mitochondria and synaptic vesicles some 30–40 nm in diameter. Some vesicles measure 40–60 nm and contain a central dense core (Grimley and Glenner 1968). These features are typical of sympathetic efferent nerve endings.

The presence of nerve endings which are apparently presynaptic to chief cells cannot be reconciled with the generally accepted view that the glossopharyngeal nerve conducts impulses predominantly centripetally from the carotid body to the petrosal ganglion of the brain. A possible explanation for this apparent contradiction has been found not in humans but in the rat by McDonald and Mitchell (1975). They examined the ultrastructure of nerve endings in the carotid bodies of rats in which selected nerves had been sectioned. They found that cutting the glossopharyngeal nerve peripheral to the petrosal ganglion resulted in degeneration of 95% of the nerve endings. Section central to the ganglion did not cause degeneration, demonstrating that the great majority of nerve endings are in fact afferent and hence sensory. Section of the sympathetic trunk demonstrated that the remaining nerves are preganglionic sympathetic efferents. All of these end as boutons, whereas the afferent sensory endings comprise a mixture of boutons and calyces. It is thus not possible to determine the nature of a nerve ending by its size or shape (McDonald and Mitchell 1975). It is possible to distinguish them by their vesicles, although measurement of their size and density may be required to achieve this. Thus sympathetic efferents contain numerous, small, densely packed synaptic vesicles with a mean diameter of 53 nm, plus many dense-core vesicles some 95–97 nm in diameter. In contrast, afferent endings contain fewer but larger synaptic vesicles 61 nm in diameter, with approximately half as many vesicles with dense cores (McDonald and Mitchell 1975). McDonald and Mitchell identified four types of afferent nerve endings in the rat. Some are presynaptic to chief cells, others are postsynaptic to their cytoplasmic processes, whereas some are postsynaptic to the main body of the cell. The fourth type is a reciprocal synapse with impulses travelling in both directions.

The direction of the impulses across synapses can be determined from high resolution electron micrographs. The two apposed membranes show cytoplasmic densities on their inner surfaces by which the direction of transmission can be deduced. The density on the presynaptic membrane takes the form of several cone-shaped projections. That on the postsynaptic membrane forms an amorphous web. In reciprocal synapses the polarity of the densities is reversed on adjacent segments of the ending. In general there are more synaptic vesicles on the presynaptic site but this may be difficult to appreciate where the synapse is reciprocal. It is not certain whether the dense-core vesicles of chief cells are involved in transmitting impulses to the postsynaptic afferent nerve endings. Certainly, they tend to be more concentrated in their vicinity but are outnumbered by much smaller, clear synaptic vesicles identical with those in nerve endings. In this respect chief cells are like neurons and it has been suggested that they act like the interneurons of the central nervous system (McDonald and Mitchell 1975). This idea is reinforced by the finding of synapses between the chief cells themselves.

Synaptic densities have not been resolved by electron microscopy of the human carotid body and hence the direction of impulses at synaptic sites cannot be determined. However, the experiments described above demonstrate that, although many nerve endings may subjectively resemble sympathetic efferents, many are in reality sensory afferent endings (see Chapter 7).

Capillaries and Pericytes

The endothelium of glomic capillaries is similar to that in other organs, consisting of a flattened layer of cytoplasm which is thickest in the region of the nucleus. In animals the endothelium is fenestrated but in humans this feature is unconspicuous (Jago et al. 1984). Its cytoplasm contains scanty mitochondria and numerous micropinocytotic vesicles. Weibel–Palade bodies are rare, these being more a feature of endothelium in larger vessels. Beneath the endothelium is a basal lamina which is usually associated with a ring of ground substance containing isolated collagen fibrils (Fig. 17.7).

Embedded within the pericapillary connective tissue are pericytes which are wrapped around the capillaries for about half of their circumference. Their nuclei are usually curved, with a broad rim

of chromatin subjacent to the nuclear membrane. The cytoplasm on either side of the nucleus extends as long, thin processes containing scanty mitochondria, free ribosomes and myofilaments of actin and myosin like those in smooth muscle. Like smooth muscle they also contain dense focal condensations within the cytoplasm and at the cell surface. The density of myofilaments and number of focal condensations is less than in smooth muscle cells in the media of glomic arterioles. Pericytes are always intimately associated with blood vessels and never send out processes that contact chief cells.

Interstitial Cells

Fibrocytes are present within the connective tissues of the carotid body. They are occasionally encountered between cell clusters and in the adventitia of glomic arterioles. They are most numerous at the edges of lobules, where they are embedded within bands of collagen fibres. Ultrastructurally they consist of highly elongated, dark nuclei with scanty cytoplasm, which is attenuated in the form of long branching streamers which ramify between the collagen fibres. Although superficially they may resemble sustentacular cells, confusion between the two cell types should not arise: their nuclei are narrower than those of sustentacular cells, they are remote from chief cells and they are not associated with axons. Elongated cells, resembling fibrocytes, are commonly found in the perineurium of nerve bundles.

Mast cells are a common finding in the perivascular connective tissue of glomic arteries and in the stromal fibrous tissue. They are rounded cells with an irregular border and contain large granular-inclusions of variable electron density.

References

Böck P (1973) The morphology of the carotid body of the mouse. Arzneimittelforschung (Drug Res) 23:1612–1613
Chen IL, Yates RD, Duncan D (1969) The effects of reserpine and hypoxia on the amine-storing granules of the hamster carotid body. J Cell Biol 42:804–816
Christie DS, Hansen JT (1983) Cytochemical evidence for the existence of norepinephrine-containing glomus cells in the rat carotid body. J Neurocytol 12:1041–1053
Dearnaley DP, Fillenz M, Woods RI (1968) The identification of dopamine in the rabbit's carotid body. Proc Roy Soc Lond [Biol] 170:195–203
Duncan D, Yates R (1967) Ultrastructure of the carotid body of the cat as revealed by various fixatives and the use of reserpine. Anat Rec 157:667–682
Grimley PM, Glenner GG (1968) Ultrastructure of the human

carotid body. A perspective of the mode of chemoreception. Circulation 37:648–665

Grönblad M (1983) Improved demonstration of exocytotic profiles in glomus cells of rat carotid body after perfusion with glutaraldehyde fixative containing a high concentration of potassium. Cell Tissue Res 229:627–637

Hansen JT (1985) Ultrastructure of the primate carotid body: a morphometric study of the glomus cells and nerve endings in the monkey (*Macaca fascicularis*). J Neurocytol 14:13–32

Höglund R (1967) An ultrastructural study of the carotid body of horse and dog. Z Zellforsch 76:576–586

Jago R, Smith P, Heath D (1984) Electron microscopy of carotid body hyperplasia. Arch Pathol Lab Med 108:717–722

Kobayashi S, Uchida T, Ohashi T et al. (1983) Immunocyto-chemical demonstration of the co-storage of noradrenaline with met-enkephalin-arg[6] -phe[7] and met-enkephalin -arg[6] -gly[7] -leu[8] in the carotid body chief cells of the dog. Arch Histol Jpn 46:713–722

Lever JD, Boyd JD (1957) Osmiophile granules in the glomus cells of the rabbit carotid body. Nature 179:1082–1083

Lever JD, Lewis PR, Boyd JD (1959) Observations on the fine structure and histochemistry of the carotid body in the cat and the rabbit. J Anat 93:478–490

McDonald DM, Mitchell RA (1975) The innervation of glomus cells, ganglion cells and blood vessels in the rat carotid body: a quantitative ultrastructural analysis. J Neurocytol 4:177–230

Morita E, Chiocchio SR, Tramezzani JH (1969) Four types of main cell in the carotid body of the cat. J Ultrastruct Res 28:399–410

Morita E, Chiocchio SR, Tramezzani JH (1970) The carotid body of the Weddell seal (*Leptonychotes weddelli*). Anat Rec 167:309–328

Ross LL (1959) Electron microscopic observations of the carotid body of the cat. J Biophys Biochem Cytol 6:253–262

Smith P (1986) Electron microscopy of the abnormal carotid body. In: Heath D (ed) Aspects of hypoxia. Liverpool University Press, Liverpool, pp 77–96

Smith P, Heath D, Fitch R, Hurst G, Moore D, Weitzenblum E (1986) Effects on the rabbit carotid body of stimulation by almitrine, natural high altitude, and experimental normo-baric hypoxia. J Pathol 149:143–153

Steele RH, Hinterberger J (1972) Catecholamines and 5-hydroxytryptamine in the carotid body in vascular, respiratory, and other diseases. J Lab Clin Med 80:63–70

18 Ultrastructure of the Hypoxic and Hyperplastic Carotid Body

In this chapter the ultrastructure of dense-core vesicles and other organelles in the glomic chief cells of hypoxic animals is discussed. The ultrastructural appearances of the human carotid body in sustentacular cell hyperplasia and dark cell proliferation are also described, with particular reference to the nature of dark and progenitor cells.

Acute Hypoxia and Dense-Core Vesicles

Changes in the structure of dense-core vesicles in response to acute episodes of hypoxia have been described, but their validity is questionable. In one such experiment, rats were subjected to severe hypoxia, in which the inspired air contained only 2.5% oxygen, for periods ranging from 5 to 20 minutes (Blümcke et al. 1967). This treatment induced a reduction in the number of dense-core vesicles, some of which were apparently discharging from the surface of the chief cells. Associated with this there was extensive cytoplasmic degeneration in the form of mitochondrial swelling and nuclear fragmentation. It is likely that the changes in the vesicles seen in this study were not physiological but a pathological response to unduly severe hypoxia. Subsequent studies have consistently failed to demonstrate exocytosis of vesicles in response to hypoxia. Cats were rendered hypoxic for 45 minutes by administering a mixture of 9% oxygen in nitrogen through an endotracheal tube (Al-Lami and Murray 1968). Although the cytoplasm of chief cells was denser than in controls,

there was no difference in the size, electron density or number of dense-core vesicles. Another study using hamsters produced similar negative findings when 3%–5% oxygen was breathed for 20 minutes (Chen et al. 1969). Electron microscopy revealed no changes in the form or number of vesicles nor in the cytoplasmic density. There is good evidence that dense-core vesicles contain catecholamines which are depleted in chief cells upon brief exposure to hypoxia (see Chapter 7). However, the ultrastructural evidence indicates that this apparent secretory activity is achieved by a total absence of exocytosis. Experiments with the potent catecholamine-releasing drug reserpine also fail to induce exocytosis or vacuolation of dense-core vesicles (see Chapter 17). Thus, dopamine and noradrenaline are discharged from chief cells by other means, perhaps by diffusing out of the vesicles through the cytoplasm. The fact that their loss does not lead to vacuolation of dense-core vesicles suggests that the structure of these organelles is due largely to their content of protein. Indeed it may be that vesicles are not directly involved in the chemosensory reflex at all (Chen et al. 1969) but are concerned with longer-term acclimatisation to hypoxia, perhaps of an endocrine nature.

Chronic Hypoxia and Dense-Core Vesicles

The term "chronic hypoxia" is often used to describe both permanent residence at high

altitude and temporary accommodation in a
decompression chamber for several weeks. While
such experimental procedures bring about a
longer exposure to hypoxia than do the short,
extreme studies described above, they are better
designated "prolonged hypoxia" rather than chronic hypoxia, which is measured in months or years.
Thus, results obtained from animals in hypobaric
chambers must be compared with those from
native highlanders only with great caution.

In contrast to acute exposure, chronic and
prolonged hypoxia induce characteristic changes
in the dense-core vesicles. The first of such studies
was that of Edwards et al. (1972) on the carotid
bodies of guinea-pigs born and bred in Cerro de
Pasco (4330 m) in the Peruvian Andes (see
Chapter 21). They showed a striking increase in
size and vacuolation of the dense-core vesicles, in
which the central core was often reduced in size
and situated eccentrically within the vesicle (see
Fig. 11.7). In some, the core was virtually absent,
producing clear vesicles up to 350 nm in diameter.
Larger vacuoles up to 2 μm in diameter were also
found and it was suggested that these may form by
progressive enlargement and fusion of the smaller,
clear vesicles (Edwards et al. 1972).

In contrast Møller et al. (1974) did not find such
vacuolation in rabbits living at a comparable
altitude on the Bolivian altiplano. Subjective
assessment suggested that there was a doubling in
the numbers of dense-core vesicles but this was not
supported by quantitative data. Blessing and Kaldeweide (1975) also found no vacuolation of
vesicles but an apparent increase in their numbers
in rats exposed to prolonged hypoxia at a simulated altitude of 7000 m. Later studies (Laidler and
Kay 1978a) found vesicular vacuolation in rats
kept in a hypobaric chamber at a pressure of
460 mmHg, simulating an altitude equivalent to
that of Cerro de Pasco. Exhaustive measures were
taken to exclude autolytic changes by perfusing
the carotid bodies with fixative in artificially ventilated, anaesthetised rats. Their glomic cells
showed enlarged, vacuolated dense-core vesicles,
as in the Peruvian guinea-pigs. Quantitative techniques were used to assess changes in population
of vesicles (Laidler and Kay 1978b). A significant
reduction in the number of vesicles per unit area
was found but was associated with a proportionately large increase in the area of each chief cell.
Thus the absolute number of dense-core vesicles
in each cell was unchanged.

More recently the carotid bodies of eight rabbits
which were born and lived in Cerro de Pasco were
compared with those from six rabbits which had
been kept in a normobaric, hypoxic chamber at an

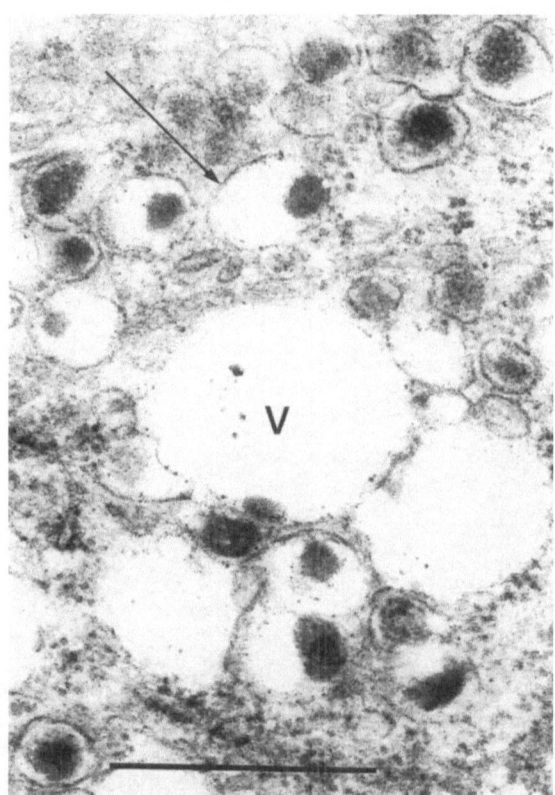

Fig. 18.1. Vacuolation of dense-core vesicles in a light cell from
the carotid body of a rabbit born and raised in Cerro de Pasco
at an altitude of 4330 m. The vesicles (*arrow*) are enlarged and
the dense cores reduced in size and situated eccentrically.
Some vesicles have enlarged sufficiently to produce small
vacuoles (*v*). Scale line = 0.5 μm.

equivalent partial pressure of oxygen for 3 or 6
months. Vacuolation of dense-core vesicles was
seen in all three groups but was confined almost
exclusively to the light variant of chief cell (Smith
et al. 1986a; Fig. 18.1). In this species light cells
contain small, circular vesicles (see Chapter 17)
and these were enlarged with reduced, eccentric
cores and correspondingly broader peripheral
haloes (Fig. 18.1). In contrast, dark chief cells with
larger, pleomorphic vesicles and indistinct haloes
were not involved (Fig. 18.2). Vacuolation of
dense-core vesicles was most prominent in the
Peruvian rabbits exposed all of their lives to
hypobaric hypoxia but, even in them, was found in
only about half of the light cells. Fewer chief cells
were affected in the rabbits which were exposed to
simulated altitude for 6 months. The changes were
scarce in rabbits made hypoxic for only 3 months.
Vacuolation of dense-core vesicles thus seems to
be dependent upon the duration of hypoxia. In
these studies individual dense-core vesicles were

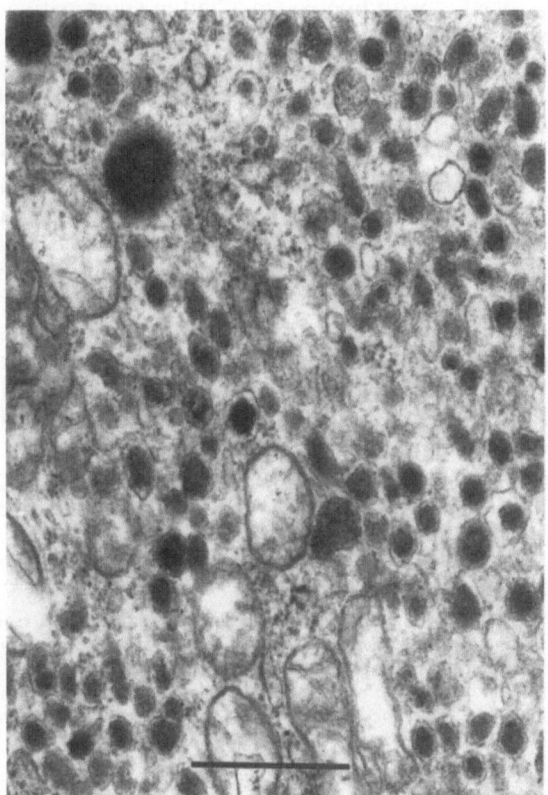

Fig. 18.2. Part of a dark chief cell from a rabbit subjected to the normobaric hypoxia of 12.7% oxygen for 3 months, simulating the same partial pressure of oxygen as in Cerro de Pasco. The cytoplasm is dense and the dense-core vesicles are very dark and of variable shape. They do not show vacuolation. Scale line = 0.5 μm.

encountered which had enlarged to the size of mitochondria (Fig. 18.1). Larger vacuoles with diameters approaching that of the nucleus were also seen, but no intermediates were found to suggest that they develop by fusion or expansion of the smaller vacuoles as suggested by Edwards et al. (1972). The large vacuoles were different in appearance from dense-core vesicles, having no limiting membrane, an irregular ragged outline and no electron-dense core. They resembled the fluid-filled vacuoles that occur during cytoplasmic degeneration. The vacuoles were at their largest and most numerous in rabbits subjected to hypoxia for 3 months, yet it was in these animals that enlargement of vesicles was scarcest.

The functional significance of vacuolation of the vesicles by prolonged or chronic hypoxia is unknown. The concept that it is associated with discharge of catecholamines in response to hypoxaemia is not supported by the length of time required for its appearance. If the process were a normal physiological response to a hypoxic stimu-lus it should be detectable acutely. What is more, depletion of catecholamines by reserpine does not cause the vesicles to vacuolate. Alternatively the increased diameter of the vesicles may represent storage of catecholamines, consistent with the increased levels of dopamine which have been demonstrated in the chief cells of animals after prolonged hypoxia (Pallot and Barer 1982). The ultrastructural changes may have nothing to do directly with chemoreception but be associated with secretion of polypeptides, hence explaining the diminution in size of the central core. The enlarged carotid bodies of spontaneously hyper-tensive rats do not show vacuolation of dense-core vesicles (Fig. 17.3; Smith et al. 1984), indicating that an elevation of blood pressure per se is not the factor responsible for its genesis. It is not known whether or not similar changes occur in humans resident at high altitude or in patients with emphysema. There is clearly a need for ultrastructural studies on the carotid bodies of hypoxic humans.

Quantitative Changes Following Prolonged Hypoxia

The ultrastructural changes in glomic cells in response to prolonged hypoxia have been assessed morphometrically in rats which had been kept in hypoxic chambers for 21–35 days and in which the carotid bodies were fixed by vascular perfusion. These investigations demonstrate that chief cells enlarge by three to four times their normal volume (Laidler and Kay 1978b; Dhillon et al. 1984; McGregor et al. 1984; Pequignot et al. 1984). Most of this enlargement is due to hypertrophy of the cytoplasm, since nuclei are either slightly reduced in size (Laidler and Kay 1978b) or only marginally enlarged (Dhillon et al. 1984). Another important contribution to enlargement of the organ is dilatation of the small blood vessels supplying the cell clusters (see Chapter 9; Laidler and Kay 1978a; Dhillon et al. 1984; McGregor et al. 1984). Indeed, it has been suggested that carotid body enlargement can be explained entirely by these two factors (McGregor et al. 1984), but Dhillon et al. (1984) found increased counts of chief cell nuclei, suggesting that hyperplasia as well as hypertrophy is responsible for the enlargement of the carotid body. In this respect there is unequivocal evidence that chief cells are capable of division during exposure to hypoxia. Numerous mitotic figures in chief cells were found by light and electron microscopy in rats treated with vincristine, which arrests mitosis in metaphase (Barer et al. 1986; Pallot 1987).

The hypertrophied and hyperplastic chief cells show quantitative changes in their organelles, particularly in mitochondria and dense-core vesicles. The mitochondria are not increased in number when expressed in terms of the unit area of cross-section of cytoplasm but in view of the increase in cytoplasmic volume their population within each cell is increased (Laidler and Kay 1978b). Dense-core vesicles show a reduced population density (Dhillon et al. 1984; Laidler and Kay 1978b) but their number per cell is normal (Laidler and Kay 1978b) or increased (Dhillon et al. 1984). Other organelles have not been quantified but qualitative assessment indicates that there is no change in the extent of the Golgi apparatus, rough endoplasmic reticulum or free ribosomes, suggesting that the synthetic capabilities of chief cells are not increased.

Ultrastructure of Sustentacular Cell Hyperplasia

This form of hyperplasia occurs in the carotid bodies of patients with pulmonary emphysema associated with hypoxaemia and in some patients with systemic hypertension. Histologically it consists of thick whorls of proliferated sustentacular cells surrounding diminished cores of chief cells and develops largely in middle-aged or elderly subjects in whom age-changes in the glomic tissue are manifest (see Chapter 10). To date the ultrastructural features have been studied in only one case, a woman of 80 years with systemic hypertension and gross left ventricular hypertrophy (Jago et al. 1984). In this case there was a proliferation of elongated cells surrounding cores of chief cells. The elongated cells were oriented at varying angles to the plane of section so that in some areas their nuclei had been cut transversely and thus appeared to be round (Fig. 18.3). They had none of the ultrastructural features of fibroblasts. Their cytoplasm contained a paucity of

Fig. 18.3. Sustentacular cell hyperplasia in a carotid body from a woman of 80 years with systemic hypertension. The nuclei (*n*) of numerous sustentacular cells have been sectioned transversely and are larger and paler than normal. Numerous nerve axons (*a*) are embedded within the cytoplasm of the sustentacular cells. Scale line = 5 μm.
▼

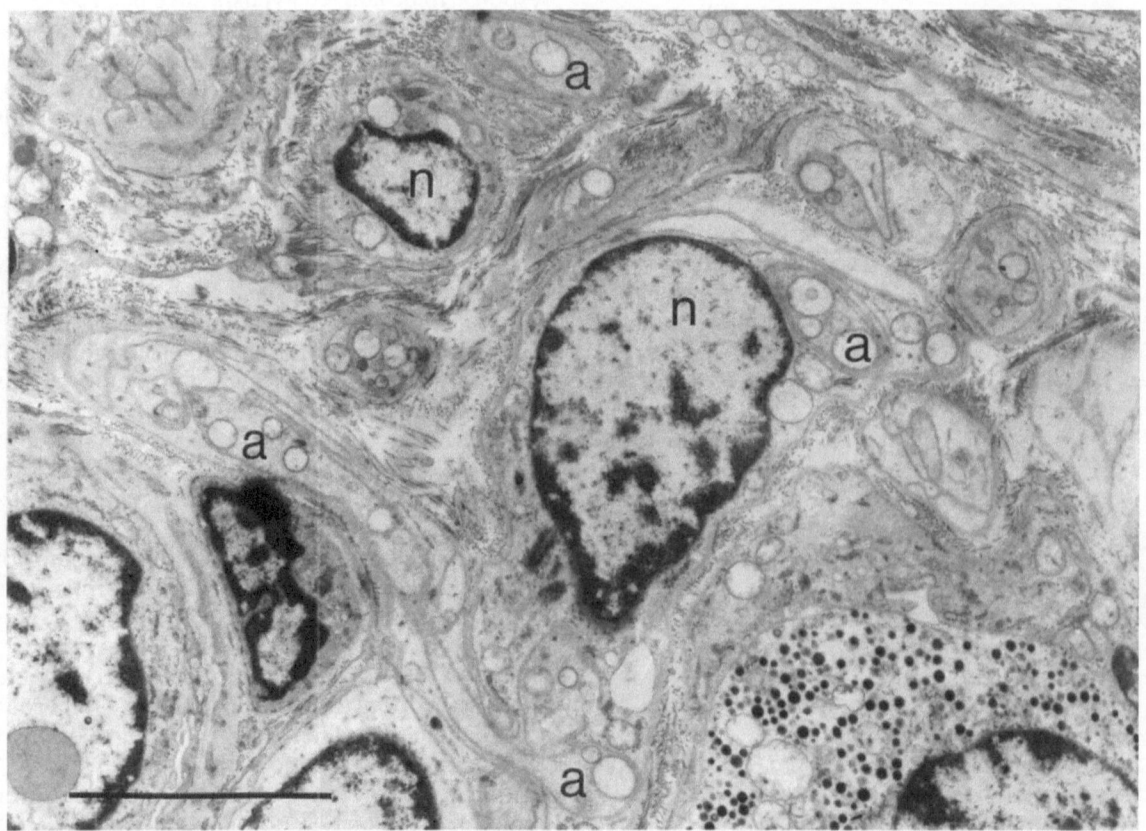

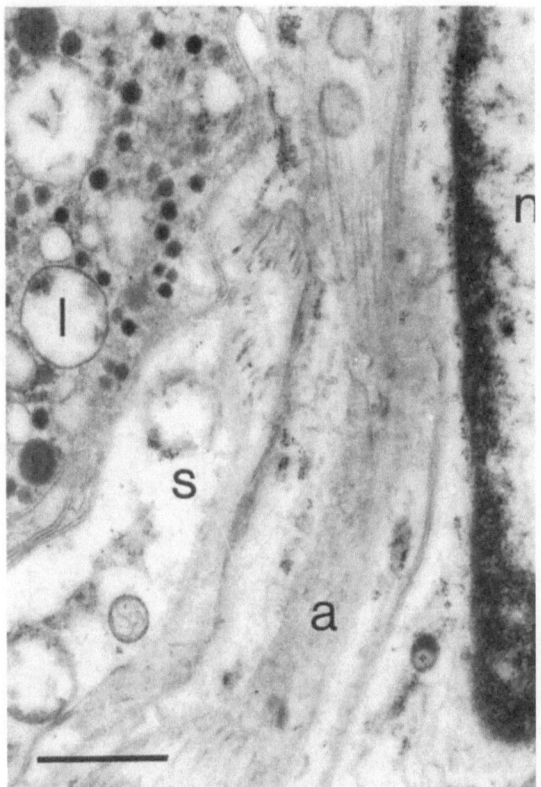

Fig. 18.4. The same case as in Fig. 18.3 showing detail of a sustentacular cell (*s*) adjacent to a light chief cell (*l*). Within the cytoplasm of the sustentacular cell is a nerve axon (*a*) containing neurofilaments. The nucleus (*n*) of the sustentacular cell is also shown. Scale line = 1 μm.

sheaths were occasionally found in cells which otherwise fulfilled the criteria for being sustentacular (Jago et al. 1984). The problem is largely one of semantics, since both types of cell probably have the same origins and perform similar rôles (see Chapter 3). However, it should be appreciated that the term "sustentacular cell hyperplasia" implies the involvement of Schwann cells.

The central cores of chief cells in this case contained an increased number of processes from sustentacular cells, with an apparent increase in the number of axons contacting chief cells, although these were too few to explain their high density at the periphery. The chief cells themselves showed no abnormality. In particular there was no vacuolation of dense-core vesicles as described above in animals. It should be borne in mind, however, that this was a case of systemic hypertension, a condition, which in the rat at least, is not associated with vacuolation of vesicles (Fig. 17.3).

A small minority of the cells involved in the hyperplasia were fibroblasts. These had narrower, dark nuclei and were confined to the periphery of the lobules forming a thin layer distant from the main mass of sustentacular cells. Many were inactive fibrocytes with scanty, finely attenuated cytoplasm but others were active fibroblasts containing extensive, dilated, rough endoplasmic reticulum (Jago et al. 1984).

organelles and small groups of intermediate filaments. Cytoplasmic extensions incorporated nerve axons (Figs. 18.3, 18.4) which were numerous so that each cell contained several axons crowded together. In fortuitous sections mesaxons could be identified (Jago et al. 1984).

The cytological appearance and the presence of axons within them identified the elongated cells as either sustentacular or Schwann cells. Many were indisputably sustentacular, since they sent out processes which enfolded both chief cells and axons. These formed the inner layers adjacent to the core of chief cells. Others at the periphery were associated with nerve bundles and myelin sheaths and were undoubted Schwann cells. Cells in an intermediate position were regarded as a mixture of the two types of cell. Difficulties in distinction between the two were exaggerated by the fact that the nuclei of sustentacular cells were broader and paler than normal and were thus indistinguishable from those of Schwann cells (Fig. 18.3). What is more, rudimentary myelin

Dark Cell Proliferation

A proliferation of the dark variant of chief cell can occur either diffusely consequent upon subacute hypoxaemia or as focal aggregates superimposed upon sustentacular cell hyperplasia (see Chapter 9). The ultrastructure of examples of both types has been described (Smith 1986; Smith et al. 1986b). In the focal aggregates of dark cells there is considerable pleomorphism in which the nuclei assume a variety of shapes with many infoldings of their membranes and show much variation in electron density. The cytoplasm contains several, large, round mitochondria and irregular cisternae of rough endoplasmic reticulum, not normally seen in dark cells. There is a wide range in diameter of dense-core vesicles, some being as large as 600 nm. There are many empty microvacuoles as large as 800 nm, apparently merging with the vesicles. The outline of the cells is very irregular, with numerous, short, branching processes. These focal aggregates of pleomorphic dark cells may develop when there is a sudden

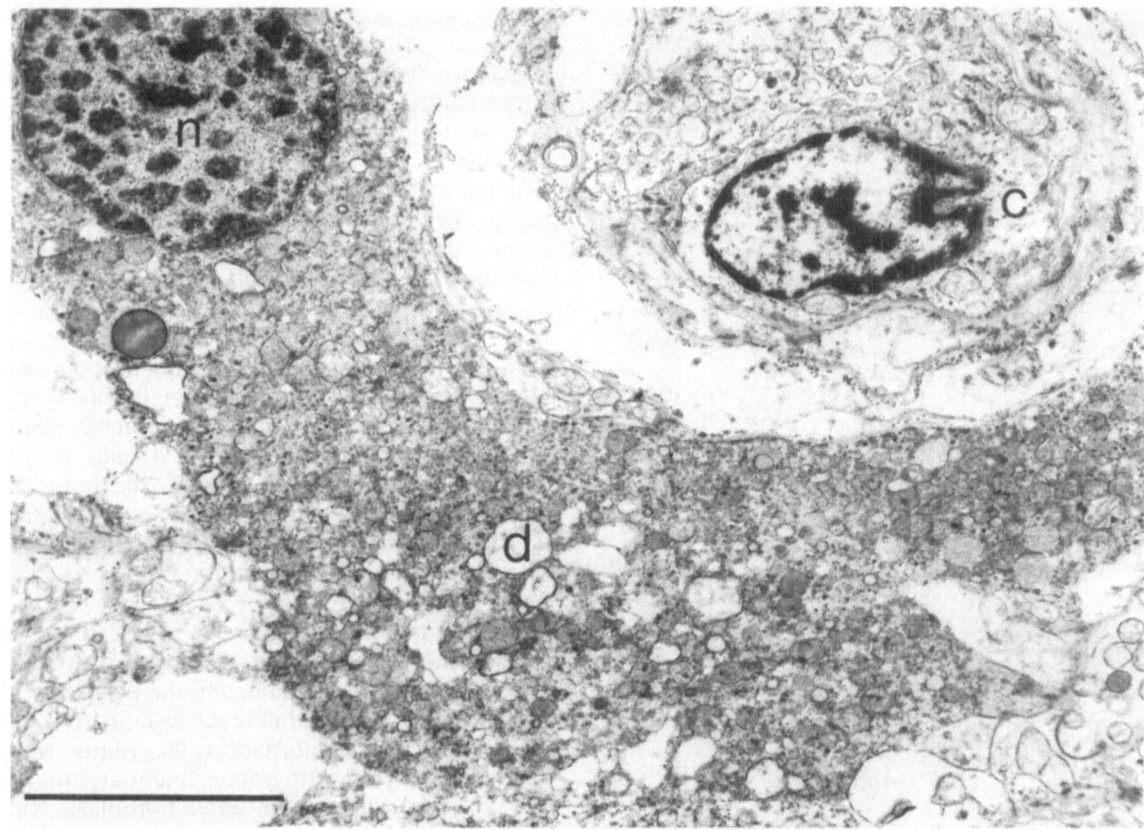

Fig. 18.5. Carotid body from a woman of 62 years with a ventricular septal defect and subacute reversal of the cardiac shunt inducing hypoxaemia for the last few months of her life. A dark chief cell (*d*) is shown with a large, round nucleus (*n*) containing clumps of heterochromatin. The cytoplasm is attenuated into a long streamer which partially surrounds a capillary (*c*). Scale line = 5 μm.

exacerbation of hypoxaemia or systemic hypertension (Heath et al. 1984).

The appearance of dark cells is different when they proliferate diffusely. The ultrastructure of a case of dark cell prominence was described in a woman of 62 years with a ventricular septal defect who died in congestive cardiac failure following subacute reversal of the shunt (Smith et al. 1986b). Dense-core vesicles were found in most chief cells but were especially numerous in the dark variety. None of the vesicles was vacuolated and this may be because the duration of hypoxaemia was too short or simply that the human carotid body does not respond to prolonged hypoxaemia in the same manner as animals. Light chief cells showed extensive cytoplasmic degeneration and contained swollen mitochondria with disrupted cristae typical of autolysis. Numerous large droplets of lipofuscin were present within them consistent with the age of the patient. In contrast, the cytoplasm of dark cells was much better preserved, denser and contained few lipofuscin droplets (Fig. 18.5). It con-

tained abundant mitochondria (Fig. 18.6) which are normally scanty in dark cells. Furthermore, many of these dark cells were bigger than normal, with large, almost circular, nuclei and copious cytoplasm, which was often elongated into streamers that partially encircled light cells, capillaries and bundles of nerves (Fig. 18.5). The ultrastructural appearances of these dark cells was suggestive of active metabolism but whether this was because they were dividing or synthesising catecholamines and polypeptides is not known.

The Nature of Dark and Progenitor Cells

The differences between the three variants of chief cell (see Chapter 3) cannot be attributed to autolysis or trauma as suggested by McDonald (1981). Thus, whilst it is possible that variations in electron density of the cytoplasmic matrix can be created by different states of hydration, the strik-

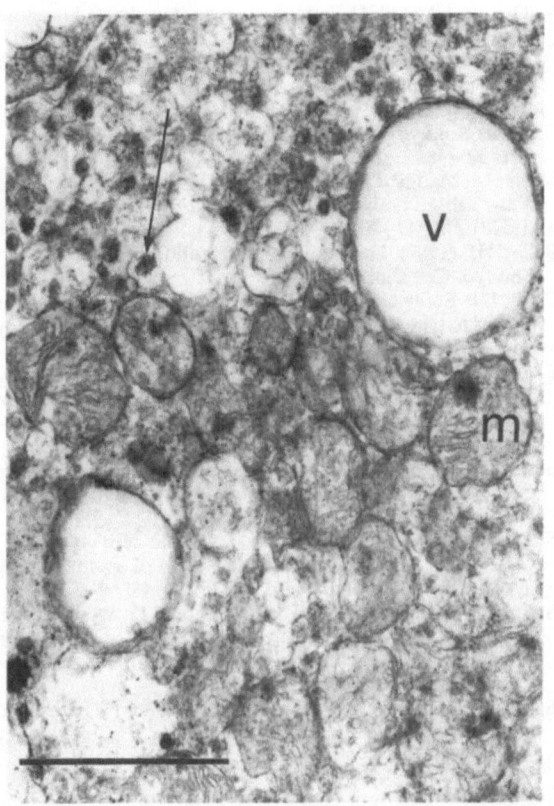

Fig. 18.6. Detail of a dark cell from the same case as in Fig. 18.5. There are numerous mitochondria (*m*) and dense-core vesicles (*arrow*). A few small vacuoles (*v*) are also present. Scale line = 1 μm.

ing numerical differences in mitochondria and dense-core vesicles is a more constant distinction, which implies functional differences between the cells. The relationship between the three variants remains unclear. Initially the light cell was regarded as the active form of chief cell, which then underwent progressive cytoplasmic condensation and nuclear contraction to form first the dark cell and finally an effete cell with a small, dense nucleus which was designated a "pyknotic cell". However, electron microscopy shows that neither dark nor progenitor cells show cytoplasmic degeneration (see Chapter 17) permitting at least two alternative hypotheses. The first supposes that dark and progenitor cells are inactive storage cells. This is based on the fact that heterochromatic nuclei are commonly associated with reduced transcription of messenger RNA and hence with little synthesis of protein. The paucity of mitochondria and large numbers of dense-core vesicles are in accord with a storage rôle. In contrast the large pale nucleus, numerous mitochondria and extensive rough endoplasmic reticulum of the light

cell is in keeping with synthetic activity and secretion of vesicles. The second alternative takes cognisance of the fact that, in most secretory organs, small cells with dark nuclei are precursors of metabolically active, mature forms. Thus, according to this view progenitor and dark variants are destined to mature into light cells.

The concept of dark and progenitor cells as precursors is supported by several observations. Dark variants contain more dense-core vesicles than the light which is contrary to what one would expect if the dark cell were an exhausted form. Dark cells contain much less lipofuscin pigment than do light variants. Since this is a pigment associated with ageing, it would be anticipated that it would be present in larger quantities in those cells which are oldest, and on this criterion it is the light, not the dark, cell which is the older of the two. The fact that dark cells always show less cytoplasmic degeneration than light cells is contrary to their being effete. Further evidence comes from studies of the effects of age on the human carotid body in which dark cells are more numerous in young people than in the elderly (Hurst et al. 1985). In the case of ventricular septal defect, dark cells were numerous and abnormally large in response to subacute hypoxaemia. In rabbits exposed to hypoxia for 3 months, dark cells, with their large pleomorphic granules (Fig. 18.2), were increased in numbers but after 6 months of hypoxia or permanent residence at high altitude they were fewer than normal (Smith et al. 1986a). Finally, as described above, dark cells in the human carotid body can proliferate focally. These studies demonstrate that, far from being inactive or effete, dark cells can multiply when stimulated to do so.

Until recently, the position of the progenitor cell in this scheme was unclear. It was regarded as either a defunct or inactive derivative of the dark cell or a more immature form of it (Smith 1986). However, the discovery of many of these cells within the developing carotid bodies of human fetuses (Heath et al. 1990) strongly suggested that they were the forerunners of the other two variants. Accordingly the term "progenitor cell" was adopted for them.

References

Al-Lami F, Murray RG (1968) Fine structure of the carotid body of normal and anoxic cats. Anat Rec 160:697–718

Barer G, Wach R, Pallot D, Bee D (1986) Almitrine, hypoxia, systemic hypertension and the carotid body. In: Heath D (ed) Aspects of hypoxia. Liverpool University Press, Liverpool, pp 113–129

Blessing MH, Kaldeweide J (1975) Light and electron micro-scopic observations on the carotid bodies of rats following adaptation to high altitude. Virchows Arch [B] 18:315–329

Blümcke S, Rode J, Niedorf HR (1967) The carotid body after oxygen deficiency. Z Zellforsch 80:52–77

Chen IL, Yates RD, Duncan D (1969) The effects of reserpine and hypoxia on the amine-storing granules of the hamster carotid body. J Cell Biol 42:804–816

Dhillon BP, Barer GR, Walsh M (1984) The enlarged carotid body of the chronically hypoxic and chronically hypoxic and hypercapnic rat: a morphometric analysis. Q J Exp Physiol 69:301–317

Edwards C, Heath D, Harris P (1972) Ultrastructure of the carotid body in high altitude guinea pigs. J Pathol 107:131–136

Heath D, Smith P, Jago R (1984) Dark cell proliferation in carotid body hyperplasia. J Pathol 142:39–49

Heath D, Khan Q, Smith P (1990) Histopathology of the carotid bodies in neonates and infants. Histopathology 17:511–520

Hurst G, Heath D, Smith P (1985) Histological changes associated with ageing of the human carotid body. J Pathol 147:181–187

Jago R, Smith P, Heath D (1984) Electron microscopy of carotid body hyperplasia. Arch Pathol Lab Med 108:717–722

Laidler P, Kay JM (1978a) Ultrastructure of carotid body in rats living at a simulated altitude of 4300 metres. J Pathol 124:27–33

Laidler P, Kay JM (1978b) A quantitative study of some ultrastructural features of the type I cells in the carotid bodies of rats living at a simulated altitude of 4300 metres. J Neurocytol 7:183–192

McDonald DM (1981) Peripheral chemoreceptors. Structure function relationships of the carotid body. In: Hornbein TF (ed) Regulation of breathing. Marcel Dekker, New York Basel, pp 105–319

McGregor KH, Gil J, Lahiri S (1984) A morphometric study of the carotid body in chronically hypoxic rats. J Appl Physiol 57:1430–1438

Møller M, Møllgård K, Sørensen SC (1974) The ultrastructure of the carotid body in chronically hypoxic rabbits. J Physiol (Lond) 238:447–453

Pallot DJ (1987) The mammalian carotid body. Adv Anat Embryol Cell Biol 102: 1–91

Pallot DJ, Barer GR (1982) Increased catecholamine content related to type I cell hyperplasia in the enlarged carotid body of chronically hypoxic animals. J Anat 135:840P

Pequignot JM, Hellström S, Johansson C (1984) Intact and sympathectomized carotid bodies of long-term hypoxic rats: a morphometric ultrastructural study. J Neurocytol 13:481–493

Smith P (1986) Electron microscopy of the abnormal carotid body. In: Heath D (ed) Aspects of hypoxia. Liverpool University Press, Liverpool, pp 77–96

Smith P, Jago R, Heath D (1984) Glomic cells and blood vessels in the hyperplastic carotid bodies of spontaneously hypertensive rats. Cardiovasc Res 18:471–482

Smith P, Heath D, Fitch R, Hurst G, Moore D, Weitzenblum E (1986a) Effects on the rabbit carotid body of stimulation by almitrine, natural high altitude, and experimental normobaric hypoxia. J Pathol 149:143–153

Smith P, Hurst G, Heath D, Drew R (1986b) The carotid bodies in a case of ventricular septal defect. Histopathology 10:831–840

19 Chemodectomas

Tumours may arise in the carotid bodies. They are true neoplasms and not hamartomas, for they have a monomorphic population of chief cells which have lost their normal relation to nerve fibrils. Furthermore, malignant change develops in a minority of them. The designation of these neoplasms remains controversial. Early authors tended to use the non-committal term of "carotid body tumour" and there is still much to recommend this. As we saw in Chapter 1, the carotid body shares identical histological features with those of other glomera scattered widely throughout the body. While neither the carotid bodies nor their closely related glomera strictly fulfil the criteria for designation as non-chromaffin paraganglia (see Chapter 1), some authorities such as Glenner and Grimley (1974) accept that tumours arising from them are non-chromaffin paragangliomas. While we accept that the individual components of the system do not meet the strict anatomical criteria for such designation, we recognise their close histological resemblance and the convenience of having a term for the group of tissues. This recognition of the system as an entity has a certain practical importance for nodules of tumour tissue composed of clusters of chief cells usually represent multicentricity of origin from the generalised system rather than metastases from a primary neoplasm in the carotid body or elsewhere (Conley 1965). Occasionally, however, when carotid body tumours become malignant, tumour masses elsewhere in the body may in reality represent metastases, although these nodules are more likely to be found in lymph nodes than in other possible sites for multicentric origin.

In 1950 Mulligan introduced the term "chemodectoma" for tumours at the base of the heart in dogs to denote a neoplasm of chemoreceptor cells. It has been accepted in the international histological classification of tumours adopted by the World Health Organization (Enzinger 1969). An objection to this term is that only the carotid and aortic bodies have been shown to be receptor organs and none of these neoplasms is functionally active as a chemoreceptor. Nevertheless, in this chapter we shall use the term "chemodectoma" for the tumour of the carotid body and "non-chromaffin paragangliomas" for tumours of the generalised system of tissues to which the carotid bodies belong.

Morbid Anatomy

In most reported series of chemodectoma the size of the tumours has ranged between 3 and 10 cm. Thus, in a collection of 37 cases Farr (1967) found 18 to be between 3 and 6 cm in diameter, 7 between 6 and 8 cm, and 12 between 8 and 10 cm. Pryse-Davies et al. (1964) reported a range of diameter of 2 to 4 cm. Earlier, Pettet et al. (1953) had found 41 resected tumours to average 4.5 cm × 4.5 cm × 3 cm. Staats and his colleagues (1966) found 11 chemodectomas to average 3.1 cm in diameter. The size of the tumour at presentation obviously depends to a large extent on the number of years it was tolerated by the patient before medical advice and surgical treatment were sought. By plotting the size of the tumour against the years of duration, Farr (1967) calculated a

growth rate of 0.5 cm/year. Hence, when neglected, chemodectomas continue their slow inex-
orable growth over the years and may eventually
assume a considerable size. Tumours as large as 10
to 20 cm have been reported by many authors over
the years, including Callison and MacKenty
(1913), Reid (1920), Brown and Fryer (1952),
Goormaghtigh and Pattyn (1954), and Yaghmai et
al. (1970). Tumours over 110 g in weight have
been reported by Szenthe and Kneiszl (1964)
(113 g), Javid et al. (1976) (120 g) and Reid (1920)
(190 g).

A small or moderately sized chemodectoma will
present as a spherical or ovoid mass situated over
the bifurcation of the common carotid artery. Fig.
19.1 shows a small chemodectoma discovered as
an incidental finding at necropsy on a woman of 74
years with systemic hypertension. Its small size
allowed its relation to these major arteries to be
demonstrated clearly. The outer surface of the
tumour is usually smooth or slightly bosselated
(Fig. 19.1). On section, the chemodectoma
presents a firm, resilient cut surface which varies in
colour from pink-grey to red-brown (Fig. 19.2).
Usually it appears to be homogeneous but it may
be irregular and mottled by areas of necrosis and

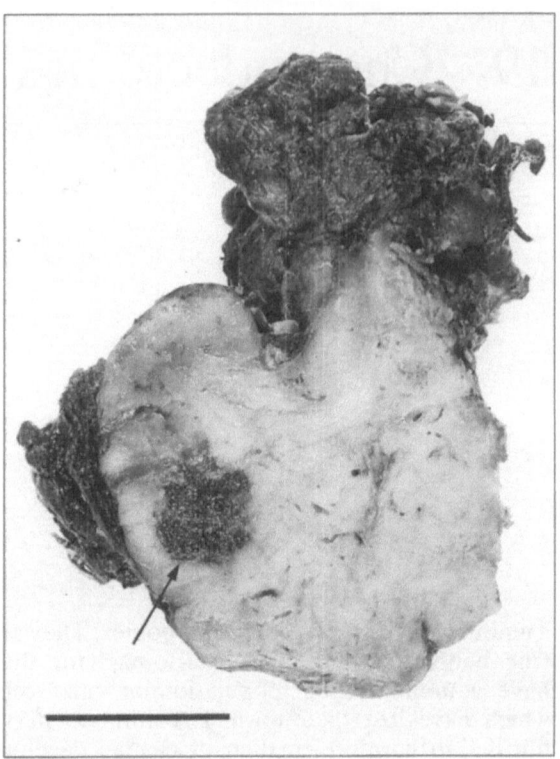

Fig. 19.2. Specimen of chemodectoma, resected at operation,
which has been sectioned. Much of the firm tumour surface is
grey in colour but there is one area of necrosis (*arrow*)
associated with both recent haemorrhage and haemosiderin
deposition. Portions of neck muscle have been removed with
the tumour. Scale line = 2 cm.

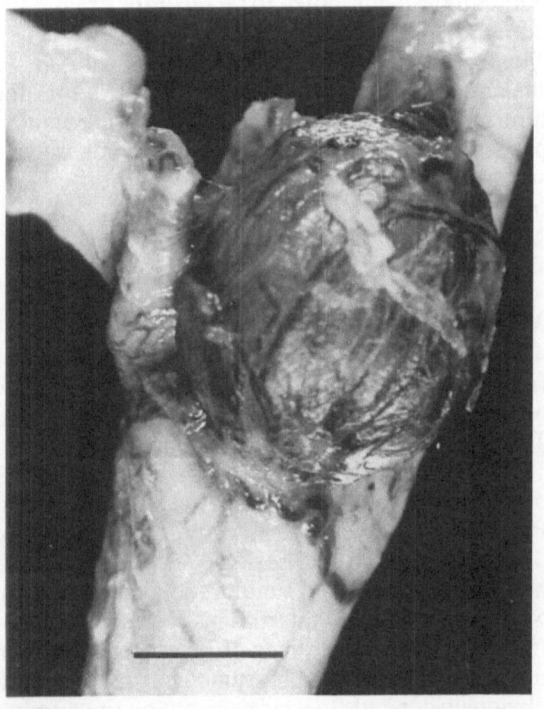

Fig. 19.1. Chemodectoma in a woman of 74 years with systemic
hypertension discovered as an incidental finding at necropsy.
Its small size allows its relation to the carotid bifurcation to be
demonstrated clearly on dissection. Scale line = 0.5 cm.

fibrosis. There may be focal purple areas due to
fresh haemorrhage (Oberman et al. 1968) or
brown areas due to deposits of haemosiderin
resulting from old haemorrhages into the substance of the tumour (Fig. 19.2). The tumour
usually has a well-defined fibrous capsule which
may be covered by a vascular plexus extending
over the carotid arteries (MacComb 1948). When
a chemodectoma is received as a surgical specimen, it will commonly show tags and adhesions of
fibrous tissue which represent the connective
tissues densely binding it to the adventitia and wall
of the carotid artery. Such a specimen will also
commonly show the grooving produced by the
closely adjacent and adherent arteries at the site of
the carotid bifurcation where it was growing
(Pryse-Davies et al. 1964). Portions of neck
muscle adherent and attached to the tumour may
also be received as part of the specimen (Fig.
19.2).

As the tumour grows to a large size, it tends to
encircle the common and internal carotid arteries
at first without compression. However, these ves-

sels become progressively embedded within the slowly growing tumours so that they are found to be compressed at operation or necropsy (Brown and Fryer 1952; Gupta et al. 1976). In some cases total occlusion of these major arteries has been reported (Turnbull 1954; Wilson 1964). In some instances occlusion of the internal carotid artery has been due to thrombosis (Brandberg 1929; McSwain and Spencer 1947). As the large chemodectoma increases in size it may engulf regional nerves such as the 10th or 12th cranial nerves and the sympathetic chain (Zak and Lawson 1982).

Histology

The classic chemodectoma mimics but does not reproduce some of the histological features of the normal carotid body described in Chapter 3. Thus the appearance of cell clusters is simulated (Fig. 19.3). They are not like those of the normal carotid body because they have a homogeneous monomorphic population of transformed chief cells (Fig. 19.4). The characteristic normal struc-

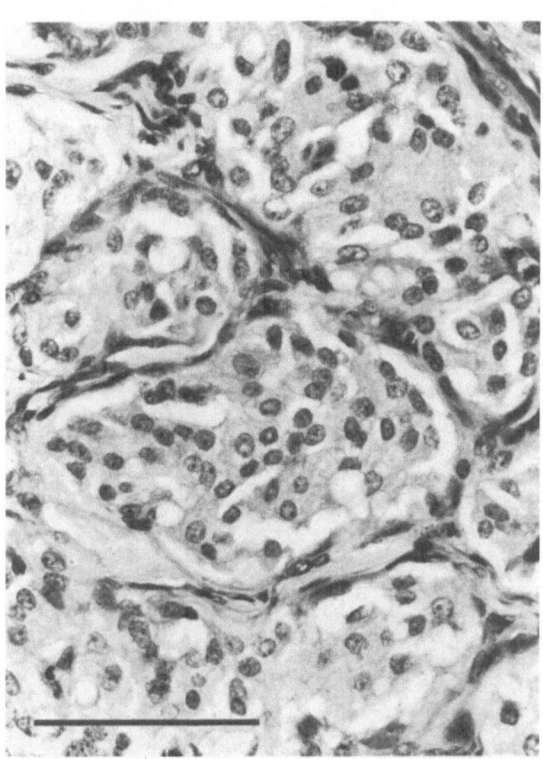

Fig. 19.4. Section of the tumour shown in Fig. 19.3 at higher magnification to show its cytological features of the "Zellballen". The appearances of these simulated cell clusters are not like those of the normal carotid body because they have a homogeneous monomorphic population of transformed chief cells. The characteristic normal structure of an inner core of chief cells composed of the customary variants and a surrounding shell of sustentacular cells is not reproduced. (HE) Scale line = 80 μm.

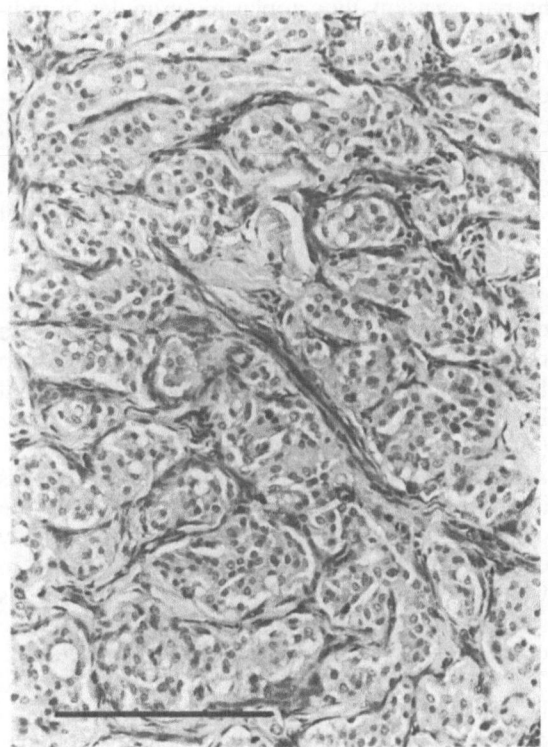

Fig. 19.3. Classic histological features of a benign chemodectoma in a man of 40 years showing the clusters of neoplastic chief cells ("Zellballen") with intervening compressed stromal cells. (Haematoxylin–eosin (HE)) Scale line = 200 μm.

ture of an inner core of chief cells and a surrounding shell of sustentacular cells is not reproduced. "Zellballen" of the neoplasm are larger than those of the normal carotid body, having a mean diameter up to 200 μm (Grimley and Glenner 1967) compared to the range of 54 to 110 μm for normal cell clusters (see Chapter 3) The tendency to form "Zellballen" is demonstrated by silver impregnation of the reticulin fibres surrounding the clusters, which presents a characteristic pattern of a network of boxes. The fibres do not penetrate between individual tumour cells.

The neoplastic cells are usually well differentiated and closely mimic the cytological features of the chief cells, as described in Chapter 3 (Fig. 19.4). However, individual tumour cells are somewhat larger, averaging 16 μm (range 8–29 μm) in diameter (Marshall and Horn 1961). Like the light variety of chief cells in the normal carotid body, the chemodectoma cells are polygonal or oval in shape and contain abundant pale eosinophilic,

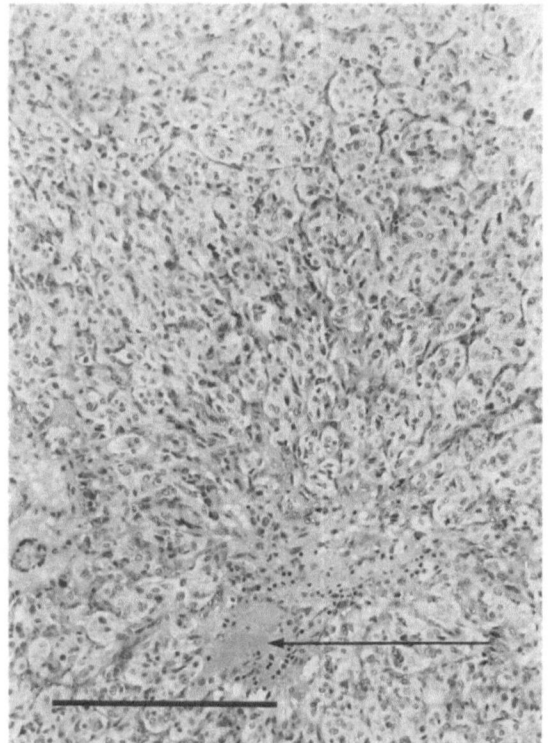

Fig. 19.5. Histological section of a chemodectoma in which the clusters are very small. In the central part of the figure the "Zellballen" are so small as to give the appearance of sheets of plump epithelial cells. Thin-walled vascular channels are present (*arrow*). They are not lined by endothelium and seem to be produced by haemorrhages into cell clusters. (HE) Scale line = 150 μm.

granular or slightly vacuolated cytoplasm. The cells have typical open round or oval nuclei, 5 to 12 μm in diameter, with a moderate chromatin content and occasional prominent nucleoli (Fig. 19.4). The margins of the cells cannot be distinguished clearly and they present a palely eosinophilic, syncytial appearance.

Glenner and Grimley (1974) hypothesised that there are differences in the cytoplasmic density in the cells of a chemodectoma which correspond to those of the light and dark variants of chief cells described in Chapter 3. They stated that this distinction can be made out, particularly when the cytoplasmic processes of adjoining light and dark cells interdigitate. Such interdigitation is seen well in thin, plastic sections of tumour fixed well in glutaraldehyde. It is, however, far more difficult to make out in routine paraffin sections.

Occasionally the clusters of neoplastic chief cells are surrounded by a rim of more elongated cells which might represent sustentacular cells (Pryse-Davies et al. 1964; Grimley and Glenner

1967), but they are greatly reduced in number compared with those in the normal carotid body (Fig. 19.3). When sheets of elongated cells are found, the possibility of malignancy must be considered, as described below. Rarely, anastomosing cords of elliptical or spherical cells are encountered, mimicking the trabecular pattern of some morphological varieties of functioning adrenal medullary phaeochromocytoma.

Not all chemodectomas show this classic alveolar pattern. In some tumours the clusters are very small (Fig. 19.5) and, if their organisation into these separate units is not recognised, could be interpreted as sheets of plump epithelial cells with abundant pale cytoplasm giving an impression of hypercellularity. Occasional well-defined and sizeable "Zellballen" may occur within these expanses of mini-clusters. Broad bands of relatively acellular fibrous tissue staining palely with eosin are characteristic. Dilated fragile vascular channels may isolate groups of cell clusters into discrete bands (Fig. 19.5). These channels may be produced by haemorrhages in cell clusters and are lined by neoplastic chief cells without intervening endothelium.

Carotid body tumours are usually non-chromaffin but occasionally cytoplasmic granules in individual or small groups of cells will exhibit chromaffinity (Pryse-Davies et al. 1964). This is usually due to catecholamines but can be due to lipochrome pigment which is sometimes present. Some chemodectoma cells will show the yellow-green colour of formalin-induced fluorescence, which is also consistent with the presence of catecholamines. Fluorescence and tissue assay for catecholamines are useful for diagnostic purposes.

Extensive study has been made of cells, in the chemodectoma, that contain granules which are reduced by the Masson-Fontana and del Rio Hortega silver methods described by Costero and Barroso-Moguel (1961), Costero and Chevez (1962) and Barroso-Moguel and Costero (1962). Argyrophilic cells were found by these authors intermingled with, and partially enclosing, the chief cells within the lobules and they probably contain one or more of the peptides described in Chapter 8. Also found were argentaffin cells situated along the boundaries of the lobules in the connective tissue surrounding the blood vessels. These cells were large and of variable shape, could be round, flattened or amoeboid, and contained a central nucleus and abundant densely granular cytoplasm. They seemed to be morphologically similar to the Kultchitsky cell of the gastro-intestinal tract, and by analogy were thought to store 5-hydroxytryptamine.

Other types of granule can be demonstrated in chemodectoma cells. As noted in Chapter 17, lipofuscin is sometimes found as an ageing pigment in the light variety of chief cells and such lipochrome pigments are also found in neoplastic chief cells. They stain with oil Red O and take up azo dyes but Schmorl's reaction is negative. Intracellular glycogen is not present and the cells do not react with the periodic acid–Schiff stain. Small deposits of haemosiderin, due to old focal haemorrhages (see Fig. 19.2), are common. Various enzyme activities of a non-specific nature were demonstrated by Pryse-Davies et al. (1964). Arylesterase ("non-specific esterase") activity of a type seen in normal paraganglia and ganglion cells may be demonstrated in most chief cells of the tumour. The tumour cells show moderate acid phosphatase activity whilst the cells lining the dilated thin-walled vessels show alkaline phosphatase activity. Lactic and succinic dehydrogenase activity is generalised in tumour and stromal cells (Pryse-Davies et al. 1964).

Usually received with a surgical specimen of chemodectoma is a variable amount of fibrous tissue beyond the confines of the tumour. Within this may be found arteries, veins, myelinated and non-myelinated nerves, ganglion cells and often a cellular element which may include lymphocytes (Pettet et al. 1953), plasma cells and mast cells. This connective tissue may form a total or partial capsule which is continuous with the adventitia of the adjacent carotid artery. Dense fibrous septa extend inwards from this capsule subdividing the tumour into lobules. Fibrous tissue further permeates the lobules, separating the parenchyma into cell clusters. While fine collagen and reticulin fibres encase the cell clusters they do not penetrate them. In some instances the fibrous stroma may undergo extensive development, reducing the tumour cells to small islands. Vascular channels may extend into the chemodectoma with the trabecular stromal tissue. Within the tumour these vessels soon come to comprise little more than an endothelial lining supported by delicate reticulin fibres and they may on occasion be so numerous as to suggest that the tumour is angiomatous in nature. The fibrous tissue within the chemodectoma may undergo hyaline and myxomatous degeneration and even calcification (Monro 1950; Pettet et al. 1953) or metaplastic bone formation (Lack et al. 1979). Harrington et al. (1941) described some cases in which hyalinised connective tissue fused with the cytoplasm of the tumour cells to form intracytoplasmic hyaline "inclusion bodies".

Nerve fibres and ganglion cells may on occasion be found within the stromal regions of the tumour but the normal relation of nerve fibres to chief cells is not seen in chemodectoma, thus providing evidence that it is truly neoplastic and not a hamartoma (Grimley and Glenner 1967) or state of hyperplasia. Reports of nerve fibrils in chemodectomas include those of Birrell (1952), Willis and Birrell (1955), Costero (1963) and Pryse-Davies et al. (1964). In many of these cases the nerve fibres were almost entirely restricted to blood vessels, without the links of nerve fibrils to glomic cells found in the normal carotid body.

Malignant Change

A characteristic histological feature of the chemodectoma is its tendency to show nuclear pleomorphism, even in tumours which are unequivocally benign on clinical grounds (Fig. 19.6). There may be considerable variation in the size and shape of chief cells. Hyperchromatic nuclei and bi- and tri-nucleate cells may be found, but mitotic figures are exceedingly rare. Even a large giant cell with a single nucleus may be discovered,

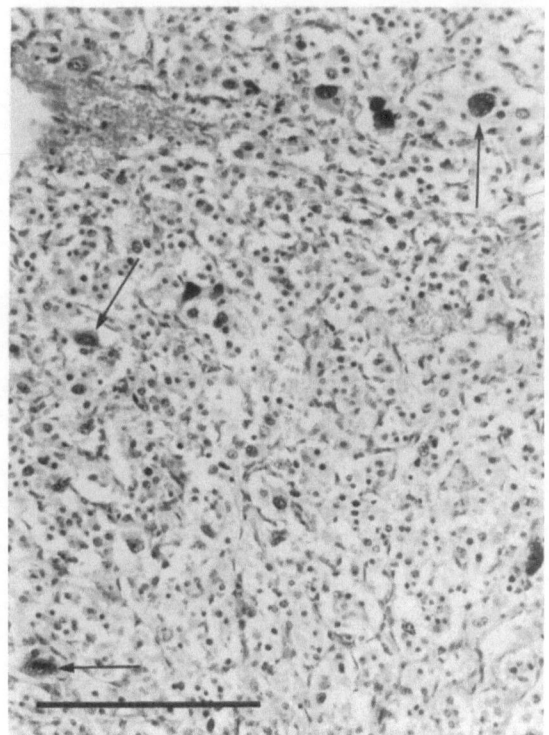

Fig. 19.6. Histological section of a chemodectoma, in a middle-aged man, which was clinically benign. A considerable degree of pleomorphism is present (*arrows*). (HE) Scale line = 200 μm.

although this may represent a surviving ganglion cell (Pryse-Davies et al. 1964). Such appearances may be disturbing but the histopathologist must be on guard against accepting such appearances as indicators of malignancy. Nevertheless, undoubted malignant change with local or distant spread does occur in the chemodectoma.

Local invasion of the wall of the carotid arterial bifurcation occurs in about 5% of cases, but metastasis to regional lymph nodes and distant viscera also takes place. Shamblin and his colleagues (1971) believe that the assertion of malignancy must rest on such spread. In 500 cases of chemodectoma reported up to 1971 they found 16 (about 3%) with local infiltration, and 16 with distant metastasis. Hence the reported incidence of malignancy was about 6%. Staats et al. (1966) had previously given the same percentages for local and distant malignancy. There is an extensive bibliography authenticating metastasis of chemodectoma to lymph nodes or viscera and a few illustrative papers are those of Donald and Crile (1948), Morfit et al. (1953), Goormaghtigh and Pattyn (1954), Romanski (1954), Turnbull (1954), Rabson and Elliott (1957), Fanning et al. (1963),

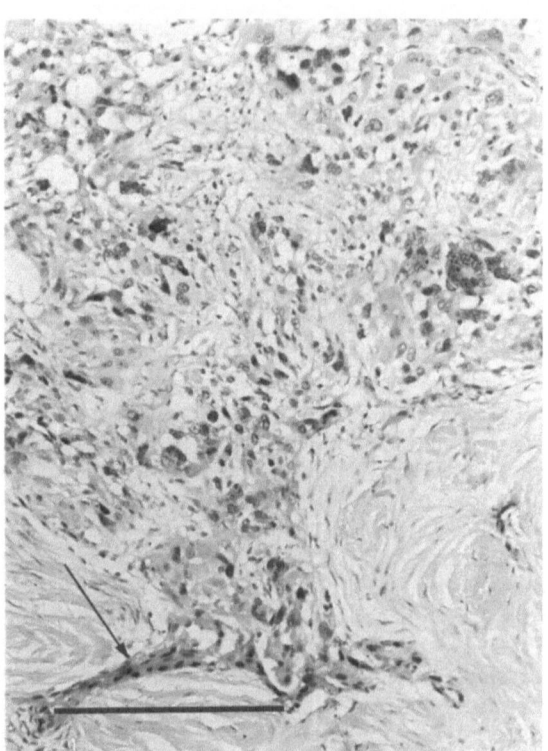

Fig. 19.8. Section of the chemodectoma illustrated in Fig. 19.7, to show frankly malignant histological features. In this field there is no simulation of cell clusters and the tumour tissue is infiltrating the surrounding tissues. In some places the tumour cells are strikingly eosinophilic and spindle-shaped (arrow). There is nuclear pleomorphism. (HE) Scale line = 200 μm.

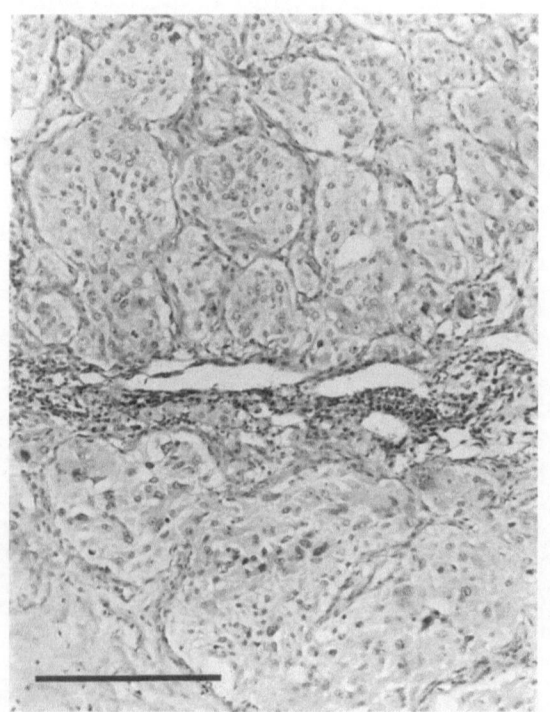

Fig. 19.7. Section of a carotid body tumour occurring in a man of 70 years. In the upper half of the figure typical "Zellballen" are seen but in the lower half the edges of the simulated cell clusters have become indistinct and the tumour tissue is hyperchromatic and pleomorphic. (HE) Scale line = 200 μm.

Reese et al. (1963), Westbury (1963) and Gustilo et al. (1965). The lungs and skeleton are especially susceptible to metastasis (Brown et al. 1967).

Prediction of the behaviour of a chemodectoma from its histological appearance is difficult, if not impossible. Cellular and nuclear pleomorphism and hypercellularity are not uncommon features of benign tumours. Glenner and Grimley (1974) examined several malignant carotid body tumours and found that they showed hyperchromatic nuclei with variation in size and shape and large nucleoli, but significantly these nuclear abnormalities appeared no more frequently than in benign lesions. Multinucleated giant cells and mitotic figures, although very rare, have been reported in carotid body tumours, but even they are not reliable indices of malignancy (Pryse-Davies et al. 1964). Shamblin et al. (1971) reported a series of cases of chemodectoma from the Mayo Clinic and this included two cases that had metastasised; in neither instance was there any histological evidence of malignancy. A patient of Lees et al. (1981) showed the histological features of a benign chemodectoma and yet subsequently exhibited widespread lymphatic metastasis. In contrast Le Compte (1951) was of the opinion that even the occurrence of chemodectoma cells within blood vessels is not a valid criterion of malignancy. Romanski (1954) carefully compared two clinically malignant chemodectomas with five benign tumours and was unable to detect histological differences from which the differing biological behaviour could have been predicted. In view of such difficulties, which may be considerable, it may be necessary for the pathologist to be very cautious when advising the surgeon or patient on prognosis.

Nevertheless, in some carotid body tumours frankly malignant histological features may supervene in a case where typical "Zellballen" were formerly present (Fig. 19.7). A striking degree of cellular and nuclear pleomorphism may develop (Fig. 19.8). The cytoplasm of some of the neoplastic cells is brightly eosinophilic and the tumour is frankly infiltrative, showing a streaming of elongated cells into the surrounding tissue (Fig. 19.8). Chemodectomas may metastasise to regional

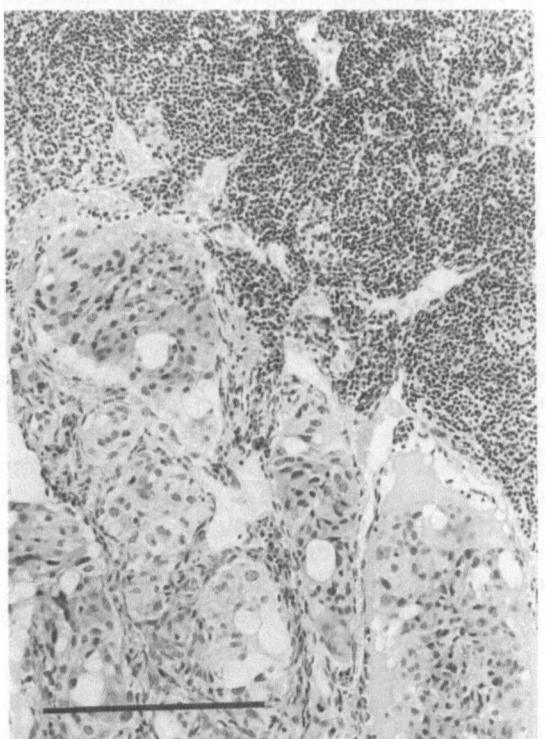

Fig. 19.9. Section of a cervical lymph node infiltrated by a metastasis from a chemodectoma in a man of 70 years. In this instance the metastatic tumour still shows a tendency to simulate cell clusters. (HE) Scale line = 200 μm.

lymph nodes. The histology of the secondary deposits may be that of infiltrative eosinophilic tumour and sheets of elongated cells or it may still simulate cell clusters (Fig. 19.9). Brown and his colleagues (1967) have pointed out that there may be a long interval of up to 35 years between observation of the primary tumour and the occurrence of metastases.

Ultrastructure

A suspected diagnosis of chemodectoma can usually be confirmed by electron microscopy, since the neoplastic cells retain many of the ultrastructural features of the normal carotid body, especially the striking dense-core vesicles (Fig. 19.10). However, the technique will not distinguish between a carotid body tumour and a vagal paraganglioma, since both have identical fine structure (Kahn 1976). Preservation of cellular detail may be poor if specific provision for immediate fixation of 1-mm^3 cubes of tumour tissue is not made in the operating theatre, with the tumour being fixed only in the customary 10% (v/v) formalin (Fig. 9.10). Ischaemia occurring during prolonged dissection may also add further damage to the vesicles. Nevertheless, it needs to be stressed that even tumour tissue fixed in 10% formalin will usually reveal dense-core vesicles and allow confirmation of the diagnosis. Fragments of chemodectomas fixed in glutaraldehyde will show much better preservation of cellular organelles together with an excellent demonstration of the vesicles (Fig. 19.11). The dense-core vesicles in the tumour cells are identical with those found in normal chief cells (Salyer et al. 1969; Fig. 19.11). Diameters have been quoted as 110 to 140 nm (Grimley and Glenner 1967), 80 to 150 nm (Macadam 1969), 140 nm (Fisher and Reidbord 1971), 40 to 100 nm (Gullotta and Helpap 1976). A variable degree of pleomorphism of these dense-core vesicles may be encountered, so that two distinct populations of them are seen, either large and small (Toker 1967; Alpert and Bochetto 1974), or tubular and dumb-bell shaped (Robertson and Cooney 1980; Fig. 19.12).

Ultrastructural study confirms that the chemodectoma has the two main components of rounded clumps of chief cells separated by a vascular stroma. The tumour differs from the normal carotid body in that there is a greater number of chief cells relative to other components. Most chief cells closely resemble those of the normal carotid body in that they are polygonal, with large,

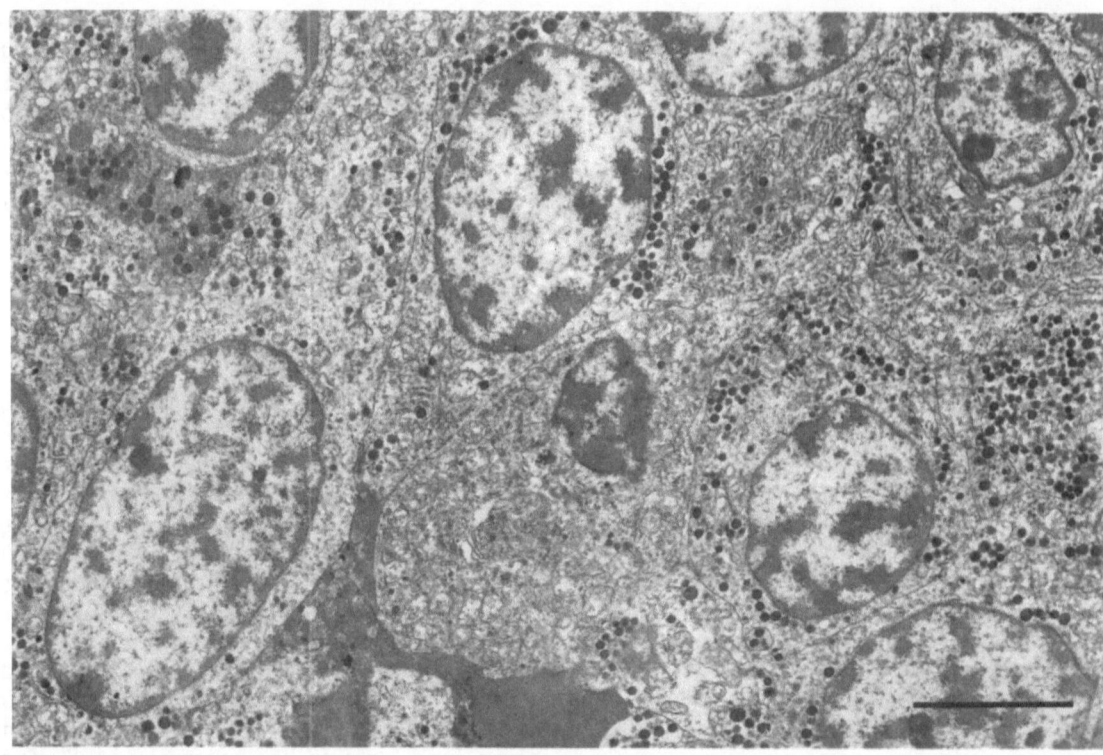

Fig. 19.10. Electron micrograph of chemodectoma, resected at operation without specific provision for appropriate fixation being available in the operating theatre, showing the characteristic round and oval nuclei. Preservation of cytoplasmic detail is poor but diagnosis of the tumour is confirmed immediately by the demonstration of dense-core vesicles, which appear as round, black bodies. Scale line = 3 μm.

pale round or oval nuclei (Fig. 19.10) and contain numerous mitochondria (Fig. 19.11) and variable amounts of smooth and rough endoplasmic reticulum (Grimley and Glenner 1967; Toker 1967; Alpert and Bochetto 1974; Robertson and Cooney 1980). The chemodectoma cells contain numerous free ribosomes and the Golgi apparatus is not infrequently prominent, hyperplastic or multicentric, features thought by Grimley and Glenner (1967) to be non-specific characteristics of neoplasia.

Interdigitation of the neoplastic chief cells is not as complex as in the normal carotid body. Cell junctions resembling desmosomes may be found occasionally between the chief cells, but in malignant chemodectomas they are extremely rare (Robertson and Cooney 1980). It has already been pointed out (see Chapter 17) that puncta adherentia are a normal feature of chief cells and hence the demonstration of desmosome-like structures should not be taken as an indication that a tumour under investigation is epidermal in nature.

In many chemodectomas light and dark variants of chief cells can be recognised largely by virtue of their cytoplasmic density (Grimley and Glenner 1967; Alpert and Bochetto 1974; Robertson and Cooney 1980). This distinction is, however, not as clear-cut as in the normal organ and a continuous spectrum of cytoplasmic density may be present instead (Robertson and Cooney 1980).

On histological examination occasionally noticeably larger "oncocytic" cells are to be found in sections of chemodectomas (Lack et al. 1979). This report is confirmed by the ultrastructural studies of several authors who describe occasional cells in which dense-core vesicles are sparse and the cytoplasm is filled instead with numerous, enlarged mitochondria reminiscent of the oncocytes of endocrine tumours (Grimley and Glenner 1967; Capella and Solcia 1971; Robertson and Cooney 1980).

Sustentacular cells and nerve axons are not features of chemodectoma. Elongated cells are frequently found at the periphery of the clumps of chief cells and have been referred to as sustentacular (Alpert and Bochetto 1974) but they often contain dense-core vesicles and are thus highly attenuated chief cells (Toker 1967; Capella and

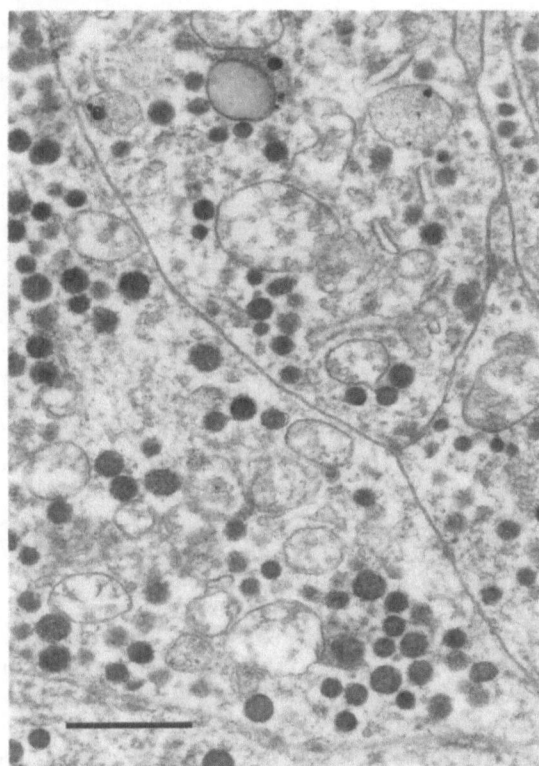

Fig. 19.11. Electron micrograph of fragments of chemodectoma fixed in 1% (w/v) glutaraldehyde. In contrast to Fig. 19.10, presentation of cellular organelles is better and mitochondria can be seen clearly. The dense-core vesicles are demonstrated. Scale line = 1 μm.

◄──────────────────────────────────

Solcia 1971; Robertson and Cooney 1980). The peripheral, elongated chief cells often contain numerous residual bodies and lipofuscin granules (Fig. 19.13). Some authors have found an abundant endoplasmic reticulum in these cells, suggesting that some of them may be fibroblasts (Grimley and Glenner 1967).

The overall impression left by ultrastructural studies is that the chemodectoma is a true neoplasm of chief cells. There is a strong propensity for them to form a monomorphic population

Fig. 19.12. Electron micrographs of fragments of chemodectoma from the same case as in Fig. 19.11. There is a considerable degree of pleomorphism in the dense-core vesicles. In the cell to the left the vesicles are small, whereas in that to the right the vesicles are larger. Scale line = 1 μm.

▼

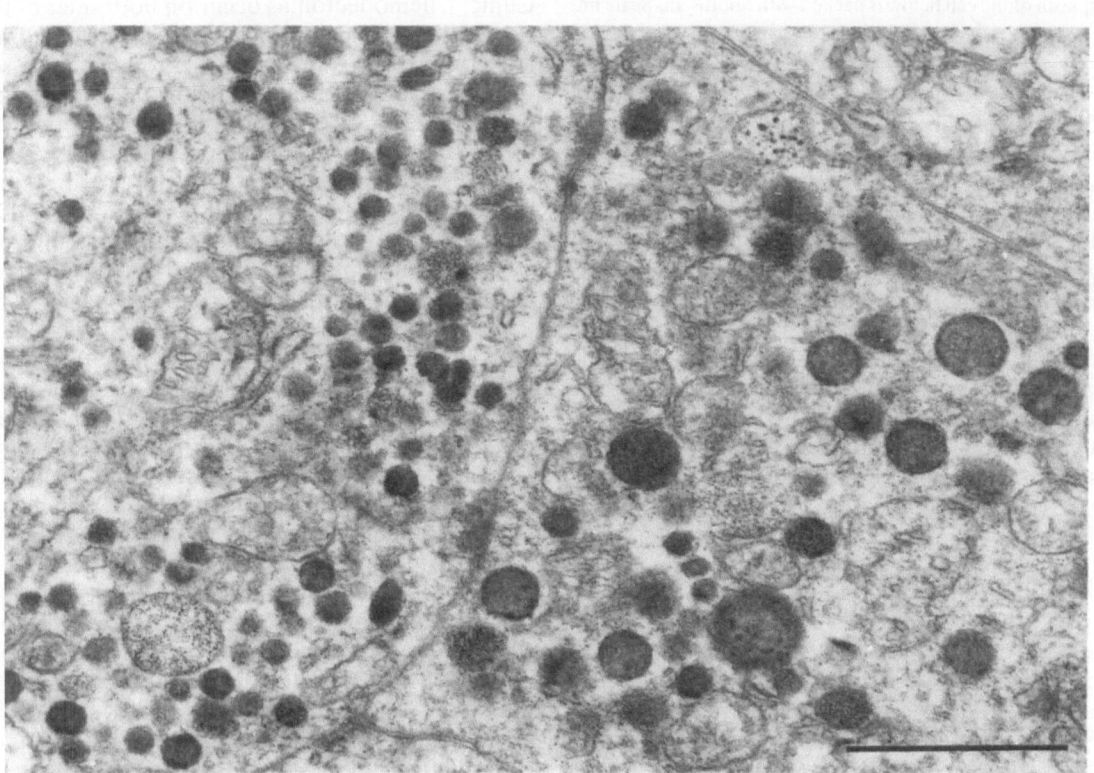

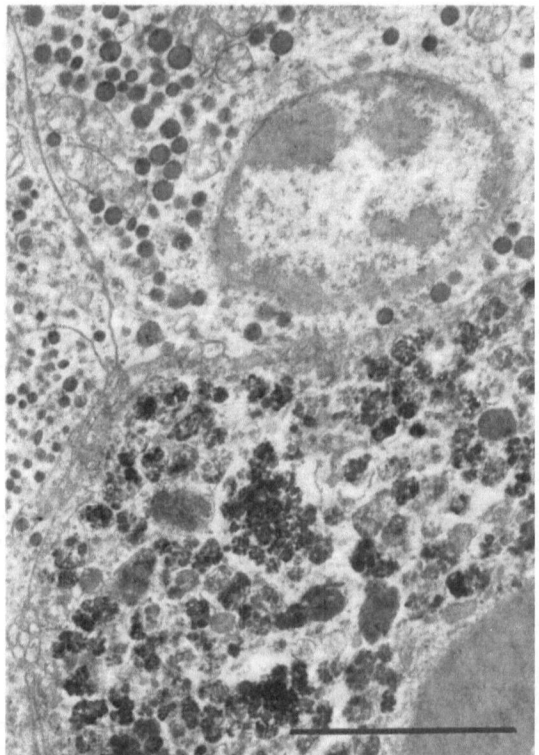

Fig. 19.13. Electron micrograph of fragments of chemodectoma from the same case. The cell above shows a typical round nucleus and contains prominent dense-core vesicles. The cytoplasm of the cell below is packed with lipofuscin. Scale line = 3 μm.

without sustentacular cells and without nerve fibrils showing an organised relation to the chief cells as in the normal carotid body.

In an ultrastructural study of a malignant chemodectoma and its cerebral metastasis, Isfort and Knoche (1966) found that many bizarre nuclear forms were present and many of the large spherical nuclei were cleft. Most chief cells had microvilli and small, compact mitochondria, often arranged in a semilunar fashion around the nucleus. The Golgi apparatus was inconspicuous. Dense-core vesicles were not found in the primary malignant tumour or in its metastasis.

Clinical Features

Chemodectomas may present at any age. In a series of 90 patients with carotid body tumours seen at the Mayo Clinic during the period 1931 to 1967, the age of presentation ranged from 12 to 63 years (Shamblin et al. 1971). However, there have

been several reports in which patients with chemodectoma coming to surgery in adulthood reported that a lump in the neck had been first noticed during childhood. For example, Oberman et al. (1968) described an adult patient with a mass in the neck since childhood which did not increase in size until 37 years later. The patient of Gustilo et al. (1965) was 8 years old at the time his lesion was found but was not operated on until the age of 30 years. At the other extreme of life was the report of a chemodectoma in a patient aged 89 years (Besznyák and Pinter 1959). The average age at diagnosis was 34 years in the 174 cases reviewed by Monro (1950) and was 41 years for 63 patients reported by McIlrath and ReMine (1963). It is of interest that Martin et al. (1973) reported a higher incidence of malignancy in younger patients, the average age at diagnosis for metastasising tumours in their series being 32 years as compared to 47 years for their patients with benign lesions.

There is no obvious sex preponderance, although Zak and Lawson (1982) were of the opinion that there is a slight predilection for women. They analysed 186 cases of chemodectoma reported by several authors and found that they included 103 females and 83 males. There was an unusually high preponderance of females with a ratio of about seven to one in the series reported by McIlrath and ReMine (1963) from the Mayo Clinic. Chemodectomas occur on both sides of the neck with equal frequency. There are two anomalous series with regard to the reporting of sex incidence and the laterality of these tumours. Saldaña et al. (1973) reported that 19 of their 22 cases among native highlanders of the Peruvian Andes were in females, with left-sided lesions being 3.2 times more frequent (see Chapter 11). Krause-Senties (1971) found that 37 of their 40 cases were in females with, once again, left-sided lesions being 2.6 times more common.

Most cases of chemodectomas reported have been in whites but there have been many reports of occurrences in negroes in the American literature (Chambers and Mahoney 1968). De Villiers (1972) described carotid chemodectomas in African negroes. A special association of carotid body tumours with Quechua Indians living at high altitude has been reported by Saldaña et al. (1973). Interpretation of this association is difficult. It could represent a racial predilection. Conversely it could be a manifestation of constant stimulation of the chemoreceptor tissue by hypobaric hypoxia due to the reduced barometric pressure (Heath and Williams 1989). This association is discussed in Chapter 11 where we consider the carotid bodies at high altitude.

The chemodectoma presents as a slowly growing firm mass in the neck. Since it is fixed to the carotid arteries it does not move with swallowing and classically is said to move laterally but not vertically. However, this clinical sign is an expression of size and, when the tumour is small, it is mobile in all directions. This sign should not be regarded as pathognomonic of chemodectoma. In the Mayo Clinic series the initial complaint of all but one of the patients was that of a mass growing in the neck. Similarly, all but two of the 41 patients reported by Lees et al. (1981) presented to a clinician on account of a mass in the neck which was increasing in size. These tumours grow very slowly as noted above. In the series of Shamblin et al. (1971) a lump in the neck had been present on average for 6.6 years before the patient presented for surgery and in one instance it had been noticed for no less than 47 years. In the case of Steimle and Steimle (1958) the tumour had been present for half a century. It is worthy of note that no fewer than 19 of the 90 patients reported that there had been a noticeable increase in size in the tumour in the 6 months before examination.

Clinically, the mass is found adjacent to the angle of the mandible and is partially covered by the sternomastoid muscle. The enlarging chemodectoma carries the common carotid artery with it laterally or anteriorly so that it or its branches can be felt on the tumour surface (Zak and Lawson 1982). Monro (1950) claimed that this lesion is the only cervical neoplasm with a large vessel pulsating in front of it. The carotid body is a highly vascular structure (see Chapter 16) and in some cases this high blood flow causes the tumour to pulsate and develop a murmur and bruit (Javid et al. 1976). This high blood flow is utilised in the diagnosis of chemodectomas through the agency of angiography, as we discuss later in this chapter. More commonly the tumour pulsates by transmission of the pulsation from the attached carotid arteries. Digital compression of the tumour or the common carotid artery may lead to a decrease in the size of the tumour (Hanford 1965). The patient may complain of a throbbing or pounding sensation in the neck (Martorell 1956; Sessions et al. 1959; Grage and Cueto 1967).

The term "potato tumour" is sometimes applied by clinicians to the chemodectoma and this has perpetuated the idea that the mass in the neck is always hard. The historical background to this term has been given by Zak and Lawson (1982). The designation "potato-like tumour" was apparently originally used by Hutchinson in 1888 to describe highly malignant fungating cervical tumours that were rapidly fatal. Many were round-cell tumours or lymphosarcomas. This term was applied to tumours of the carotid body by Gilford and Davis (1904) some 16 years later and this introduced the idea that chemodectomas are always hard. In reality the firmness of the tumour depends on its histological composition. When there is a large vascular angiomatous component the tumour will be soft and apparently cystic, but when there is a big fibrous stromal content the tumour will be hard.

The carotid body tumour may be smooth, nodular or lobulated. Although it is usually ovoid or cylindrical it is sometimes hour-glass shaped, producing a pharyngeal bulge with displacement of the tonsil, soft palate and uvula to the opposite side if the tumour is large (Zak and Lawson 1982). If the pharyngeal component of a large tumour is significant, it may lead to spontaneous pharyngeal bleeding from dilated mucosal blood vessels lining the pharyngeal bulge. Shamblin et al. (1971) reported 15 out of 90 chemodectomas to have both cervical and pharyngeal components.

As a general rule chemodectomas are not tender on palpation and pain from them is exceptional. However, if there is pressure or infiltration of the nerves of the cervical and brachial plexuses by a large tumour there may be local tenderness or pain radiating to the head, neck and shoulders. Zak and Lawson (1982) found pain to be associated with 19% of 254 carotid chemodectomas reported in the decade up to 1982. They reviewed a wide range of different types of pain found in association with carotid body tumours. These include pain on exertion, or pain referred to the ear and temporal area, to the face, jaws or occipital area. The patient may complain of frontal, occipital or temporal headache or even toothache.

A wide variety of other symptoms may be produced by carotid body tumours. Bulky tumours may lead to dysphagia due to compression of the oesophagus or impingement on the hypopharynx or by involvement of the vagus or hypoglossal nerves (Zak and Lawson 1982). Its incidence with these tumours has been estimated at 10% by Rush (1962). The carotid sinus syndrome is occasionally found in association with a chemodectoma (see Chapter 20).

Hoarseness is usually the result of a vocal cord paralysis secondary to involvement of the vagus nerve or its recurrent laryngeal branch, by pressure or infiltration of a chemodectoma. Stridor may be caused by narrowing of the upper airway by a pharyngeal component of the tumour. Zak and Lawson (1982) estimated the incidence of this to be 8% after a review of 254 cases of chemodectoma. Pressure of a carotid body tumour on the larynx, trachea or vagus nerve may lead to a cough which may be episodic or chronic.

Hormonal Activity

The chief cells of the carotid body have a high content of biogenic amines including adrenaline, noradrenaline, 5-hydroxytryptamine and especially dopamine (see Chapter 7). It might be anticipated that on neoplastic transformation and proliferation of these cells there would be an outpouring of such amines, with systemic hypertension and a clinical picture not unlike that of a phaeochromocytoma. In fact this occurs in only a small number of chemodectomas. Thus, a tumour was reported to have developed in the left side of the neck in a normotensive boy of 12 years (Glenner et al. 1962). Its massage led to systemic hypertension, sweating and pallor and its removal resulted in intractable hypotension and death. This boy had increased concentrations of urinary noradrenaline and its metabolites, and fluorometric and chromatographic assays of the neoplasm demonstrated high levels of noradrenaline (1.5 mg/g of tumour). Histologically the mass was consistent with being a carotid body tumour with clusters of polygonal cells which were in the main non-chromaffin. A similar neoplasm was reported in a 66-year-old normotensive man by Berdal et al. (1962). As in the previous case manipulation of it led to systemic hypertension and assay of the neoplasm revealed both noradrenaline (550 µg/g) and adrenaline (33 µg/g). Other functionally active carotid body tumours have been reported by Hamberger et al. (1967), Hörtnagl et al. (1973) and Newell (1971). Grimley and Glenner (1967) observed that, although most of these tumours are at or near the carotid bifurcation, they are also commonly intimately attached to the cervical sympathetic chain near the superior cervical ganglion. It would seem that about 1% of carotid body tumours manifest hormonal activity with the release of catecholamines (Zak and Lawson 1982). In contradistinction to this, most phaeochromocytomas show such functional activity. Functioning carotid body tumours tend to declare themselves in childhood or in early adult life. They show no sexual predilection. When a patient with a chemodectoma complains of repeated headache or attacks of palpitation or sweating the possibility of its becoming active hormonally should be considered. Elevation of systemic blood pressure or changes in cardiac rhythm during palpation of the mass should greatly raise diagnostic suspicion.

Preoperative determination of urinary catecholamines or their metabolites is indicated in these cases because of the attendant morbidity and mortality in the removal of functioning tumours without the administration of adrenergic blocking drugs (Zak and Lawson 1982).

The chief cells are also rich in peptides such as leucine- and methionine-enkephalins, substance P and vasoactive intestinal polypeptide (see Chapter 8). We are not aware of the content of these peptides in carotid body tumours or of any clinical effects produced by them.

Familial Incidence

It has long been appreciated that the chemodectoma tends to show a familial incidence (Chase 1933; McNealy and Hedin 1939; Lewison and Weinberg 1950). The genetic transmission is most consistent with autosomal dominance (Glenner and Grimley 1974). The largest family group involved was that described by Sprong and Kirby (1949), with 11 siblings. Tumours of the neck occurred in nine members of this family and were surgically resected in six of them. Katz (1964) reported a family in which three members had bilateral carotid body tumours. Families with a tendency to develop chemodectomas also show an increased incidence of glomus jugulare tumours. Zak and Lawson (1982) provided an exhaustive catalogue of references from 1916 to 1977 concerning 63 families with a tendency to develop non-chromaffin paragangliomas, in which 209 members had been affected. Chemodectomas occurred in 166 subjects, and tumours of the glomus jugulare in 42, of the vagus in 12, of the aortic body in 1, and of the retroperitoneal area in 2. In 33 of the 63 families there was only a chemodectoma involved. In 6 families only glomus jugulare tumours were transmitted. However, in 24 families there were various combinations of tumours of the carotid body, vagus, aortic body, glomus jugulare and retroperitoneum (Zak and Lawson 1982). Multicentricity is thus a striking feature of heredofamilial glomic tumours. Zak and Lawson (1982) found from their review of reports of familial glomus tumours referred to above that no fewer than a third (69 of 209) of the patients had multicentric tumours. A quarter (50 of 209) of them had bilateral chemodectomas. The remaining subjects showed some combination of chemodectoma, and tumours of the vagus, aortic body and glomus jugulare. In 10 cases (5%) three glomic neoplasms were present. The biological behaviour of familial glomus tumours is otherwise identical, with random cases, including malignant change, in a small minority (Dent et al. 1976). There is no convincing evidence to support sex-linked transmission of the carotid body tumour trait.

Multicentricity

Irrespective of the familial tendency, non-chromaffin paragangliomas as a group frequently have a multicentric origin. This characteristic is usually manifested as bilateral carotid body tumours, but Zak and Lawson (1982) provided an exhaustive list of references exemplifying a wide variety of combinations of tumours of various glomera. Bilateral chemodectomas in the neck, therefore, may be associated with unilateral or bilateral glomus jugulare tumours. Bilateral glomus jugulare tumours may occur on their own or with a unilateral carotid body tumour. A unilateral tumour of the carotid body may occur with one of the glomus jugulare. A variety of other multicentric non-chromaffin paragangliomas in the orbit, larynx, thyroid and so on may be accompanied by carotid body tumours. Bilateral tumours of the vagal body are sometimes found.

Zak and Lawson (1982) reviewed 249 cases of multicentric glomus tumours reported by many authors in the literature. They found that 207 involved 2 sites, 33 involved 3 sites, 7 involved 4 sites and in 2 cases 5 tumours were present. The commonest combination, as noted above, is the bilateral carotid body tumour which accounts for between 3% and 14% of multicentric tumours (Phelps et al. 1937; Rush 1963; Shamblin et al. 1971; Irons et al. 1977). Zak and Lawson (1982) pointed out that a distinction must be drawn between the sporadic and familial types of bilateral carotid body tumours. In patients with a positive familial history, bilateral tumours are found in as many as 26%–33% of cases (Rush 1963; Pratt 1973).

Chemodectomas and Hypoxaemia

It has been suggested that the basis for the multicentricity of the tumours of the carotid body and other glomera is a chronic reactive hyperplasia to a sustained as yet unidentified stimulus. In the chemoreceptors this stimulus could well be hypoxaemia (see Chapters 10 and 11). In support of this view are reports in the literature of chemodectomas in patients with chronic obstructive lung disease. Chedid and Jao (1974) reported the

occurrence of 11 tumours of the carotid body and one non-chromaffin paraganglioma of the ganglion nodosum of the right vagus nerve in six members of two consecutive generations of a family. All affected members had bilateral tumours of the carotid bodies with but a single exception. No fewer than four of these six patients had associated chronic obstructive lung disease with a persistently low arterial oxygen tension and a sustained hypercarbia.

There is a further interesting association between chemodectomas and hypoxaemia in the frequency of these tumours in Quechua Indians of the Andes of Central Peru chronically exposed to the hypobaric hypoxia of high altitude. This association is described and discussed in Chapter 11.

Diagnosis

Most chemodectomas can be diagnosed from the clinical history, signs and symptoms but there are many modern investigations which can be employed to confirm these clinical impressions and above all to substantiate the diagnosis at an early stage of the growth of the neoplasm when its surgical extirpation can be carried out to avoid complications to the cerebral circulation or to cranial nerves.

Fine-Needle Aspiration

This technique is now being used in the diagnosis of a wide range of tumours and it has been tried in the case of chemodectomas, with limited success it has to be said and not without danger. In the patient of Chen and Hwang (1989), a man of 21 years, the aspirate from a carotid body tumour included a large amount of fresh, bright red blood, as though the needle were in an artery. The authors thought this to be such a characteristic finding as to constitute evidence for diagnosis of a chemodectoma. Kapoor et al. (1989) carried out fine-needle aspiration in a woman of 38 years with a swelling in the neck and also found that it yielded largely blood, with only sparse groups of tumour cells. They concluded this investigation should be contemplated cautiously in the diagnosis of chemodectomas, an opinion expressed earlier by Jacobs and Waisman (1987). Engzell et al. (1971) aspirated a suspected right carotid body tumour in a man aged 33 years. Subsequently his right common carotid artery became obstructed and he became hemiplegic and died 3 days postoperatively. At necropsy there was a thromboembolus in the right cerebral artery.

The cytological details reported as having been found in fine-needle aspirates reflect accurately the histological features of chief cells described in Chapter 3. They have been reported as epithelioid, round or polygonal in shape, and sometimes elongated, with copious finely granular, slightly eosinophilic cytoplasm containing small vacuoles and having an indistinct border (Mincione and Urso 1989). The nuclei, as in histological sections, are round or oval, and have a single prominent nucleolus and unevenly distributed clumped chromatin. In the nucleus clear areas may be seen around chromatin aggregates and at the nuclear periphery where they assume a semilunar shape (Mincione and Urso 1989). The tumour cells may be isolated or arranged in acini, clusters or rosettes. Cells enveloping nearby fellows with their cytoplasm ("cell embracing") may sometimes be seen. Other workers have reported very similar cytological findings (Hood et al. 1983; Chen and Hwang 1989; Kapoor et al. 1989). González-Cámpora et al. (1988) carried out fine-needle aspiration cytology in six patients with paragangliomas, including two carotid body tumours from women aged 29 and 51 years, respectively. The fine-needle aspirates were similar in all the paragangliomas.

Jacobs and Waisman (1987) in an aspirate from a chemodectoma in a 56-year-old man found intranuclear vacuoles, previously thought to be characteristic of papillary thyroid adenocarcinoma, although they have also been found in malignant melanoma, endometrial adenocarcinoma, hepatocellular carcinoma and meningioma. Intranuclear inclusions were also found in paragangliomas by González-Cámpora et al. (1988) but were found by them to be most prominent in phaeochromocytomas.

Angiography

In the past angiography has been the mainstay of diagnosis of carotid body tumours, since these tumours are highly vascular (see Chapter 16). In recent years computed tomography (CT) has displaced angiography from this prime position in diagnosis but high quality angiograms are commonly necessary before surgery for chemodectomas to define accurately the arteries supplying blood to the tumours. Conventional radiography of the neck provides little information of diagnostic value about carotid body tumours except for demonstrating a soft-tissue mass or displacement

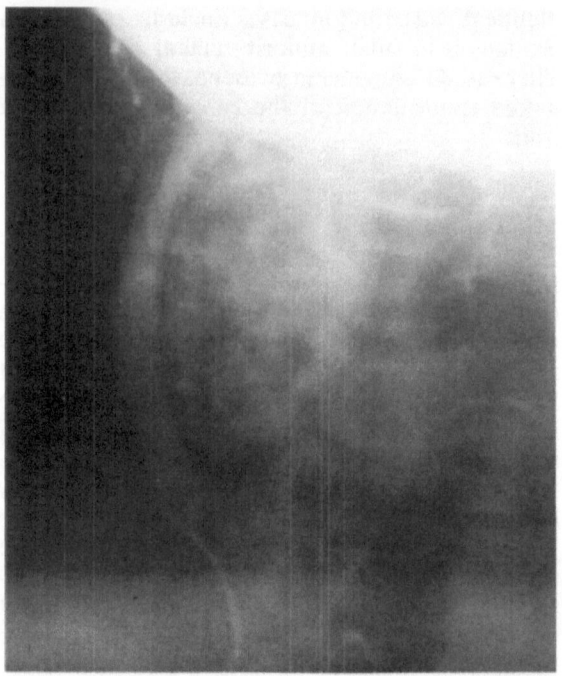

Fig. 19.14. Carotid angiogram demonstrating the "vascular blush" of a chemodectoma situated in the bifurcation of the common carotid artery.

of the larynx, trachea or oesophagus by a large tumour. In contrast, angiography is diagnostic. In the lateral projection there is a spreading of the carotid bifurcation to produce what Zak and Lawson (1982) termed a "lyre-shaped configuration", with the internal carotid artery usually pushed laterally and posteriorly while the external carotid artery is displaced anteriorly and laterally or medially. In the early phases of angiography numerous small and large blood vessels are seen about the tumour site. Later the tumour itself is picked out intensely as the contrast medium pools in the extensive vascular network within it (Fig. 19.14). Chemodectomas have the capacity to develop a collateral circulation from the regional vessels, including the external carotid artery, the subclavian artery via the thyrocervical trunk, the vertebral arteries, the internal carotid artery and the glomic artery arising from the carotid bifurcation (Zak and Lawson 1982). In addition to establishing the diagnosis, angiography provides information about the position of the tumour in relation to the carotid bifurcation, its size and its vascular supply. It demonstrates the patency and course of the carotid arteries. The contrast medium may show the tumour to surround and deform the internal and external carotid arteries thus indicating difficulties and potential dangers of surgical excision. It permits preoperative differential diagnosis from such lesions as vagal body tumours, which appear as vascular neoplasms causing lateral displacement of the carotid vessels without involvement of the carotid bifurcation, and aneurysms within the carotid arterial tree. Circulation times during angiography reveal a rapid arterial phase lasting under 1.5 seconds, immediately followed by a more uniform tissue accumulation or parenchymatous phase of 1.5–6.0 seconds' duration (Tihansky and Porter 1989).

Computed Tomography and Magnetic Resonance Imaging

Ancillary techniques to angiography such as CT scans have assumed increasing importance in the diagnosis of chemodectomas. They can determine both tumour extent and metastases and the findings in four patients with bilateral carotid body tumours reported by Casselman et al. (1987) are typical of the results obtained by this technique. Casselman et al. found well-circumscribed and strongly contrast-enhancing masses, situated at the bifurcation of the common carotid artery, displacing the adjacent structures. Scanning demonstrated the exact size of the tumours. Their experience convinced them that CT was an excellent, non-invasive examination technique for the diagnosis of chemodectoma. More recently the role of magnetic resonance imaging in the diagnosis of chemodectomas has been evaluated. Olsen et al. (1987) used the technique in 15 paragangliomas in 10 patients and compared the results with those of CT scanning in 13 tumours from 8 patients. They found that all the tumours were detected by both techniques, with the exception of one small glomus tympanicum tumour that was seen only in retrospect with magnetic resonance imaging. CT was found to be better for demonstrating subtle osseous changes of the skull base and the relation of paragangliomas to middle-ear structures, while magnetic resonance imaging was superior for showing the relation of such tumours to the adjacent carotid artery and internal jugular vein. Vogl and his associates (1989) also commented on the ability of magnetic resonance imaging to demarcate carotid body tumours from the carotid bifurcation. Olsen et al. (1987) found that the appearance of paragangliomas on magnetic resonance imaging was characteristic and based on their vascularity. Such appearances differed from meningiomas, neuromas and metastatic disease of the base of the skull.

Ultrasound has occasionally been used to assess chemodectomas (Gooding 1979; Lewis et al. 1980) but it does not seem to be reliable for differentiating between solid and cystic masses in the neck (Steinke et al. 1989). Doppler colour flow imaging detects blood flow in small tumours even in the absence of sonographically identifiable wall structures, and may improve the diagnosis of chemodectomas at a very early stage. Steinke et al. (1989) described the use of the technique in two cases of carotid body tumour, one in a 38-year-old man and the other in a 69-year-old woman. These new techniques offer the opportunity to detect chemodectomas at a very early stage when they are very small. They should improve our knowledge of the prevalence and natural history of chemodectomas, which are largely unknown at present because of the difficulties of early diagnosis and follow-up by non-invasive means. The greatest advantage will be to offer the patient and surgeon the option of early surgical intervention when the tumour, and the difficulties and complications of removal, are small.

Treatment

Whether or not to remove a carotid body tumour presents the surgeon with a classic dilemma. On the one hand, these tumours bleed torrentially due to the rich vascularity of the neoplasm both within its substance and in its capsule. The organ receives an enormous blood flow for its size (see Chapter 16). As we have noted above, even fine-needle aspiration can result in haemorrhage, yielding a specimen sometimes composed almost entirely of pure arterial blood. Chemodectomas are intimately attached to the carotid bifurcation and to the internal carotid artery, which conduct the major blood supply to the brain and these vessels may be readily torn during surgery. In view of these facts many surgeons believe it best to leave these tumours well alone. There is no doubt that surgery for carotid body tumours even today carries a high rate of significant complications.

Shamblin and his associates (1971) reported the results of surgery for chemodectoma performed at the Mayo Clinic over the period 1931 to 1967. Their report illustrates that in the early years before the advent of modern vascular surgical techniques there was a significant postoperative morbidity associated with the removal of these tumours, including damage to the carotid arterial tree with the onset of postoperative cerebrovascular accidents including transient or permanent hemiplegia or aphasia. Paralysis involving the 9th to 12th cranial nerves was also common. Sometimes this was produced by the tumour but often followed surgery (Van Asperen de Boer et al. 1981). In more recent times the incidence of cerebrovascular complications has fallen to a much lower level. Thus, in a recent review of 153 cases operated on during the half century up to 1985, Hallett et al. (1988) found no perioperative deaths and morbidity due to strokes fell to only 2.7% during the last decade. However, at the same time the morbidity due to injury of the lower cranial nerves remained unchanged over the last 50 years, remaining at 40%. This persistent risk to the hypoglossal or vagus nerves resulting in an awkward tongue or a husky voice is directly related to the size of the carotid body tumour being operated on and emphasises yet again the importance of early diagnosis.

To achieve and maintain such a low risk of cerebrovascular complications it is essential to obtain a clear demonstration of the blood supply to the tumour and to have an experienced vascular surgeon present to resect and repair damaged areas of the carotid arterial tree. Local invasion of the carotid arterial wall occurs in about 5% and in such cases the tumour has to be removed "en bloc". For this to be done safely the cerebral circulation has to be maintained with a shunt and the excised vessel replaced with a reversed autogenous graft of saphenous vein. The impossibility of predicting whether or not this major procedure will be necessary makes it necessary for a vascular surgeon to be present.

Tight control of diagnosis and surgical treatment to minimise the risk of postoperative stroke is seldom achieved. The chemodectoma is often misdiagnosed as an enlarged cervical lymph node and discovered by a young, inexperienced surgeon attempting a biopsy, sometimes under local anaesthesia (Browse 1982). According to Browse:

No surgeon should attempt to remove a carotid body tumour unless he is experienced in the surgical technique used for carotid artery surgery for atheromatous disease. This means that any surgeon who finds such a tumour unexpectedly should stop the operation, without a biopsy, and arrange for the patient to have further investigations and reexploration by an experienced vascular surgeon. To proceed unprepared with excision without the correct vascular instruments and surgical skill is hazardous and might cause a stroke, an unacceptable complication of what the patient had probably been told was to be a simple biopsy of a node.

With this imposing catalogue of difficulties and dangers associated with removal of carotid body tumours it might be thought preferable to leave them well alone. This viewpoint would be totally acceptable if the chemodectoma was always slowly growing and benign. However, as noted above, carotid body tumours carry a 6% risk of malignancy comprising 3% local invasion and 3% distant metastasis. This has led some authors to recommend a more vigorous surgical approach (Javid et al. 1976) but the all-important condition for this is that the tumour is treated early in its life history when it is small and can be removed easily by operating within the adventitia. Even then the dissection is always bloody but the oozing stops as soon as the tumour is free, hence the importance of the early diagnosis of these tumours when they are small by the modern techniques of investigation described above.

Radiotherapy

Chemodectomas are not very sensitive to X-radiation but occasionally radiotherapy may be the first line of treatment in advanced tumours and when the patient is a poor surgical risk (Bujko and Kawecki 1987). It may also be used as an adjuvant to incomplete resection and in cases with recurrent tumour after previous surgery. The response to irradiation may be slow. Farr (1967) reported no appreciable benefit following radiotherapy in six cases of chemodectoma. Oberman et al. (1968) reported failure of radiotherapy to modify growth of an advanced paraganglioma involving the base of the skull. No regression was observed and the patient died one year later showing symptoms of increased intracranial pressure. Shamblin and his colleagues (1971) noted that carotid body tumours had become smaller in only three of 16 irradiated patients. In the study of Lack et al. (1977) one patient was treated by chemotherapy alone and died 16 years later of unrelated causes but at necropsy proved to have residual disease in the neck.

References

Alpert LI, Bochetto JF (1974) Carotid body tumor: ultrastructural observations. Cancer 34:564–573

Barroso-Moguel R, Costero I (1962) Argentaffin cells of the carotid body tumour. Am J Pathol 41:389–403

Berdal P, Braaten M, Cappelen C, Mylino EA, Walaas O (1962) Noradrenaline–adrenaline producing nonchromaffin paraganglioma. Acta Med Scand 172:249–257

Besznyák I, Pinter E (1959) Karotisknötchentumor (Chemodectom). Zentralbl Allg Pathol 99:575–582

Birrell JH (1952) Carotid body tumours. A study of three cases including a bilateral example. Aust NZ J Surg 22:123–135

Brandberg R (1929) A case of tumour of the carotid body with thrombosis of the arteria carotis interna. Acta Chir Scand 65:464–474

Brown JB, Fryer MP (1952) Carotid body tumors: report of

removal of tumor thought to be largest recorded. Carotid removal, hemiplegia, recovery. Surgery 32:997–1002

Brown JW, Burton RC, Dahlin DC (1967) Chemodectoma with skeletal metastasis: report of two cases. Mayo Clin Proc 42:551–555

Browse NL (1982) Carotid body tumours. Br Med J 284:1507–1508

Bujko K, Kawecki A (1987) Radiotherapy of neck chemodectomas: long term results. Radiobiol Radiother (Berl) 28:749–753

Callison JG, MacKenty JE (1913) Tumors of the carotid body. Ann Surg 58:741–765

Capella C, Solcia E (1971) Optical and electron microscopical study of cytoplasmic granules in human carotid body, carotid body tumours and glomus jugulare tumours. Virchows Arch [B] 7:37–53

Casselman JW, Wilms GE, Baert AL (1987) Computed tomography of bilateral carotid body tumours. ROFO 146:381–386

Chambers RG, Mahoney WD (1968) Carotid body tumors. Am J Surg 116:554–558

Chase WH (1933) Familial and bilateral tumors of the carotid body. J Pathol Bacteriol 36:1–12

Chedid A, Jao W (1974) Hereditary tumours of the carotid bodies and chronic obstructive pulmonary disease. Cancer 33:1635–1644

Chen LT, Hwang WS (1989) Fine needle aspiration of carotid body paraganglioma. Acta Cytol 33:681–682

Conley JJ (1965) The carotid body tumor: a review of 29 cases. Arch Otolaryngol 81:187–193

Costero I (1963) Recent advances in the knowledge concerning chemodectomas. Lab Invest 12:270–284

Costero I, Barroso-Moguel R (1961) Structure of carotid body tumor. Am J Pathol 38:127–141

Costero I, Chevez AZ (1962) Carotid body tumor in tissue culture. Am J Pathol 40:337–357

Dent TL, Thompson NW, Fry WJ (1976) Carotid body tumors. Surgery 80:365–372

De Villiers DM (1972) Patologiese en chirurgiese aspekte van maligne nie-chromaffien paragangliome (chemodektome). S Afr Med J 46:1238–1240

Donald RA, Crile G (1948) Carotid body tumors. Am J Surg 75:435–440

Engzell U, Franzén S, Zajicek J (1971) Aspiration biopsy of tumors of the neck: cytologic findings in 13 cases of carotid body tumor. Acta Cytol 15:25–30

Enzinger FM (1969) Histological typing of soft tissue tumours. International histological classification of tumours no. 3. World Health Organization, Geneva

Fanning JP, Woods FM, Christian HJ (1963) Metastatic carotid body tumor. JAMA 185:49–50

Farr NW (1967) Carotid body tumors. A thirty year experience at Memorial Hospital. Am J Surg 114:614–619

Fisher ER, Reidbord H (1971) Electron microscopic evidence suggesting the myogenous derivation of the so-called alveolar soft-part sarcoma. Cancer 27:150–159

Gilford H, Davis KLH (1904) 'Potato tumours' of the neck and their origin as endotheliomata of the carotid body, with an account of three cases. Practitioner 73:729–739

Glenner GG, Grimley PM (1974) Tumors of the extra-adrenal paraganglion system (including chemoreceptors). Armed Forces Institute of Pathology, Washington, DC

Glenner GG, Crout JR, Roberts WC (1962) A functional carotid-body-like tumour secreting levarteronol. Arch Pathol Lab Med 73:230–240

González-Cámpora R, Otal-Salaverri C, Panea-Flores P, Lerma-Puertas E, Galera-Davidson H (1988) Fine needle aspiration cytology of para ganglionic tumors. Acta Cytol 32:386–390

Gooding GAW (1979) Gray-scale ultrasound detection of carotid body tumors. Radiology 132:409–410

Goormaghtigh N, Pattyn S (1954) A presumably benign tumor and a proved malignant tumor of the carotid body. Am J Pathol 30:679–693

Grage TB, Cueto J (1967) Surgical management of carotid body tumors. Surgery 62:742–749

Grimley PM, Glenner GG (1967) Histology and ultrastructure of carotid body paragangliomas. Comparison with the normal gland. Cancer 20:1473–1488

Gullotta F, Helpap B (1976) Tissue culture, electron microscopic and enzyme histochemical investigations of extra-adrenal paragangliomas. Pathol Eur 11:257–264

Gupta S, Gupta OP, Verma DN, Rastogi BL (1976) Carotid body tumour: excision with resection of carotid arteries. J Laryngol Otol 90:305–310

Gustilo RB, Lober PH, Salovich EL (1965) Chemodectoma (carotid body tumor) metastasizing to bone. J Bone Joint Surg [Am] 47:155–160

Hallett JW, Nora JD, Hollier IH, Cherry KJ, Pairolero PC (1988) Trends in neurovascular complications of surgical management for carotid body and cervical paragangliomas: a fifty-year experience with 153 tumors. J Vasc Surg 7:284–291

Hamberger CA, Hamberger CB, Wersäll J, Wågermark J (1967) Malignant catecholamine-producing tumour of the carotid body. Acta Pathol Microbiol Scand 69:489–492

Hanford JM (1965) Carotid body tumors: non chromaffin paragangliomas of the cervical region. Am J Surg 110:398–404

Harrington SW, Clagett OT, Dockerty MB (1941) Tumors of the carotid body. Clinical and pathological considerations of twenty tumors affecting nineteen patients (one bilateral). Ann Surg 114:820–833

Heath D, Williams DR (1989) High-altitude medicine and pathology. Butterworths, London

Hood IC, Qizilbash AH, Young JE, Archibald SD (1983) Fine needle aspiration biopsy cytology and ultrastructural studies of three cases. Acta Cytol 27:651–657

Hörtnagl H, Hörtnagl H, Propst A, Schwingshakl H, Weiser G, Winkler H (1973) Catecholamine storage in liver metastases of a malignant carotid body tumour. Virchows Arch [B] 12:330–337

Hutchinson J (1888) Sarcomatous tumors under the upper part of the sternomastoid (potato-like tumors). Lond Illust News 1:50

Irons GB, Weiland LH, Brown WL (1977) Paragangliomas of the neck: clinical and pathological analysis of 116 cases. Surg Clin North Am 57:575–583

Isfort A, Knoche H (1966) Tumoren des Glomus Caroticum. Brun's Beitr Klin Chir 212:417–440

Jacobs DM, Waisman J (1987) Cervical paranganglioma with intranuclear vacuoles in a fine needle aspirate. Acta Cytol 31:29–32

Javid H, Chawla SK, Dye WS, Hunter JA, Najafi H, Goldin MD, Serry C (1976) Carotid body tumor: resection or reflection. Arch Surg 111:344–347

Kahn LB (1976) Vagal body tumor (non chromaffin paraganglioma, chemodectoma, and carotid body-like tumor) with cervical node metastasis and familial association. Ultrastructural study and review. Cancer 38:2367–2377

Kapoor R, Saha MM, Das DK, Gupta AK, Tyagi S (1989) Carotid body tumor initially diagnosed by fine needle aspiration cytology. Acta Cytol 33:682–683

Katz AD (1964) Carotid body tumors in a large family group. Am J Surg 108:570–573

Krause-Senties LG (1971) Tumores del cuerpo carotideo. Arch Invest Med (Mex) 2:25–30

Lack EE, Cubilla AL, Woodruff JM, Farr HW (1977) Paragangliomas of the head and neck region. Cancer 39:397–405

Lack EE, Cubilla AL, Woodruff JM (1979) Paragangliomas of the head and neck region. A pathologic study of tumors from 71 patients. Hum Pathol 10:191–218

Le Compte PH (1951) Tumors of the carotid body and related structures. Fascicle 16. Atlas of tumor pathology. Armed Forces Institute of Pathology, Washington, DC

Lees CD, Levine HL, Beven EK, Tucker HM (1981) Tumors of the carotid body. Experiences with 41 operative cases. Am J Surg 142:362–365

Lewis RR, Beasley MG, Coghlan BA, Yates AK, Gosling RG (1980) Demonstration of a carotid body tumour by ultrasound. Br J Radiol 53:368–371

Lewison EF, Weinberg T (1950) Carotid Body tumors; case report of bilateral carotid body tumors with unusual family incidence. Surgery 27:437–447

Macadam RF (1969) The fine structure of a human carotid body tumour. J Pathol 99:101–104

MacComb WS (1948) Carotid body tumours. Ann Surg 127:269–277

Marshall RB, Horn RC (1961) Non chromaffin paraganglioma, a comparative study. Cancer 14:779–787

Martin CE, Rosenfeld L, McSwain B (1973) Carotid body tumors: a 16-year follow-up of seven malignant cases. South Med J 66:1236–1243

Martorell F (1956) Tumor of the carotid body. Angiology 7:228–232

McIlrath DC, ReMine WH (1963) Parapharyngeal tumors. Surg Clin North Am 43:1041–1047

McNealy RW, Hedin RF (1939) Surgery of carotid body tumours. J Int Coll Surg 2:285–294

McSwain B, Spencer FC (1947) Carotid body tumor in association with carotid sinus syndrome. Report of two cases. Surgery 22:222–229

Mincione G, Urso C (1989) Fine needle aspiration cytologic findings in a case of carotid body paraganglioma (chemodectoma). Acta Cytol 33:679–681

Monro RS (1950) The natural history of carotid body tumours and their diagnosis and treatment. With a report of five cases. Br J Surg 37:445–453

Morfit HM, Swan H, Taylor ER (1953) Carotid body tumors; report of twelve cases including one case with proven visceral dissemination. Arch Surg 67:194–213

Mulligan RM (1950) Chemodectoma in the dog. Am J Pathol 26:680–681

Newell R (1971) Diagnosis and treatment of glomus jugulare and carotid body tumors. Am Acad Ophthalmol Otol Course 506

Oberman HA, Holtz F, Sheffer LA, Magielski JE (1968) Chemodectomas (non chromaffin paragangliomas) of the head and neck. Cancer 21:838–851

Olsen WL, Dillon WP, Kelly WM, Norman D, Brant-Zawadzki M, Newton TH (1987) MR imaging of paragangliomas. Am J Roentgenol 148:201–204

Pettet JR, Woolner LB, Judd ES (1953) Carotid body tumors (chemodectomas). Ann Surg 137:465–477

Phelps FW, Case SW, Snyder GA (1937) Primary tumors of the carotid body. Review of 159 histologically verified cases. Report of a case. West J Surg Obstet Gynec 45:42–46

Pratt LW (1973) Familial carotid body tumours. Arch Otolaryngol 97:334–336

Pryse-Davies JI, Dawson IMP, Westbury G (1964) Some morphologic, histochemical, and chemical observations on chemodectomas and the normal carotid body, including a study of the chromaffin reaction and possible ganglion cell elements. Cancer 17:195–202

Rabson AS, Elliott JL (1957) Carotid body tumors with regional lymph node involvement with report of a case. Surgery 42:381–385

Reese HE, Lucas RN, Bergman PA (1963) Malignant carotid body tumors. Ann Surg 157:232–243

Reid MR (1920) Adenomata of the carotid gland. Bull Johns Hopkins Hosp 31:177–184

Robertson DI, Cooney TP (1980) Malignant carotid body paraganglioma: light and electron microscopic study of the tumor and its metastases. Cancer 46:2623–2633

Romanski R (1954) Chemodectoma (non-chromaffinic) paraganglionic of the carotid body with distinct metastasis with illustrative case. Am J Pathol 30:1–9

Rush BF (1962) Current concepts in treatment of carotid body tumors. Surgery 52:679–684

Rush BF (1963) Familial bilateral carotid body tumors. Ann Surg 157:633–636

Saldaña MJ, Salem JE, Travezan R (1973) High altitude hypoxia and chemodectomas. Hum Pathol 4:251–263

Salyer KE, Ketchum LD, Robinson DW, Masters FW (1969) Surgical management of cervical paragangliomata. Arch Surg 98:572–578

Sessions RT, McSwain B, Carlson RI, Scott HW (1959) Surgical experiences with tumors of the carotid body, glomus jugulare and retroperitoneal nonchromaffin paraganglia. Ann Surg 150:808–823

Shamblin WR, Remine WH, Sheps SG, Harrison EG (1971) Carotid body tumor (chemodectoma). Clinicopathologic analysis of ninety cases. Am J Surg 122:732–739

Sprong DH, Kirby FG (1949) Familial carotid body tumors: report of nine cases in 11 siblings. Ann West Med Surg 3:241–242

Staats EF, Brown RL, Smith RR (1966) Carotid body tumors, benign and malignant. Laryngoscope 76:907–916

Steimle P, Steimle R (1958) Les chémodectomas ou paragangliomes nonchromaffines. J Chir (Paris) 76:559–572

Steinke W, Hennerici M, Aulich A (1989) Doppler color flow imaging of carotid body tumors. Stroke 20:1574–1577

Szenthe L, Kneiszl F (1964) Die Chemodektome Bruns. Beitr Klin Chir 209:363–380

Tihansky DP, Porter PS (1989) Pulsed Doppler-ultrasonic diagnosis of carotid body tumor. NY State J Med 89:580–582

Toker C (1967) Ultrastructure of a chemodectoma. Cancer 20:271–280

Turnbull FM (1954) Malignancy in carotid body tumors with presentation of a proved case of metastases to the lung. West J Surg Obstet Gynec 62:382–390

Van Asperen de Boer FRS, Terpstra JL, Vink M (1981) Diagnosis, treatment and operative complications of carotid body tumours. Br J Surg 68:433–438

Vogl T, Bruning R, Schedel H, Kang K, Grevers G, Hahn D, Lissner J (1989) Paragangliomas of the jugular bulb and carotid body: MR imaging with short sequences and Gd-DTPA enhancement. Am J Roentgenol 153:583–587

Westbury G (1963) Carotid body tumours. Curr Med Drugs 4:3–5

Willis AG, Birrell JH (1955) The structure of a carotid body tumor. Acta Anat (Basel) 25:220–265

Wilson H (1964) Carotid body tumors: surgical management. Ann Surg 159:959–966

Yaghmai I, Shariat S, Shamloo M (1970) Carotid body tumors. Radiology 97:559–563

Zak FG, Lawson W (1982) The paraganglionic chemoreceptor system. Physiology, pathology and clinical medicine. Springer Verlag, Berlin Heidelberg New York

20 Carotid Sinus

So far in this book we have been concerned with the main chemoreceptor in humans, the carotid body, which is situated in the fork of the carotid bifurcation. Closely adjacent in the first part of the internal carotid artery arising from the bifurcation is the main baroreceptor of the human body.

The Discovery of the Baroreceptor

The realisation that the first part of the internal carotid artery is dilated is usually credited to Burns (1811). Right from the outset he noted that this dilatation had a predilection for disease and he pondered as to whether it might provide the basis for aneurysm formation. Over the following half century only passing reference was made to this dilatation until Luschka (1862, p 333) redescribed it, once again referring to it as "the normal prototype of a fusiform aneurism". Some four years later Patrick Manson, Assistant Medical Officer in the Durham County Asylum, carried out 17 post mortems on subjects who died from such conditions as general paralysis of the insane. He found what he called a "peculiar affection of the internal carotid artery in connexion with disease of the brain". This comprised a dilatation at the origin of this vessel. Thus, in his first publication (Fig. 20.1) at the age of only 22 years he gave one of the first clear descriptions of what we now call the carotid sinus. Both Manson (1866) and Schäfer (1877) found that the structure was not associated with increasing age and the latter author noted that it was to be found in children who had been healthy during life. Binswanger

(1879) called attention to the fact that even in such children free of vascular disease there were areas of marked thinning of the wall of the carotid sinus, especially in its ventromedial part.

The dilatation of the first part of the internal carotid artery was ascribed by Stahel (1886) to abnormal haemodynamics in the region of the bifurcation. He thought, for example, that the internal carotid artery was unable to dilate to accommodate the pulsating flow of blood through the bifurcation and thus dilated in its proximal part. He considered that the vascular dilatation dampened down the pulse wave protecting the delicate brain substance. It was also thought likely that the circumscribed areas of thinning, particularly of the ventromedial aspect of the sinus, were secondary to haemodynamic stresses. A century later, studies of the haemodynamics within the carotid bifurcation and sinus were still being carried out (Zarins et al. 1983) as we describe below.

Much later the alternative view that these areas of thinning are primary had to be considered. Views on the nature and function of the first part of the internal carotid artery changed dramatically in 1923 when Hering drew attention to the bradycardia and fall of systemic blood pressure which result from direct pressure on the carotid bifurcation. These observations led him to conceive of a sinus reflex (Hering 1924a) which was mediated by a sinus nerve which was a branch of the glossopharyngeal (Hering 1924b). In this way the carotid sinus came to be recognised as the major baroreceptor in the human body. However, as early as 1925 Drüner expressed the opinion that the carotid body itself also had the capacity to

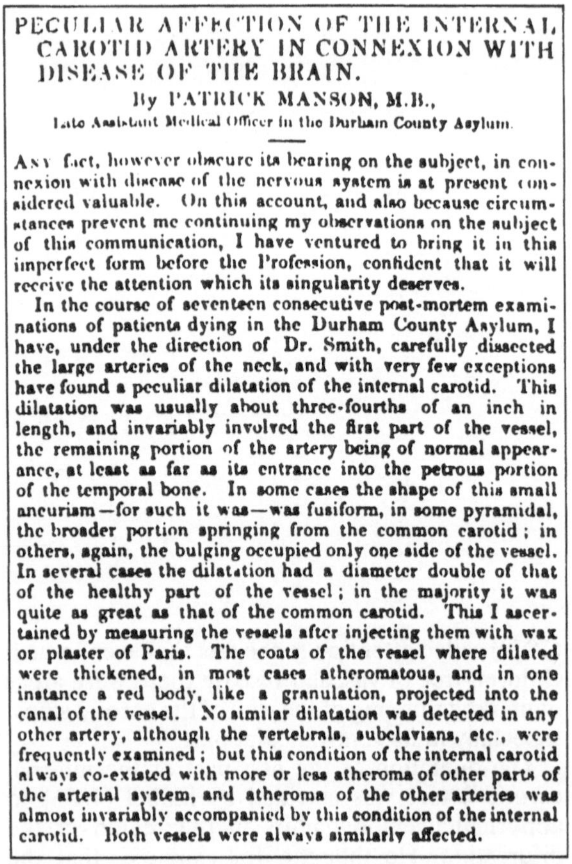

PECULIAR AFFECTION OF THE INTERNAL CAROTID ARTERY IN CONNEXION WITH DISEASE OF THE BRAIN.

By PATRICK MANSON, M.B.,

Late Assistant Medical Officer in the Durham County Asylum.

ANY fact, however obscure its bearing on the subject, in connexion with disease of the nervous system is at present considered valuable. On this account, and also because circumstances prevent me continuing my observations on the subject of this communication, I have ventured to bring it in this imperfect form before the Profession, confident that it will receive the attention which its singularity deserves.

In the course of seventeen consecutive post-mortem examinations of patients dying in the Durham County Asylum, I have, under the direction of Dr. Smith, carefully dissected the large arteries of the neck, and with very few exceptions have found a peculiar dilatation of the internal carotid. This dilatation was usually about three-fourths of an inch in length, and invariably involved the first part of the vessel, the remaining portion of the artery being of normal appearance, at least as far as its entrance into the petrous portion of the temporal bone. In some cases the shape of this small aneurism—for such it was—was fusiform, in some pyramidal, the broader portion springing from the common carotid; in others, again, the bulging occupied only one side of the vessel. In several cases the dilatation had a diameter double of that of the healthy part of the vessel; in the majority it was quite as great as that of the common carotid. This I ascertained by measuring the vessels after injecting them with wax or plaster of Paris. The coats of the vessel where dilated were thickened, in most cases atheromatous, and in one instance a red body, like a granulation, projected into the canal of the vessel. No similar dilatation was detected in any other artery, although the vertebrals, subclavians, etc., were frequently examined; but this condition of the internal carotid always co-existed with more or less atheroma of other parts of the arterial system, and atheroma of the other arteries was almost invariably accompanied by this condition of the internal carotid. Both vessels were always similarly affected.

Fig. 20.1. The beginning of the paper by Manson (1866) which provided one of the first descriptions of the carotid sinus.

function as a baroreceptor. As we have seen in Chapter 16, the small interlobular glomic arteries of the carotid body have an elastic structure inappropriate for vessels of such small dimensions but similar in histological appearance to the carotid sinus. Such observations are entirely in line with Drüner's (1925) early view that the carotid body as well as the sinus may subserve baroreception.

Indeed, the special innervation of the carotid sinus is derived from the intercarotid plexus, either directly, or through the carotid body (de Castro 1926). The nerve endings from this plexus end in the adventitia of the carotid sinus and particularly just outside the external elastic lamina. At first thought to be vagal, these nerve fibrils were later recognised to be branches of the glossopharyngeal. While the carotid sinus shows areas of thinning of the media, due to loss of muscle and replacement by closely applied elastic laminae as we shall see later, the adventitia becomes thicker over these thinned areas. The nerve endings are found to lie in an annular band, increasing in width as it passes around the sinus, and fading away gradually above and below the band.

Development of the Carotid Sinus

Although it has been suggested from the end of the last century that the carotid sinus represents the transverse part of the third aortic arch, the development of the region has been considered almost entirely from the point of view of the carotid body. Kastschenko (1887), Marchand (1891) and Paltauf (1892) all found that the wall of the developing internal carotid artery, destined to become the carotid sinus, becomes markedly thickened, although the adult sinus is characterised by ventromedial thinning as we describe below. All three authors associated the thickening of the carotid artery with the formation of the carotid body. In this century Hammar (1934) and Boyd (1937) confirmed this thickening in human fetuses, while Ito (1950) found it in the rabbit. Boyd (1937) found what he called a primary condensation of mesodermal cells lying like a cuff around the commencement of the third aortic arch, corresponding to the segment where the carotid sinus develops. He considered this mesodermal condensation to be the primary anlage of the carotid body but other workers have thought it more likely that it represents a stage in the development of the carotid sinus. Kohn (1900), for example, described the same thickening in pig embryos but he claimed that it was not connected with the development of the carotid body. More recently Nielsen (1968) found in dogs a very thick cellular coat around the developing internal carotid artery at the site of the future carotid sinus. This did not diminish until around birth when the carotid sinus attained its characteristic mural thinning, described later in this chapter. Hence there remains controversy as to whether this thick peri-arterial sheath is related to the developing carotid body or carotid sinus.

At the time of writing, we are engaged in a study of this peri-arterial thickening. So far we have investigated the histological appearances of this cellular sheath but positive identification of its components will require electron microscopy and this is being carried out. The vascular smooth muscle cells of the developing internal carotid artery show considerable mitotic activity. Their cytoplasm shows a bubbly appearance and many of these small clear areas show a faint basophilia.

In many sites they are enlarged and clear. The outer limit of the media is ill defined, as elongated and strap-like muscle cells are seen apparently migrating into the peri-arterial sheath. It seems to us that this is a significant finding, for it suggests that the sheath tissues are derived from vascular smooth muscle. The sheath itself is much looser in texture and includes a considerable variety of cells. Strap-like cells with a faintly eosinophilic cytoplasm and with cytological features consistent with origin from vascular smooth muscle cells show elongated, tapering cytoplasmic extensions. Commonly they are adjacent to clear areas larger than those found in the compact media and showing in their walls linear structures with a network of small round prominences. Other cells have narrow fusiform nuclei with elongated filamentous cytoplasmic extensions. Many of these extensions are sharply angled. It is clear from these preliminary histological studies that the structure of this peri-arterial sheath is complex and requires electron microscopy for its definitive elucidation. However, the observations so far are consistent with the views expressed by Kohn (1900) and Nielsen (1968) that this cellular sheath may be associated with the developing carotid sinus.

Macroscopic Features

The carotid sinus is often easy to delineate from the portions of the common and internal carotid arteries lying either side of it. It is commonly fusiform or triangular in shape but bulges more on its dorsolateral aspect. Early observations (Schäfer 1877) suggested that the transverse diameter of the sinus is usually less than twice that of the internal carotid artery, and is very rarely more than 2.5 times that diameter. Stahel (1886) found all diameters for the carotid sinus up to a maximum of 10.9 mm, compared to 8.9 mm for the common carotid and 5.9 mm for the rest of the internal carotid artery. We have found a range of 5.3–9.3 mm for the diameter of the carotid sinus, with a mean value of 7.6 mm (Table 20.1). While the carotid sinus involves only the first part of the internal carotid artery, not infrequently it incorporates the termination of the common carotid artery as well. It is often virtually impossible to measure the length of the carotid sinus as its extremities merge imperceptibly into the adjacent arteries with which it is continuous. Krause (1879) felt confident enough to quote a length of between 7 and 10 mm.

Histological Structure

Qualitative and quantitative studies of the histology of the walls of the vessels forming the bifurcation of the common carotid artery have confirmed that the carotid sinus has a specialised structure (Heath et al. 1973). The parent vessel of the sinus, the common carotid artery, is an elastic conducting artery with a structure similar to that of the aorta. Its thick media contains many crenated elastic laminae running roughly parallel with one another. The bulk of the media is composed of

Table 20.1. Quantitative histological features of the carotid sinus and common and internal carotid arteries (after Heath et al. 1973)

Subject	Common carotid artery		Carotid sinus		Internal carotid artery	
	Range	Mean	Range	Mean	Range	Mean
Greatest transverse diameter on fresh specimen (mm)	6.0–9.3	8.0	5.3–9.3	7.6	3.0–4.3	4.7
Greatest transverse diameter on histological examination[a] (mm)	4.4–7.6	6.4	4.4–7.7	6.3	3.0–5.0	4.0
Absolute medial thickness (μm)	626–1034	767	164–588	319	318–590	463
Medial thickness as % of external diameter	8.3–16.7	12.1	2.5–8.7	5.2	10.1–15.8	12.4
Thickness of intimal proliferation as % of external diameter	1.0–9.6	4.5	2.1–19.7	10.0	0–2.4	1.2
% of lumen occluded by atherosclerosis	—	—	12.8–69.0	37.4	—	—

[a] It will be noted that the external diameters of the fixed arteries as determined by histological measurements are smaller than the transverse diameters measured on fresh specimens. This is because the external diameters measured between diametrically opposite points on the external elastic laminae do not include the adventitia, as do measurements on photographic prints of the fresh specimens. Furthermore, histological measurements are made on tissue inevitably shrunken as a result of fixation and processing.

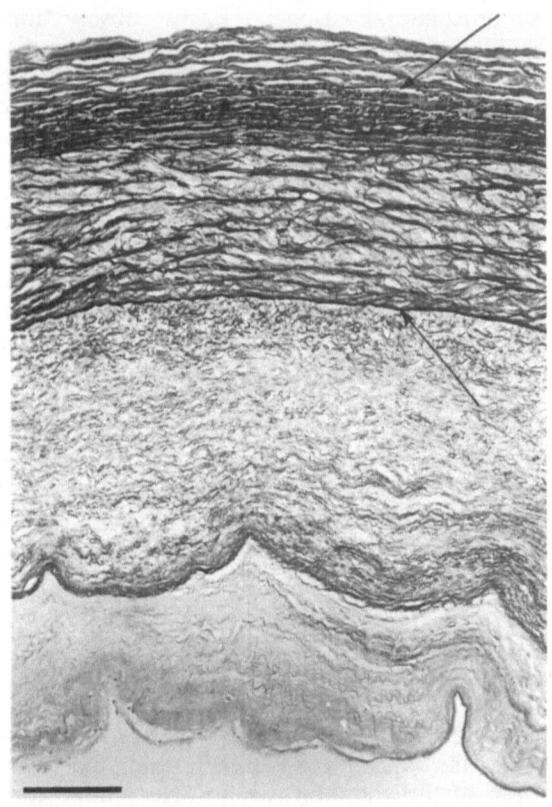

Fig. 20.2. Arc of transverse section of carotid sinus from a man of 61 years. There is a pronounced thinning of the media, the limits of which are indicated by *arrows*. The elastic fibrils in the outer media are tightly packed together. There is severe intimal fibroelastosis. (Elastic-Van Gieson (EVG)) Scale line = 200 μm.

ously thicker than that of the sinus, although lying distal to it. Its internal elastic lamina is thick and well defined, with elastic fibrils in the subjacent zone of the media. There is a compact zone of tightly packed parallel elastic fibrils in the zone of demarcation between the outer media and adventitia.

The medial thickness of the carotid sinus is much less than that of the carotid artery on either side of the sinus. The range and mean values of medial thickness expressed in absolute terms and as a percentage of the external diameter of the vessel are shown in Table 20.1. The data confirm that the carotid sinus is a definite anatomical entity characterised by an abrupt thinning of the media, which is just as rapidly replaced by the normal muscularity of the media on distal progression in the internal carotid artery (Table 20.1). In fact the medial thickness of the carotid sinus in circumscribed areas is only some 5% of the external diameter of the artery, a figure more appropriate to the pulmonary than to the systemic circulation.

The Circumscribed Nature of Wall-Thinning

The anatomical thinning of the wall of the carotid sinus is restricted to circumscribed areas. According to Adams (1958), Meyer (1876) was the first to observe that, on opening the carotid artery and holding it up to the light, circumscribed areas of thinning can be seen. On distending the sinus he was able to produce small ectasias of the wall. There is a particularly thin part of the wall of the sinus ventromedially, opposite the maximum bulge of the posterolateral wall. This thin area is close to and faces the carotid body. Adams (1958) noted that the wall of the carotid sinus just below its equator has been found to be extremely thin, extending as a belt around the vessel. de Castro (1926) related this thinning to the innervation of the sinus for it represents the point at which the nerves first reach the vessel and from which they fan out around it in an annular belt.

fibres of smooth muscle with a few collagen fibres but in its outer zone there may be areas where the elastic fibres are crowded together with little or no intervening muscle. In contrast, in the carotid sinus there is an abrupt thinning of the media (Fig. 20.2). Its inner zone still has the structure of a conducting elastic artery but, in the outer zone adjacent to the adventitia rich in nerve fibrils, the elastic fibrils are packed tightly together with no muscle fibres interposed (Fig. 20.2). In some subjects the media of the carotid sinus is very thin and composed largely of a compact layer of elastic fibrils. With increasing age the inner portion of the media becomes fibrous.

On the distal side of the carotid sinus the internal carotid artery reverts rapidly into the structure of a typical muscular artery in which the media is largely composed of tightly packed smooth muscle with but few small elastic fibrils. The media of the internal carotid artery is obvi-

The Adventitia and Nerve Network

The adventitia of the carotid sinus is thicker and more elastic than that of the adjacent arteries and it contains most of the terminations of the afferent nerves passing from the sinus. In the very thin part of the wall of the carotid sinus the highly elastic media merges into the elastic adventitia. These thinned areas show rich vascularisation and the constituent vessels effect communication with those around the carotid body.

The innervation which is largely restricted to the adventitia appears to be of vagal and glossopharyngeal origin, directly or by way of the intercarotid plexus, but there may be a contribution from the sympathetic to the media (Adams 1958). The orthodox view is that the adventitial nerve endings all belong to medullated fibres from which telodendria spread out into the deeper part of the adventitia between the lamellae of collagen and elastic fibres. The nerve endings comprise both neurofibrillar nets and end-bulbs resembling Meissner's corpuscles (Adams 1958).

Age-Change and Atherosclerosis

Like all human blood vessels the carotid sinus is subject to age-change taking the form of proliferation of myofibroblasts in the intima leading to fibrosis or fibroelastosis there. Although the sinus is a normal anatomical structure, it has the general configuration of a small aneurysm as was immediately apparent to early observers (Burns 1811; Luschka 1862). As a result it is subjected to increased haemodynamic stresses that result in exaggerated age-changes and the development of atherosclerosis. The relation of these stresses to the propensity for the occurrence of atheroma in different areas of the carotid arterial tree has been studied by Zarins et al. (1983). They found that intimal thickening and atherosclerosis were most likely to be severe in the outer, lateral wall of the carotid sinus. This may be related to the fact that this is subjected to a relatively low wall shear stress with an alteration in the blood flow. The usual unidirectional flow aligned to the axis of the carotid vessels is replaced by complex secondary flow patterns including helical trajectories which may actually cause local reversals of axial flow in the carotid sinus, leading to exaggeration of atherosclerosis. In contrast, along the inner wall of the carotid sinus and in the external carotid artery, rapid laminar flow is maintained and intimal thickening is delayed. Beyond the carotid sinus the two dissimilar flows "reattach" to form a rapid axial flow (Zarins et al. 1983) and this carries on into the distal carotid artery, which then becomes like the external carotid artery to be exposed to a rapid unidirectional and axially aligned flow subjecting the walls to moderate-to-high shear stress forces which tend to delay the onset of atherosclerosis, according to Zarins and his group.

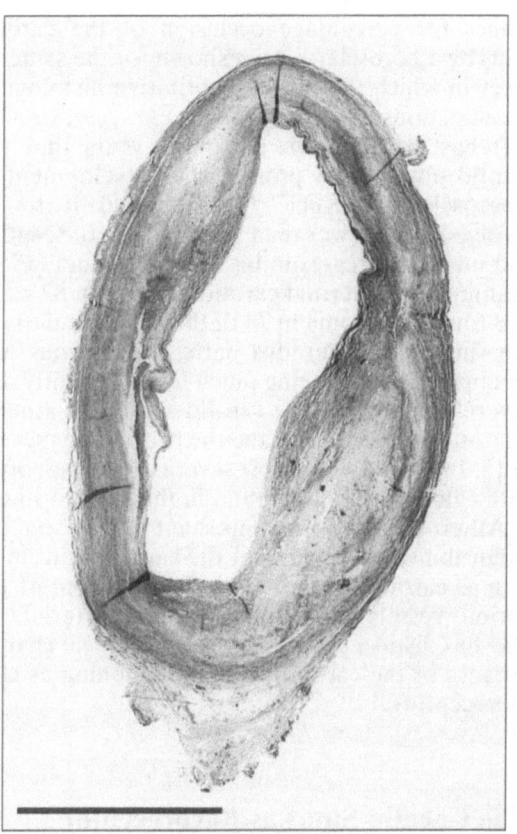

Fig. 20.3. Transverse section of right carotid sinus from a man of 63 years with bronchial carcinoma and duodenal ulcer. It shows severe atherosclerosis leading to an appreciable diminution in the lumen and fibrosis of the media. (EVG) Scale line = 3 mm.

Thus from the very outset the carotid sinus has a form that almost inevitably will lead to pathological changes likely to interfere with its function. When the sinus of a subject beyond early adult life is cut open, it commonly shows severe atherosclerosis with degeneration, rigidity and often calcification of much of its wall thickness (Fig. 20.3). There is frequently thickening, puckering and scarring of its lining so that the sinus is contracted. On histological examination the thick layer of atherosclerosis commonly extends into the underlying media, fragmenting its elastic fibrils and leading to a "moth-eaten appearance". On transverse sections of the carotid sinus from 38 subjects coming to necropsy we measured the areas occupied by the remaining patent lumen, intimal proliferation and media (Heath et al. 1973). It thus proved possible to measure the percentage blockage of the lumen of the carotid sinus by intimal proliferation. In Table 20.1 the mean values and ranges for percentage occlusion of the carotid sinus by atherosclerosis are shown for the same 13 cases in which the other quantitative histological investigations had been made.

It has been known for many years that the carotid sinus is very prone to the development of atherosclerosis. Keele (1933) found it to be involved in no fewer than 50 of the 55 consecutive and unselected cases in his series. Samuel (1956) examined the internal carotid arteries in 82 cases and found atheroma in 74 of them, particularly in the sinus and cavernous parts, the petrous and cerebral segments being much less frequently and severely involved. The carotid sinus was atherosclerotic in 60 instances and the process was severe in 15. In his study the most severe atheroma, often with calcification, was found in the carotid sinus.

Atherosclerosis has important effects on the extensibility and rigidity of the carotid sinus in its rôle as carotid baroreceptor. Involvement of the various vessels comprising the carotid arterial tree also has distinct effects on the histological characteristics of the carotid bodies functioning as chemoreceptors.

The Carotid Sinus as Baroreceptor

As we have noted above, more than half a century ago it was demonstrated in animals that pressure on the walls of the carotid sinus leads to slowing of the heart and a fall in systemic blood pressure (Hering 1927; Heymans 1929). This would appear to be the physiological function associated with the characteristic thinning of the media of the carotid sinus described above. Such a structure suggests that the young and healthy sinus might well be more distensible than its neighbouring common and internal carotid arteries to allow stimulation of the branches of the glossopharyngeal nerve in the adventitia and the initiation of baroreceptor reflexes. One has to be careful, however, in applying the results of studies in animals uncritically to those in humans, thinking of the carotid sinus as a thin-walled, highly elastic segment of the carotid arterial bifurcation. In humans, as we have seen, the histological adaptation of the wall of the carotid sinus to subserve a baroreceptor rôle is heavily modified by age-change intimal fibrosis and atherosclerosis. Indeed, in the elderly the carotid sinus may become hardly more than a fibrous or even partly calcified sac. For this reason we carried out a study of the extensibility of the human carotid sinus (Winson et al. 1974).

Extensibility

We measured the extensibility of the carotid sinus in 30 subjects coming to necropsy between the ages of 16 to 77 years, using the apparatus we had employed previously to measure the extensibility of the pulmonary trunk in cases of congenital heart disease (Harris et al. 1965). This consisted of a central Perspex chamber in which a 1 cm length of circumferential strip of the sinus was stretched between two vertical clamps. The lower one was attached to a platform which could be moved up and down between runners. The upper one was suspended vertically above the lower clamp by means of a stiff steel wire attached to one arm of a balance, the opposite arm of which acted as a pointer against a circular scale which had been previously calibrated between 0 and 100 g weight. Thus by moving the platform up and down the strip of artery could be subjected to increasing extensible loads the magnitude of which could be read from the scale on the balance. The length of the strip at any load was measured by means of a Vernier microscope. In view of the variation in initial length the extension produced by each extensile load was expressed as a percentage of the original unstretched length, given by the formula:

$$\frac{\text{stretched length} - \text{original length}}{\text{original length}} \times 100$$

When the results were expressed graphically it was found that there was a curved relation between extensile load and percentage extension produced by that load, so that with increasing extensile loads the extensibility decreased (Fig. 20.4). This relation is similar to that described previously in systemic arteries elsewhere in the body (Burton 1954) and in the human pulmonary trunk (Harris et al. 1965). Roach and Burton (1957) ascribed the curved shape of the length–tension diagram of the walls of arteries to the elastic fibres having a shorter relaxed length relative to the collagen fibres. Thus, the first and more horizontal part of the graph was thought to be due to the extension of the elastic fibres alone, and the more vertical part of the graph to gradual recruitment of collagen fibres.

The extensibility of the carotid sinus decreases with age (Fig. 20.4). We found that at an extensile load of 100 g the carotid sinus of a boy of 16 years extended by as much as 64% of its original circumference, whereas in a woman of 75 years it extended by only 16%. Such a fall in extensibility has been described in both systemic and pulmonary arteries (Burton 1954; Harris et al. 1965) and is

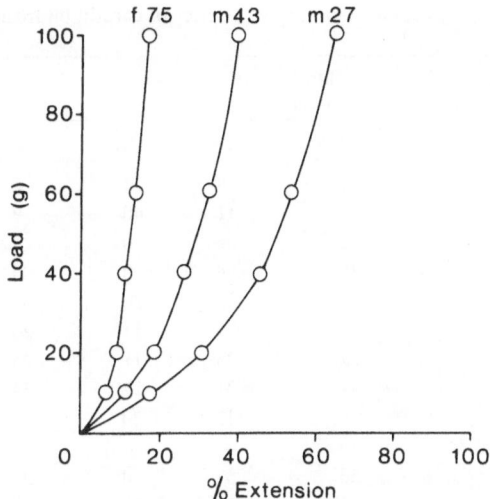

Fig. 20.4. Circumferential extensibility of the right carotid sinus in two men (m) aged 43 and 27 years old and in a woman (f) aged 75 years. The extensibility falls progressively with age.

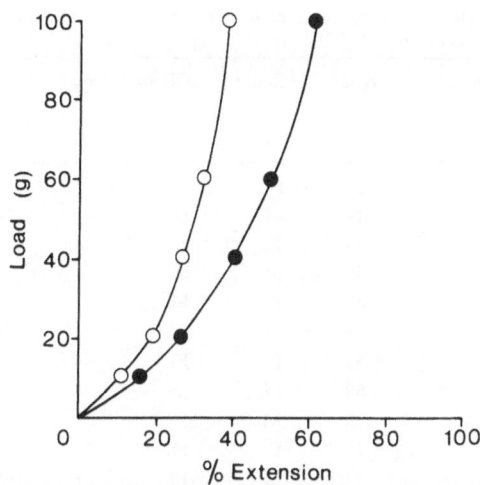

Fig. 20.5. Circumferential extensibility of the right carotid sinus (open circles) and right common carotid artery (closed circles) in a 43-year-old man. The carotid sinus is less extensible than the common carotid artery.

almost certainly attributable to the increase of collagen within the arterial walls. In the case of the carotid sinus the loss of circumferential extensibility is certainly due to the pronounced age-change intimal fibroelastosis and atherosclerosis described above. As a result of its predilection to age-change intimal fibrosis and atherosclerosis, the carotid sinus is more extensible than the common carotid artery only in the very young and at loads below 40 g. From early adulthood onwards and with greater loads it is less extensible and soon the carotid sinus becomes less extensible than the common carotid artery (Fig. 20.5).

Baroreflex Insensitivity

Clinical physiologists have confirmed that the loss of arterial distensibility which occurs both with ageing and with systemic hypertension reduces the efficiency of the carotid sinus. Gribbin et al. (1971) studied baroreflex sensitivity in 61 male and 20 female patients, aged 19–66 years, whose mean systemic arterial pressure ranged from 70 to 150 mmHg. The index of sensitivity that they used was the increase in pulse interval in milliseconds which occurs reflexly in response to each mmHg rise in systolic pressure induced by the intravenous injection of phenylephrine. They found that increasing age and systemic arterial pressure act independently to reduce baroreflex sensitivity. Sleight (1971) found that a linear relation is obtained when the systolic pressure of a pulse beat

is plotted against the pulse interval that follows. Thus the slope of this line may be employed as an index of the baroreflex arc. He thought that increased intravascular pressure may inactivate the baroreceptors by damaging the nerve endings or splint them in the stiff, fibrous walls of the sinus. The effects of atherosclerosis on baroreflex sensitivity must depend to some extent on the orientation of the receptors in the arterial wall. If these are pulled circumferentially then an increased rigidity of the arterial wall should lead to a decreased baroreceptor response. If, on the other hand, they are compressed from within outwards, an increased rigidity of the vessel wall may not have such an effect. It is known that digital compression of a stiff carotid sinus will still stimulate receptors.

Carotid Artery Atherosclerosis and Carotid Body Histological Age-Change

We have been able to define characteristic histological changes in the human carotid body associated with increasing age and we give an account of these in Chapter 4. We wondered if these changes could be ascribed to ageing of the glomic tissues per se or whether they might be related to ischaemia brought about by progressive blockage by atherosclerosis of the carotid arteries from which the minute glomic arteries arise (see Chapter 16). Accordingly we related histological "age-changes" in the carotid bodies to the percentage

Table 20.2. Histological appearances of carotid body related to age and percentage blockage of arteries of carotid bifurcation (after Lowe et al. 1987)

Case no.	Age	Sex	CB histology	% of CB occupied by glomic tissue	% blockage of arteries of carotid bifurcation			
					CCA	CS	ICA	ECA
1	29	F	N	50	4	8	14	6
2	37	M	D	42	6	11	40	9
3	54	F	N	61	18	32	0	0
4	54	M	N	30	25	58	0	0
5	68	F	Fi	28	45	33	26	58
6	68	M	S	46	30	27	11	26
7	69	F	Fi	45	27	25	11	26
8	69	M	S	26	—	31	27	34
9	70	M	S	17	—	12	20	46
10	72	F	Fi	26	—	—	—	44
11	73	M	Fi	29	36	53	0	24
12	77	M	Fi	39	23	56	20	7
13	78	F	S	20	13	3	33	39
14	78	M	S	13	35	53	54	47
15	80	F	S	22	41	63	69	64
16	80	M	Fi	23	60	63	39	39
17	85	F	Fi	40	40	51	50	65
18	85	F	S	20	21	19	39	26
19	100	F	Fi	17	30	—	29	34

CB, carotid body; CCA, common carotid artery; ICA, internal carotid artery; ECA, external carotid artery; CS, carotid sinus; N, normal; D, dark cell prominence; S, sustentacular cell hyperplasia; Fi fibrosis; M, male; F, female.

blockage of the various arteries of the carotid bifurcation in 19 subjects (Table 20.2). The relations between the mean percentage blockage by atherosclerosis in the right and left carotid sinus, age, and histological appearances in the carotid body are shown in Fig. 20.6. The relations between the mean percentage blockage by atherosclerosis in the right and left external carotid arteries, age and histological appearances in the carotid body are shown in Fig. 20.7. The results of our study shown in these two figures and in Table 20.2 are consistent with the view that age-related histological changes in the carotid body and the associated fall in the percentage of glomic tissue in the carotid body are an expression of the ischaemia brought about by progressive occlusion of the arteries of the carotid bifurcation.

However, the effects on the carotid body depend on which component of the carotid arterial tree is blocked. Even severe occlusive atherosclerosis of the carotid sinus does not appear to be a marker of any deleterious effect on the carotid body. In cases 3 and 4 in Table 20.2 extensive obliteration of the lumen of the sinus had taken place, while the carotid body maintained a normal histological appearance, its blood supply being apparently uncompromised. On the other hand in the four cases with histologically normal carotid

bodies (cases 1 to 4) there was insignificant reduction in the lumen of the external carotid. From a practical standpoint it would seem that a knife-cut through the external carotid artery at necropsy is more likely to give an indication of the histological structure of the carotid body than one through the sinus. The reason for this is that the lateral wall of the carotid sinus, which is the part of the carotid arterial tree most susceptible to atherosclerosis, is most remote from the origin of the glomic artery supplying the carotid body. In conclusion, the histological "age-changes" in the human carotid body may well be an expression of ischaemia brought about by the unusual susceptibility of the arteries of the carotid arterial tree to atherosclerosis. In younger subjects the degree of patency of the external carotid artery is likely to be a much better marker of the histological normality of the carotid body than that of the carotid sinus, which is peculiarly susceptible to atherosclerosis on account of its shape which predisposes to abnormal flow patterns within it.

Carotid Sinus Syndrome

Weiss and Baker (1933) suggested that, in sharp contrast to baroreflex insensitivity discussed

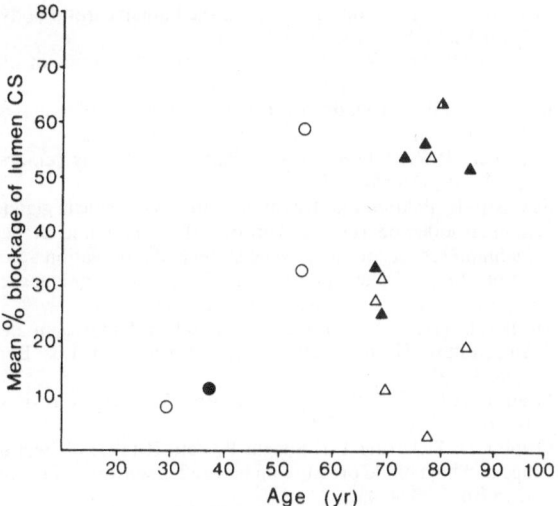

Fig. 20.6. Relation between the mean percentage blockage by atherosclerosis in the right and left carotid sinus (CS), age, and histological appearances in the carotid bodies. (○) Normal histological appearance; (●) dark cell prominence; (△) sustentacular cell hyperplasia; (▲) fibrosis with focal accumulation of lymphocytes.

above, occasionally the reflex arising in the human carotid sinus is abnormally sensitive and may be responsible for attacks of unconsciousness and even convulsions. They reported 15 such cases and showed that the attacks could be induced by pressure over one carotid sinus. The cause of syncope was considered to be cerebral anoxaemia resulting from a sudden fall in systemic blood

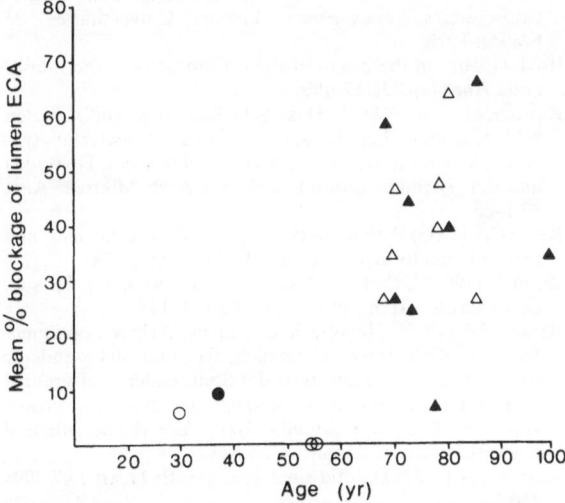

Fig. 20.7. Relation between the mean percentage blockage by atherosclerosis in the right and left external carotid arteries (ECA), age, and histological appearances in the carotid bodies.

pressure or even cardiac asystole. This clinical syndrome was described in further detail in a subsequent paper by Weiss and his colleagues in 1936. They were convinced that a wide variety of signs and symptoms may result from this so-called carotid sinus syndrome. These included fainting, dizziness and weakness, hyperpnoea, facial pallor, bradycardia, numbness and tingling of the extremities, drowsiness and convulsions. Weiss et al. (1936) stated that "The diagnosis of carotid sinus syncope is applied to short attacks of syncope and related manifestations which usually develop when the patient is in the upright position and which can be reproduced with regularity by means of mechanical stimulation of the carotid 'sinus'". They found that the syndrome involved middle-aged and elderly subjects in whom there is likely to be atherosclerosis of the carotid sinus with rigidity and diminished extensibility of its walls. Mechanical stimulation of the carotid sinus does not give rise to symptoms in young subjects and increasing pressure in the sinus leads only to a slight and temporary fall in systemic blood pressure. However, stimulation of the diseased carotid sinus in the middle-aged or elderly may induce symptoms as well as a characteristic fall in blood pressure. Weiss and his colleagues were so convinced of the validity of the syndrome as a clinical entity that they recognised three types: the vagal, the depressor and the cerebral. In the vagal form the dizziness and fainting were considered to result from cardiac asystole itself, thought to be due to either sinoatrial or atrioventricular block, which induces acute cerebral anoxaemia.

The carotid sinus syndrome is sometimes associated with a chemodectoma. In these cases attacks of fainting brought about by reflex bradycardia and systemic hypotension as described above may be accompanied by nausea, blurred vision, trembling and excessive perspiration (Gratiot 1943). Such episodes may occur spontaneously on movement of the head, or be brought about by pressure on the tumour. Estimates of the incidence of the carotid sinus syndrome with chemodectomas vary widely. Some authors believe the incidence to be as low as 2% (Pettet et al. 1953) or 3% (Gratiot 1943). McSwain and Spencer (1947) found only four definite cases of carotid sinus syndrome associated with carotid body tumours on review of the English literature to that date. In contrast, others have claimed the incidence to be as high as 5 of 16 cases by Sessions et al. (1959) and 18 of 38 cases by Chambers and Mahoney (1968). It is clear from these figures that different authors have very different ideas as to what constitutes the carotid sinus syndrome.

The structure of the human glomic vasculature described in Chapter 16 provides an additional basis for understanding why pressure on a normal carotid body or a chemodectoma may stimulate baroreceptors and lead to the symptoms listed above. The main glomic artery and its branches, despite their small size, retain the structure of an elastic vessel which is more typical of the carotid sinus or large systemic arteries. In the normal carotid body even the interlobular arteries which permeate the substance of the glomus have histological appearances which are reminiscent of those of the carotid sinus, suggesting that the glomic artery may subserve a similar baroreceptor function. Pressure of an enlarged neoplastic carotid body on the carotid sinus or glomic arteries might well produce the carotid sinus syndrome.

Further cases of the syndrome were but rarely reported after the initial papers by Weiss and his associates, until Morley et al. (1982) described no fewer than 70 patients who had been treated by artificial pacing over a period of four years. Leatham (1982) found this wealth of cases so surprising that he felt it raised the whole issue of the definition of the carotid sinus syndrome. Leatham thought that reflex slowing of sinoatrial rate from carotid sinus massage was an almost universal finding and that it was more pronounced in older subjects and was more noticeable in patients with disease of the conducting tissue. He thought that cardiac slowing in the cases reported by Morley and his colleagues (1982) might be primarily sinoatrial disease rather than an abnormality of the wall of the carotid sinus. Leatham did not accept as valid the statement that none of their patients had overt sick sinus syndrome on the resting or 24 hour electrocardiogram, because he felt that patients with episodes of even extreme bradycardia may have weeks or even months of normality. Thus, Leatham concluded that most, if not all, of the patients described by Morley and his colleagues had sinoatrial disease. He felt that further work was required to decide whether primary carotid sinus syncope is other than an extremely rare medical curiosity. He regarded carotid sinus stimulation more as a means of unmasking conduction disease.

References

Adams WE (1958) The comparative morphology of the carotid body and carotid sinus. Charles C. Thomas, Springfield, IL

Binswanger O (1879) Anatomische Untersuchungen über die Ursprungsstelle und den Anfangstheil der Carotis interna. Arch Psychiatr Berl 9:351–368

Boyd JD (1937) The development of the human carotid body. Contr Embryol Carneg Inst 26:1–31

Burns A (1811) Observations on the surgical anatomy of the head and neck. Thomas Bryce, Edinburgh

Burton AC (1954) Relation of structure to function of tissues of the wall of blood vessels. Physiol Rev 34:619–642

Chambers RG, Mahoney WD (1968) Carotid body tumors. Am J Surg 116:554–558

de Castro F (1926) Sur la structure et l'innervation de la glande intercarotidienne (glomus caroticum) de l'homme et des mammifères, et sur un nouveau système d'innervation autonome du nerf glossopharyngien. Études anatomiques et expérimentales. Trav Lab Rech Biol 24:365–432

Drüner L (1925) Ueber die anatomischen Unterlagen der Sinusreflexe Herings. Dtsch Med Wochenschr Lpz Berl 51:559–560

Gratiot JH (1943) Carotid-body tumors. A collective review. Internat Abstr Surg 77:177–186

Gribbin B, Pickering TG, Sleight P, Peto R (1971) Effect of age and high blood pressure on baroreflex sensitivity in man. Circ Res 29:424–431

Hammar JA (1934) Konstitutionsanatomische Studien über die Neurotisierung des Menschenembryos. 1. Zur Bildungsgeschicte des Glomus Caroticum. Z Mikrosk Anat Forsch 35:602–630

Harris P, Heath D, Apostolopoulos A (1965) Extensibility of the human pulmonary trunk. Br Heart J 27:651–659

Heath D, Smith P, Harris P, Winson M (1973) The atherosclerotic human carotid sinus. J Pathol 110:49–58

Hering HE (1923) Der Karotisdruckversuch. München Med Wochenschr 70:1287–1290

Hering HE (1924a) Der Sinus caroticus an der Ursprungsstelle der Carotis interna als Ausgangsort eines hemmenden Herzreflexes und eines depressorischen Gefässreflexes. (Gleichzeitig II. Mitteilung über den Karotisdruckversuch.) München Med Wochenschr 71:701–704

Hering HE (1924b) Sinus reflexe vom Sinus caroticus werden durch einen Nerven (Sinusnerv) vermittelt, der ein Ast des Nervus glossopharyngeus ist. München Med Wochenschr 71:1265–1266

Hering HE (1927) Die Karotissinusreflexe auf Herz und Gefässe. Theodor Steinkopff, Dresden

Heymans C (1929) Le sinus carotidien et les autres zones vasosensibles réflexogènes. Presses Universitaires de France, Paris

Ito T (1950) On the origin of the carotid body in the rabbit. Folia Anat Jap 23:117–130

Kastschenko N (1887) Das Schicksal der embryonalen Schlundspalten bei Saugetieren. (Zur Entwicklungsgeschicte des mittleren und ausseren. Ohres, der Thyroidea und der Thymus. Carotidenanlage.) Arch Mikrosk Anat 30:1–26

Keele CA (1933) Pathological changes in the carotid sinus and their relation to hypertension. Q J Med 2:213–220

Kohn A (1900) Ueber den Bau und die Entwicklung der sog. Carotisdrüse. Arch Mikrosk Anat 56:81–148

Krause W (1879) Handbuch der menschlichen Anatomie. Durchaus nach eigenen Untersuchungen und mit besonderer Rücksicht auf das Bedürfuiss der Studirenden, der praktischen Aerzte und Wundärzte und der Gerichtsärzte, verfasst von C. F. T. Krause, 3rd edit. Halm. Bd. II Specielle und macroskopische Anatomie, Hanover

Leatham A (1982) Carotid sinus syncope. Br Heart J 47:409–410

Lowe P, Heath D, Smith P (1987) Relation between histological age-changes in the carotid body and atherosclerosis in the carotid arteries. J Laryngol Otol 101:1271–1275

Luschka H (1862) Die Anatomie des Menschen in Rücksicht

auf die Bedürfnisse der praktischen Heilkunde. Bd 1, Abt 1 Der Hals, H. Laupp, Tubingen

Manson P (1866) Peculiar affection of the internal carotid artery in connexion with disease of the brain. Med Times Gaz Lond 53:336–337

Marchand F (1891) Beitrage zur Kenntnis der normalen und pathologischen Anatomie der Glandula carotica und der Nebennieren. Int Beitr Wiss Med 1:535–581

McSwain B, Spencer FC (1947) Carotid body tumor in association with carotid sinus syndrome. Report of two cases. Surgery 22:222–229

Meyer L (1876) Ueber aneurysmatische Veränderungen der Carotis interna Geisterskranker. Arch Psychiatr Berl 6:84–109

Morley CA, Perrins EJ, Grant P, Chan SL, McBrien DJ, Sutton R (1982) Carotid sinus syncope treated by pacing. Analysis of persistent symptoms and role of atrioventricular sequential pacing. Br Heart J 47:411–418

Nielsen EH (1968) Sinus caroticus. Stougaard, Copenhagen

Paltauf R (1892) Ueber Geschwulste der Glandula carotica nebst einem Beitrage zur Histologie und Entwickelungs geschicte derselben. Beitr Pathol Anat 11:289–300

Pettet JR, Woolner LB, Judd ES (1953) Carotid body tumors (chemodectomas). Ann Surg 137:465–477

Roach MR, Burton AC (1957) The reason for the shape of the distensibility curves of arteries. Can J Biochem Physiol 35:681–690

Samuel KC (1956) Atherosclerosis and occlusion of the internal carotid artery. J Pathol Bacteriol 71:391–401

Schäfer EA (1877) Ueber die aneurysmatische Erweiterung der Carotis interna an ihrem Ursprung. Allg Z Psychiatr 34:438–451

Sessions RT, McSwain B, Carlson RI, Scott HW (1959) Surgical experiences with tumors of the carotid body, glomus jugulare and retroperitoneal nonchromaffin paraganglia. Ann Surg 150:808–823

Sleight P (1971) What is hypertension? Recent studies in neurogenic hypertension. Br Heart J 33 [suppl]:109

Stahel H (1886) Ueber Arterienspindeln und über die Beziehung der Wanddicke der Arterien zum Blutdruck. Abhandlung II. Arch Anat Physiol Lpz Anat Abt 309–334

Weiss S, Baker JP (1933) The carotid sinus reflex in health and disease: its role in the causation of fainting and convulsions. Medicine (Baltimore) 12:297–354

Weiss S, Capps RB, Ferris EB Jr, Munro D (1936) Syncope and convulsions due to a hyperactive carotid sinus reflex: diagnosis and treatment. Arch Intern Med 58:407–417

Winson M, Heath D, Smith P (1974) Extensibility of the human carotid sinus. Cardiovasc Res 8:58–64

Zarins CK, Giddens DP, Bharadvaj BK, Sottjurai VS, Mabon RF, Glagov S (1983) Carotid bifurcation atherosclerosis. Quantitative correlation of plaque localization with flow velocity profiles and wall shear stress. Circ Res 53:502–514

21 Comparative Histopathology

Most mammals have to adjust to the hypobaric hypoxia of high altitude, when they ascend into mountains, by the complex processes of acclimatisation which is described in detail elsewhere (Heath and Williams 1989). Acclimatisation is a reversible, non-inheritable change in the anatomy or physiology of an organism which enables it to survive in an alien environment. Humans have to acclimatise in this manner. The carotid body has an important rôle as a chemoreceptor in initiating and maintaining an enhanced ventilatory response to hypoxia and in humans shows histological changes including prominence of the dark variant of chief cells (see Chapter 11), which is rich in peptides such as enkephalins (see Chapter 8).

Indigenous Mountain Species

In sharp contrast, some indigenous mountain species have lived for countless millenia at high altitude and have, through a process of natural selection, become fully genetically adapted to the hypoxia of high altitude. Adaptation is the development of biochemical, physiological and anatomical features which are heritable and of genetic basis, enabling the species to explore the environment to its best advantage. Mammals adapted to high altitude include the yak in Himalaya (Heath et al. 1984b), the Tibetan snow-pig in the Tibetan plateau (Sun et al. 1989), and the mountain viscacha of the Andes (Heath et al. 1981). Probably the best-known group of indigenous mountain species, however, comprises the high-altitude camelids of the Andes. This group includes the llama, alpaca, vicuña and guanaco (Harris et al.

1982). It is of interest to study the histology of the carotid body from such species for the appearances are those of chemoreceptor tissue biologically adapted rather than acclimatised to high altitude.

The carotid bodies of high-altitude camelids are small. Those of the alpaca are composed of clusters of chief cells surrounded by sustentacular cells (Heath et al. 1985). The clusters are some 35 μm in diameter and are composed of chief cells, 10 μm in diameter, with pale eosinophilic cytoplasm and ill-defined borders, containing large vacuoles (Fig. 21.1). The nuclei of the chief cells are ovoid and have an open chromatin pattern. Surrounding the clusters are elongated sustentacular cells with faintly eosinophilic cytoplasm and nuclei measuring 7 μm $\times$ 3 μm (Fig. 21.1). Differential cell counts reveal a high proportion of these sheath cells (Table 21.1). An insignificant number of cells with compact, chromatin-packed nuclei, and a darkly staining cytoplasm are seen (Heath et al. 1984a). They are probably comparable to the dark variants of chief cells seen in the human carotid body. The entire appearance is one of quiescence, with no suggestion of hyperplasia of either type of cell or evidence of any pathological process. Presumably such quiescent histological appearances represent biological adaptation rather than acclimatisation to the hypobaric hypoxia of high altitude.

Cattle

Cattle are not adapted to high altitude but instead have to undergo the complex processes of acclima-

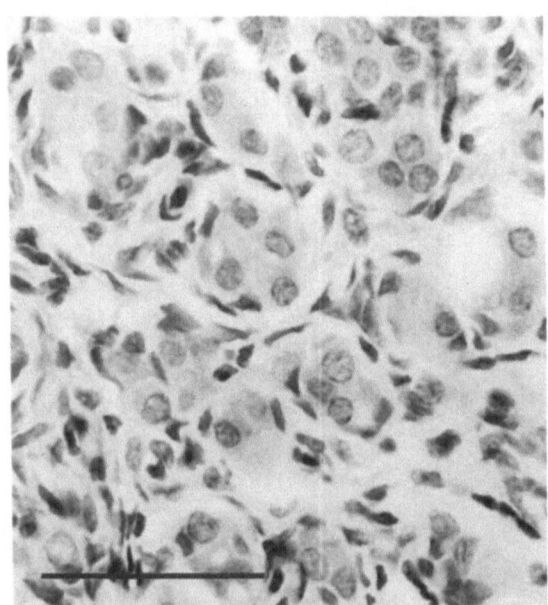

Fig. 21.1. Carotid body of alpaca from La Raya (4200 m), Peru, showing clusters of chief cells with round prominent nuclei, separated by intervening sustentacular cells with elongated nuclei. (Haematoxylin–eosin (HE)) Scale line = 75 μm.

tisation to adjust to life in the mountains just like humans. Indeed, calves may have difficulty in initiating successful acclimatisation and susceptible animals in the Wasatch mountains in Utah may die from "brisket disease" due to the onset of significant pulmonary hypertension and conges-

tive cardiac failure. In cattle at high altitude there may be considerable enlargement of the carotid bodies, this appearance being in striking contrast to the small carotid bodies of the high-altitude camelids described above. Histologically the glomic tissue of cattle at high altitude looks hyperplastic and somewhat disorganised so that the margins of the cell clusters are indistinct. In this way it differs from the undisturbed appearance of the carotid body of the alpaca. Many of the cells arranged in a rather haphazard fashion are light chief cells with large, round vesicular nuclei, 7 μm in diameter, and very pale eosinophilic cytoplasm showing pronounced vacuolation. There are in addition many prominent focal collections of dark cells which are larger than those seen in cattle from sea level (Fig. 21.2). These cells have compact nuclei, 5.5 μm in diameter, and dark basophilic cytoplasm which shows some degree of vacuolation. The maturation of dark cells into light cells can be picked out, the nuclei with plentiful heterochromatin and intervening clear areas becoming clearer with enlargement of these oval areas so that the nucleus has only remaining dots and nodules of chromatin connected by thick bars (Fig. 21.2). Later the nucleoplasm becomes even clearer, with thin attenuated strands connected to dots of chromatin in the periphery of the nucleus. At the same time there is progressive lightening and vacuolation of the cytoplasm. Such appearances are consistent with the view that dark cells are precursors of the light variant (see Chapter 3). The focal proliferations of dark cells account for a

Table 21.1. Differential cell counts in the carotid bodies of different species (after Heath et al. 1985)

Species and strain	Functional status	% cells			
		Light	Dark	Progenitor	Sustentacular
(a) Alpaca	H	26	0	0	74
(b) Llama	H	54	2	0	44
(a) Cattle	H	49	29	1	21
	CH	62	21	0	17
(b) Cattle	E	39	28	1	32
(a) Guinea-pigs	H	39	31	1	29
(b) Guinea-pigs	E	31	32	4	33
(a) Rabbits	H	25	47	1	27
(b) Rabbits	E	41	21	<1	37
(a) Dogs	H	28	36	1	35
(b) Dogs	E	31	17	1	51
(a) Wistar rat	E	←——— 46 ———→			54
(b) Okamoto-Aoki rat	S	←——— 47 ———→			53
Humans					
(a) Healthy control	E	54	5	2	39
(b) Pulmonary emphysema	H	32	4	2	62
(c) Systemic hypertension	S	40	4	2	54

E, euxoxic; H, hypoxic; S, systemic hypertensive; CH, chemodectoma.

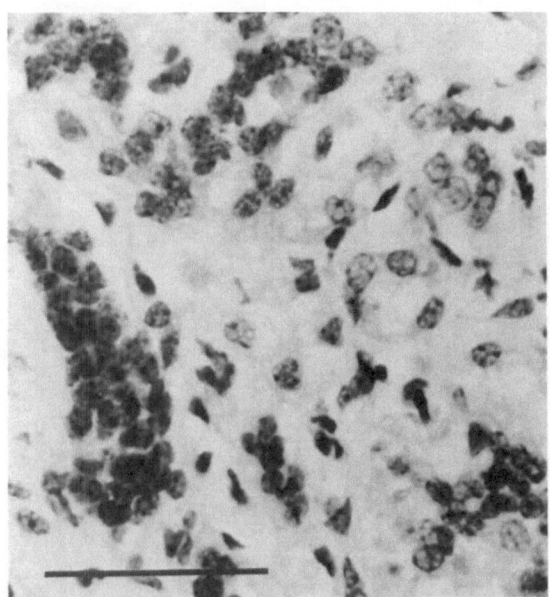

Fig. 21.2. Carotid body of cow from Cerro de Pasco (4300 m) in the Peruvian Andes. Large groups of the dark variants of chief cells are present. There appears to be a transition between them and the light variant of chief cells. (HE). Scale line = 100 μm.

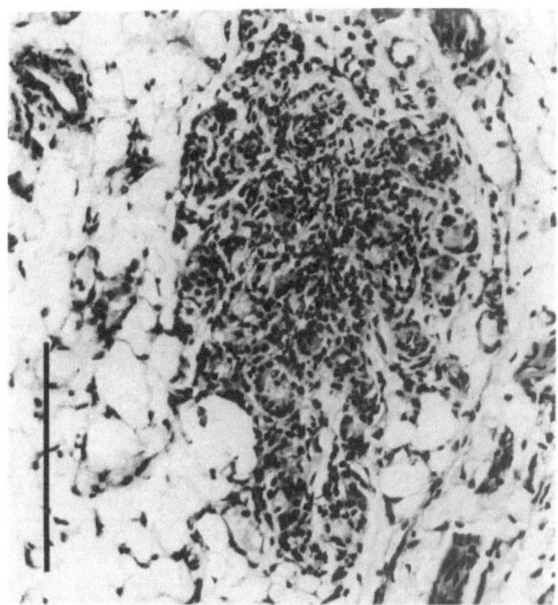

Fig. 21.3. Carotid body of low-altitude guinea-pig consisting of a compact, fusiform collection of glomic cells situated in the adipose tissue between the internal and external carotid arteries. (HE) Scale line = 200 μm.

pronounced shift in the differential cell count at the expense of the light variant of chief cells and of the sustentacular cells compared with that of the high-altitude camelids (Table 21.1). Such prominence of dark cells in the carotid body of cattle at high altitude is comparable to that found in the native highlanders of Ladakh, as described in Chapter 11, where its significance is discussed.

In one animal that we studied, the carotid bodies were noticeably larger, with several large lobes. The histological appearances differed in presenting multiple small clusters of the light variety of chief cells with strikingly fewer dark cells. Such appearances are very reminiscent of the "Zellballen" of the chemodectoma (see Chapter 19). Arias-Stella and Bustos (1976) also found that as many as 40% of cattle living at an elevation of 4300 m in the vicinity of Cerro de Pasco in the Andes of Central Peru developed histological changes in their carotid bodies suggestive of chemodectoma.

Guinea-pigs

The carotid bodies of guinea-pigs exposed to the hypobaric hypoxia of high altitude are also enlarged but they show histological changes different from those found in cattle. The main dis-

tinction is that in guinea-pigs there is a widespread prominence of greatly vacuolated light cells rather than dark cells and appearances suggestive of early chemodectoma formation as seen in cattle. In low altitude guinea-pigs the carotid body consists of a compact, fusiform collection of glomic cells situated in the adipose tissue between the internal and external carotid arteries (Fig. 21.3). Light and dark variants of chief and sustentacular cells are apparent, if we employ the same criteria as were described for cattle in the preceding section. Considerable numbers of dark cells are present (Table 21.1) and the apparent syncytium formed by them gives the whole section a darkly stained appearance (Fig. 21.4). The cytoplasm of the light cells is faintly eosinophilic and vacuolated, the vacuoles varying in size and number between cells but not being striking in appearance. The sustentacular cells are elongated.

The carotid bodies of high-altitude guinea-pigs can be shown by methods of tissue morphometry to be larger than those of low-altitude animals (Edwards et al. 1971). Even when viewed at low power the histological sections show strikingly different appearances. As well as being much larger, sections of the carotid body present a much clearer appearance (Fig. 21.5). On closer inspection at high power this pallor is due to some extent to the increase in the number of the light variant of chief cells (Fig. 21.6). However, much of it is a

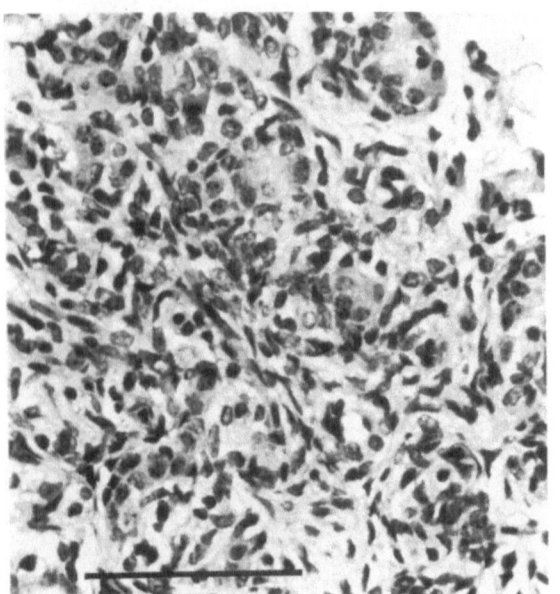

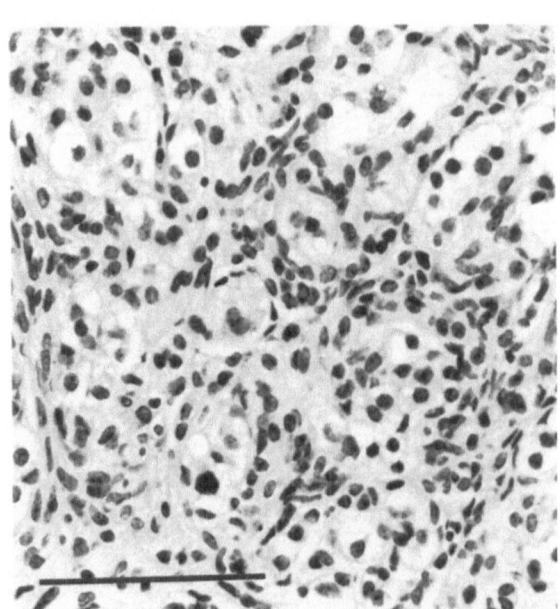

Fig. 21.4. Carotid body of low-altitude guinea-pig. An apparent syncytial appearance due to groups of the dark variant of chief cells. (HE) Scale line = 80 μm.

Fig. 21.6. Carotid body of high-altitude guinea-pig from the same town as in Fig. 21.5. Compared to the carotid body in Fig. 21.4 the staining is paler due to a relative hyperplasia of the light variant of chief cells with cytoplasmic vacuolation. (HE) Scale line = 80 μm.

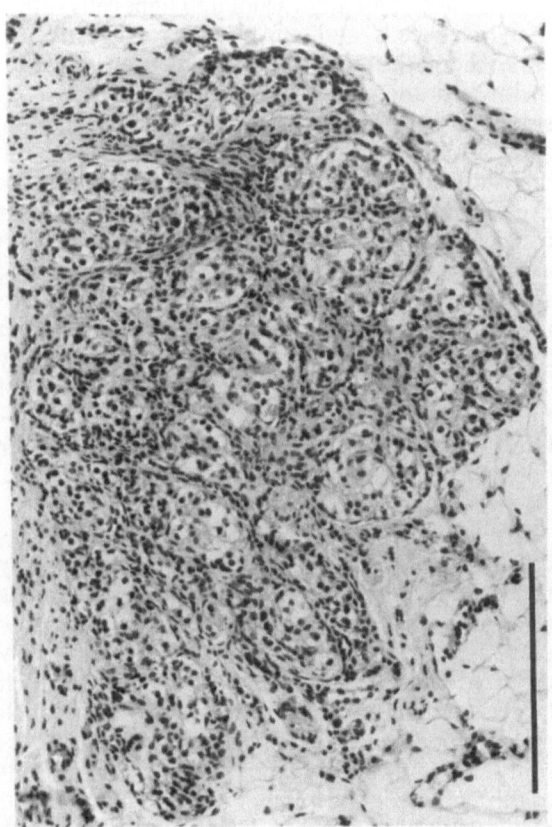

consequence of striking vacuolation of the chief cells (Fig. 21.6). In Chapter 9 we presented experimental evidence that in rabbits the initial response of the glomic tissue to hypoxia after a period of 3 months is a proliferation of dark cells rich in peptides. Later, after a period of 6 months, the dark cells have been lost to be replaced by light cells. It seems likely that the same phenomenon occurs in guinea-pigs so that the resulting appearance of sustained response to continuation of hypobaric hypoxia is one of predominant light cells. The vacuolation is likely to be due to enlargement of dense-core vesicles (Edwards et al. 1972). They show swelling of the original clear halo to form a microvacuole (see Chapter 11). At the same time there is progressive pallor and shrinkage of the central, osmiophilic core, which commonly moves eccentrically (see Chapter 11). The precise biochemical and physiological significance of this microvacuolation remains obscure but perhaps there is liberation of biogenic amines and peptides.

Fig. 21.5. Carotid body of high-altitude guinea-pig from Cerro de Pasco (4300 m) in the Peruvian Andes. Compared to that shown in Fig. 21.3 the glomus is enlarged and presents a paler appearance. (HE) Scale line = 200 μm.

Rabbits

The glomic tissue of the carotid body of the low-altitude rabbit is similar to that of humans in its micro-anatomical arrangement: lobules of parenchymal cells are separated by intervening strands and septa of fibrous tissue. The cellular tissue forming the lobules appears to comprise two main variants. In one the nucleus is compact and shows small, clear areas between fairly compact chromatin. In the other the nucleus is large, round and clear with a sparse vesicular pattern. These types of cell seem to correspond to the dark and light variants seen in humans. The cytoplasm of the chief cells has an ill-defined outline and contains small vacuoles. Even on low-power light microscopy the differences in the histological appearances of the carotid bodies of high-altitude rabbits from their sea-level counterparts are apparent. We confirmed by tissue morphometry that the carotid bodies are enlarged in high-altitude rabbits (Edwards et al. 1971). There is more cellularity with less obvious fibrous tissue within the confines of the glomus. On staining with haematoxylin and eosin the glomic tissue appears lighter in colour due to pronounced vacuolation in the cytoplasm of the cells, as we describe above for guinea-pigs. In one study (Edwards el al. 1971) a considerable increase was found in the differential cell count of dark cells from 21% to 47% (Table 21.1) and hence much of the pallor of sections must be held to be due to vacuolation in dark cells rather than hyperplasia of the light variant of chief cells. The arrangement of glomic cells in clusters is more obvious in rabbits living at high altitude. There is a difference in the pattern of the reticulin network. In low-altitude rabbits, individual clusters of glomic cells are delineated by reticulin. In high-altitude rabbits the clusters of glomic cells are more numerous and grouped together so that the collections of cells contained within the respective reticulin envelopes are larger.

Dogs

The carotid bodies of dogs from low and high altitude present appearances similar to those seen in rabbits. The light variant of the chief cells has a large, ovoid nucleus and a lightly stained cytoplasm with vacuoles and an ill-defined outline. The dark cell has a relatively small, ovoid, hyperchromatic nucleus with areas of very densely stained heterochromatin and some small clear areas. The cytoplasm of these cells is similar to that of the light variety of chief cell. The carotid bodies of high-altitude dogs have greater mean linear dimensions and volumes than those of their sea-level counterparts (Edwards et al. 1971). There is an increase in the proportion of dark chief cells, with a corresponding fall in the number of sustentacular cells.

Rats

The glomic tissue of the rat carotid body forms a compact, non-lobulated structure between the arms of the internal and external carotid arteries. It is subdivided into several clusters of chief cells, each surrounded by elongated sustentacular cells. The clusters are roughly circular in cross-section at the periphery of the organ and more elongated deep within its substance.

The characteristic histological features of many disease-states of this organ in humans are not reflected in the tissues of the rat chemoreceptor. As early as 1926 de Castro was able to distinguish two types of chief cell, one with a large pale-staining nucleus (which we designate the light cell) and one with a small densely staining nucleus (which we would subdivide into dark and progenitor cells). In electron microscopic studies of the carotid bodies of several species light and dark cells have been distinguished on the basis of the electron density of their cytoplasm. The distinction between light and dark cells has been made by various authors in the rabbit (Lever el at. 1959), horse and dog (Höglund 1967), seal (Morita et al. 1970), cat (Abbott et al. 1972), and hamster (Chen et al. 1969). The distinction has even been made in the rat (Niedorf 1970). Nevertheless, McDonald (1981, pp 121 – 123) considered the density of staining of the cytoplasm to be an unreliable criterion for distinguishing subtypes of chief cells in the carotid body. He found light and dark cells in rat carotid bodies that were manipulated using surgical procedures before fixation, or were removed from the animal immediately after fixation by vascular perfusion. He found that dark cells were not present in carotid bodies of normal rats that were left untouched for at least 1 hour after the perfusion.

In our experience the chief cells of the rat carotid body comprise a homogeneous population so that, although there is some variation in nuclear diameter and staining intensity, there is no clear distinction between light and dark variants as in the human carotid body. We have also found no distinctive progenitor cells in the murine carotid body. For this reason we feel that the rat is unsuitable as a model for the study of carotid body pathology in humans. The nuclei of the uniform chief cells are slightly ovoid, with a moderately haematoxyphilic stippled chromatin pattern. The cytoplasm of all chief cells is uniformly palely eosinophilic and finely vacuolated. Sustentacular cells have darkly haematoxyphilic, elongated nuclei and pale cytoplasm which merges imperceptibly with the surrounding Schwann cells and connective tissue. They are more elongated than their human counterparts. There are no differences in this cellular pattern between Wistar and Okamoto-Aoki rats. The carotid bodies of rats undergo rapid and reversible enlargement due to acute vascular engorgement on exposure to simulated high altitude in a decompression chamber (Heath et al. 1973; see Chapter 9).

In spontaneously hypertensive rats of the Okamoto-Aoki strain, organic enlargement of the carotid bodies due to cellular proliferation occurs. One study (Smith et al. 1984) found that in 21 control Wistar albino rats (Table 21.1) 46% of the glomic cells were chief cells of unspecified type and 54% were sustentacular. In 20 spontaneously hypertensive rats of the Okamoto-Aoki strain the enlarged carotid bodies showed 47% unspecified chief cells and 53% sustentacular cells. In other words, both types of cell shared equally in the proliferation so that neither cell predominated (Smith et al. 1984). In one Okamoto-Aoki rat, however, there were isolated groups of chief cells in which the nuclei were much darker and smaller than normal, and resembled the foci of dark cell hyperplasia described in Chapter 9. The differential cell counts in patients with systemic hypertension and in rats with spontaneous hypertension are virtually identical. It should be noted, that in hypertensive humans the high percentage of sustentacular cells (54%) is reached by hyperplasia of a much lower percentage (39%) (Table 21.1). In rats the percentage, while equally high (53%) starts from an already high proportion in control Wistar albino rats (54%) (Table 21.1). In other words, in rats there is no sustentacular hyperplasia as in man.

Conclusions

The observations made in this chapter make it clear that the histological appearances of the enlarged carotid body are not uniform but appear rather to be characteristic of the species (Table 21.2) and its biological adjustment to environmen-

Table 21.2. Summary of histological changes in the carotid bodies of various animal species exposed to hypobaric hypoxia of high altitude

Animal	Histology of carotid body	Association
High-altitude camelids (e.g. llama, alpaca)	Quiescent appearance of chief and sustentacular cells	Genetic adaptation to hypobaric hypoxia
Cattle	Dark cell prominence	Peptide-rich cells appearing in initial response to hypoxia
	Later, appearance suggestive of chemodectoma	Possibly neoplasia associated with continuous hypoxic stimulation
Guinea-pigs	Pallor of vacuolated chief cells	Sustained response to hypoxia
Rabbits	Dark cell prominence	As above
	Pallor of vacuolated chief cells	As above
Dogs	Dark cell prominence	As above
	Pallor of vacuolated chief cells	As above
Rats	Homogeneous population of chief cells	

tal or pathological hypoxia. Thus in one Peruvian town such as Cerro de Pasco in the Andes (4330 m) the carotid bodies in llamas, cattle, and guinea-pigs may show totally different histological appearances although exposed to an identical partial pressure of oxygen.

References

Abbott CP, Daly M de B, Howe A (1972) Early ultrastructural changes in the carotid body after degenerative section of the carotid sinus nerve in the cat. Acta Anat 83:161–185

Arias-Stella J, Bustos F (1976) Chronic hypoxia and chemodectomas in bovines at high altitudes. Arch Pathol Lab Med 100:636–639

Chen I, Yates RD, Duncan D (1969) The effects of reserpine and hypoxia on the amine-storing granules of the hamster carotid body. J Cell Biol 42:804–816

de Castro F (1926) Sur la structure et l'innervation de la glande intercarotidienne (glomus caroticum) de l'homme et des mammifères, et sur un nouveau système d'innervation autonome du nerf glossopharyngien. Études anatomiques et expérimentales. Trav Lab Rech Biol 24:365–432

Edwards C, Heath D, Harris P, Castillo Y, Krüger H, Arias-

Stella J (1971) The carotid body in animals at high altitude. J Pathol 104:231–238

Edwards C, Heath D, Harris P (1972) Ultrastructure of the carotid body in high altitude guinea pigs. J Pathol 107:131–136

Harris P, Heath D, Smith P, Williams DR, Ramirez A, Krüger H, Jones DM (1982) Pulmonary circulation of the llama at high and low altitudes. Thorax 37:38–45

Heath D, Williams DR (1989) Adaptation to hypobaric hypoxia. In: High-altitude medicine and pathology. Butterworths, London, pp 315–326

Heath D, Edwards C, Winson M, Smith P (1973) Effects on the right ventricle, pulmonary vasculature and carotid bodies of the rat of exposure to, and recovery from, simulated high altitude. Thorax 28:24–28

Heath D, Williams D, Harris P, Smith P, Krüger H, Ramirez A (1981) The pulmonary vasculature of the mountain-viscacha (*Lagidium peruanum*). The concept of adapted and acclimatized vascular smooth muscle. J Comp Pathol 91:293–301

Heath D, Smith P, Jago R (1984a) Dark cell proliferation in carotid body hyperplasia. J Pathol 142:39–49

Heath D, Williams D, Dickinson J (1984b) The pulmonary arteries of the yak. Cardiovasc Res 18:133–139

Heath D, Smith P, Fitch R, Harris P (1985) Comparative pathology of the enlarged carotid body. J Comp Pathol 95:259–271

Höglund R (1967) An ultrastructural study of the carotid body of horse and dog. Z Zellforsch 76:568–576

Lever JD, Lewis PR, Boyd JD (1959) Observations of the fine structure and histochemistry of the carotid body in the cat and rabbit. J Anat 93:478–490

McDonald DM (1981) Peripheral chemoreceptors. Structure–function relationships of the carotid body. In: Hornbein TF (ed) Regulation of breathing. Marcel Dekker Inc, New York Basel, pp 105–319

Morita E, Chiocchio SR, Tramezzani JH (1970) The carotid body of the Weddell seal (*Leptonychotes weddelli*). Anat Rec 167:309–328

Niedorf HR (1970) Die normale und pathologische Anatomie des Glomus caroticum. Med Welt 21:251–257

Smith P, Jago R, Heath D (1984) Glomic cells and blood vessels in the hyperplastic carotid bodies of spontaneously hypertensive rats. Cardiovasc Res 18:471–482

Sun SF, Sui GJ, Liu YH, Cheng XS, Anand IS, Harris P, Heath D (1989) The pulmonary circulation of the Tibetan snow pig (*Marmota himalayana*). J Zool 217:85–91

Subject Index